Executive Politics and Governance

Series Editors
Martin Lodge, London School of Economics and Political Science,
London, UK
Kai Wegrich, Hertie School of Governance, Berlin, Germany

The Executive Politics and Governance series focuses on central government, its organisation and its instruments. It is particularly concerned with how the changing conditions of contemporary governing affect perennial questions in political science and public administration. Executive Politics and Governance is therefore centrally interested in questions such as how politics interacts with bureaucracies, how issues rise and fall on political agendas, and how public organisations and services are designed and operated. This book series encourages a closer engagement with the role of politics in shaping executive structures, and how administration shapes politics and policy-making. In addition, this series also wishes to engage with the scholarship that focuses on the organisational aspects of politics, such as government formation and legislative institutions. The series welcomes high quality research-led monographs with comparative appeal. Edited volumes that provide in-depth analysis and critical insights into the field of Executive Politics and Governance are also encouraged.

More information about this series at
http://www.palgrave.com/gp/series/14980

Andrea Mennicken · Robert Salais
Editors

The New Politics of Numbers

Utopia, Evidence and Democracy

Editors
Andrea Mennicken
Department of Accounting and
Centre for Analysis
of Risk and Regulation
London School of Economics and
Political Science
London, UK

Robert Salais
Institutions et dynamiques historiques
de l'économie et de la Sociètè
(IDHES)
École Normale Supérieure
Paris-Saclay
Gif-sur-Yvette, France

Executive Politics and Governance
ISBN 978-3-030-78200-9   ISBN 978-3-030-78201-6   (eBook)
https://doi.org/10.1007/978-3-030-78201-6

In Memory of Alain Desrosières and Anthony Hopwood

Foreword: What Numbers Do

Numbers do things. They highlight and obscure. They integrate and disaggregate. They mark and measure. They represent and intervene. They tame and inflame. They structure people's interactions. They create new objects and new kinds of people. They possess a power that hides itself. They are rhetoric that is anti-rhetorical. What all of these features of numbers share is that they express a certain agency. They perform.

The agency of numbers is not the same as human agency. It is not the creative human action that invents numbers and finds new uses and contexts for them. But it is agency, nonetheless, because numbers order and make possible specific kinds of cognition and action and preclude others. For example, numbers make it almost impossible not to compare the entities that share a scale. In doing so, numbers give rise to extraordinary amounts and kinds of comparisons. Our understanding of what is shared and what is unique requires comparison. Numbers make it possible to reduce complex, diverse information into a 'sense-able' sequence. Once we know a number, it is difficult not to think in terms of quantity, whether this is time, distance, price, or some other unit (an exception constitute phone numbers, or street addresses). Numbers produce hierarchy where more is typically better than less. Difference is in degrees, not kind.

We form relationships through and to numbers. Those that took the same standardized test now share a tie with one another, one that is abstract but sometimes fateful. And people become invested in particular numbers or even particular kinds of numbers. Witness the fixation on the country risk index in Argentina that de Santos (2009) describes. Or the way that measures of sexual behavior shaped the gay rights movement in the United States where the 10% became the widely accepted estimate of the gay population (Espeland and Michaels 2018). Numbers dehumanize, too, as the tattoos of Jews during the Holocaust attest. But they can also humanize, as being counted in a national census offers a sense of inclusion and potency (Rodríguez-Muñiz 2017).

This volume takes the agency of numbers seriously. In different domains, the chapters it contains address how distinctive combinations of human agency and the agentic properties of numbers give rise to particular modes of coordination, administration and governance. Coordination, the art of breaking down complex tasks into actionable (and complementary) units, depends heavily on numbers. We use metrics to synchronize our machines. We use accounting to assess what we have, what is owed, and what worked in the past. We use numbers to project ourselves into uncertain futures when we analyse risk. And we hold leaders accountable with numbers: crime rates, budgets, votes, cost-benefit ratios, rankings, economic indicators, and returns on investments. As the seminal work of Miller and Rose (1990) established, numbers permit states to govern from a distance. But exactly how this governing happens in different institutional contexts is less established. This volume helps to fill that empirical and theoretical gap.

Numbers are like any symbol in that they are a form that systematically includes, excludes, and organizes information. What is distinctive about numbers is how scrupulously they edit the world and the nature of their authority. We associate numbers with precision, rigor, objectivity, and rationality. Numbers are firm while words are malleable. We believe that the rules for constructing measures are constraining such that numbers are less easy to manipulate, less open to multiple interpretations, than other kinds of information. The discipline needed to make numbers is enough to tame the self-interest or politics that threaten their objectivity.

But numbers are rarely as innocent as they seem. They make possible new forms of surveillance that inform and control. As Ota De Leonardis demonstrates, the quantification of poverty translates inequalities into distance, something which mutes discussions of political power. Similarly, Ousmane Sidibé describes how the advent of benchmarking with performance indicators has contributed to undermining public politics around development in Africa. Tong Lam examines how with the advent of big data, algorithms now construct profiles of individuals that informs their credit, their politics, and their consumption. Robert Salais shows how the transformation to governance by numbers, with its emphasis on efficiency, neglects the 'fair balance' between the state and its citizens.

But numbers can also constrain power. In times of political upheaval, measurement has often been a pressing issue, bound up with issues of equity and fairness. The Magna Carta, for example, declared that there should be 'one measure of wine throughout our whole realm, and one measure of ale and one measure of corn [...]' in order to protect common people from being cheated with the use of arbitrary measures. And the French Revolution famously introduced the meter and the metric system to the world (Alder 2002). Witold Kula (1986) sees conflict over measures as a fundamental part of class struggle. Isabelle Bruno, Emmanuel Didier, and Tommaso Vitale analyse forms of political action that mobilized statistics in struggles of various kinds, which they term statactivism (Bruno et al. 2014) (see also Boris Samuel's contribution in this book).

It is easy to underestimate the work and infrastructure that is required to produce numbers that others find credible. As Laurent Thévenot shows, quantification depends on prior processes of formatting and codifying, all of which require discipline and standardizing. For example, coding, a process for turning disparate qualitative information into standardized 'data' is a complex intervention that requires often hundreds of decisions about whether something is or isn't an instance of some preestablished category. Coding is governed by established, written protocols that must be learned by those making the coding decisions. And all this transformation takes resources, time, training, and imagination—in other words, administrative capacity. Corine Eyraud describes the developmental stages of the indicators used to measure education in France showing how these operate at different analytical levels. And Thomas Amossé shows how France's statistical infrastructure has shifted over time, and how its representative household survey, biographical investigation, and matched panel are statistical practices that articulate specific ways of revealing the world, social science theory, methods and conceptions of public action.

Particularly for English-speaking readers, this volume is a welcome introduction to some of the most important theorists of quantification. Together the original essays it contains offer sophisticated and detailed analyses that reveal the administrative capacity that is required to make and use numbers and their power to reshape our core institutions. They allow us to see how decisions to make and use numbers set in motion consequences that were never envisioned. They reveal how numbers hide their power. After reading this volume, it is impossible to take numbers at face value, a worthy accomplishment.

Wendy Nelson Espeland
Department of Sociology
Northwestern University
Evanston, USA

References

Alder, K. (2002). The measure of all things: The seven-year Odyssey and hidden error that transformed the world. The Free Press.

Bruno, I., Didier, E., & Vitale, T. (2014). Statactivism: Forms of action between disclosure and affirmation. *PaCo: Partecipazione & Conflitto, 7*(2), 198–220.

de Santos, M. (2009). Fact-totems and the statistical imagination: The public life of a statistic in Argentina. *Sociological Theory, 27*(4), 466–489.

Espeland, W. N., & Michaels, S. (2018). The history of 10%: Measures of sexual behavior and the gay rights movement in the U.S. Paper presented at the Swedish Collegium for Advanced Study, Uppsala University, Uppsala, Sweden,

Kula, W. (1986). Measures and men. Princeton University Press.

Miller, P., & Rose, N. (1990). Governing economic life. *Economy and Society, 19*(1), 1–31.

Rodríguez-Muñiz, M. (2017). Cultivating consent: Nonstate leaders and the orchestration of state legibility. *American Journal of Sociology, 123*, 1–41.

Wendy Nelson Espeland is Professor of Sociology at Northwestern University. She works in the fields of culture, organizations, law, and knowledge production, with an emphasis on quantification and accountability. She has written about the effects of quantification on indigenous populations (*The Struggle for Water: Politics, Rationality, and Identity in the American Southwest*, University of Chicago Press, 1998) and, with Michael Sauder, on the effects of university rankings on higher education (*Engines of Anxiety: Academic Rankings, Reputation, and Accountability*, Russell Sage Foundation, 2016). She is also conducting research, with Stuart Michaels, on the relationship between measures of sexual behavior and the gay rights movement. Her articles have appeared in the *American Journal of Sociology*, the *American Sociological Review*, *Law & Society Review*, the *Annual Review of Sociology*, the *European Journal of Sociology*, and the *Annual Review of Law and Social Science*. She has received fellowships from the Russell Sage Foundation, the Australian National University, the Radcliffe Institute for Advanced Study, the Wissenschaftskolleg zu Berlin, and the Swedish Collegium for Advanced Study.

Acknowledgments

The idea for this book originated at a workshop held at the Wissenschaft-skolleg zu Berlin in April 2014. The workshop brought together the Wissenschaftskolleg's Quantification Focus Group and a group of scholars working on quantification selected by the Nantes Institute for Advanced Study (IAS-Nantes) in France. Alain Supiot, who was then the Director of the IAS-Nantes, had proposed this meeting given that in France, under the initiative of Alain Desrosières and other scholars, who were located mainly within the milieu of public statisticians (especially the INSEE—Institut National de la Statistique et des Etudes Economiques), a vibrant field of research on quantification had developed studying statistical tools and their history, conventions of measurement, categories, and classification.

The Quantification Focus Group was composed of the following 2013–2014 Fellows: Bruce Carruthers (Northwestern University, USA), John Carson (University of Michigan, USA), Lorraine Daston (Permanent Fellow at the Wissenschaftskolleg and Director of the Max Planck Institute for the History of Science, Germany), Emmanuel Didier (CNRS, France), Wendy Nelson Espeland (Convenor of the Group, Northwestern University, USA), Tong Lam (University of Toronto, Canada), Andrea Mennicken (London School of Economics and Political Science, UK), Jahnavi Phalkey (King's College London, UK), Theodore Porter (University of California, Los Angeles, USA). The Nantes Group consisted of Thomas Amossé (Centre d'Etudes de l'Emploi), Ota De Leonardis

(Università degli Studi di Milano-Bicocca, Italy), Corine Eyraud (Aix-Marseille University), Samuel Jubé (IAS-Nantes and University of Nantes), Martine Mespoulet (University of Nantes), Robert Salais (Ecole Normale Supérieure de Cachan), Ousmane Sidibé (Bamako University, Mali), Alan Supiot (University of Nantes, IAS-Nantes), Laurent Thévenot (Ecole des Hautes Etudes en Sciences Sociales, EHESS, Paris).

Considering the fecund and lively nature of the exchanges at the workshop, the two groups decided to further intensify their collaboration and engage in the production of a jointly edited book. Andrea Mennicken (then Fellow at the Wissenschaftskolleg) and Robert Salais (Fellow at the IAS-Nantes in 2011–2012 and Fellow at the Wissenschaftskolleg in 2005–2006) were selected as coordinators of the project—a project which eventually resulted in the publication of this volume.

Two further workshops facilitated the development of the book; one of these was held at the IAS-Nantes in April 2015 and one at the ENS Cachan (now ENS Paris-Saclay) in April 2016. We gratefully acknowledge the financial and logistical support we received from the Fellow Forum of the Wissenschaftskolleg, the IAS-Nantes, and ENS Cachan for the organization of these meetings, where earlier versions of the works presented in this edited volume were discussed.

From the Wissenschaftskolleg, four Fellows of the 2013–2014 Quantification Focus Group contribute to this volume: Emmanuel Didier, Wendy Nelson Espeland, Tong Lam, and Andrea Mennicken. From the Nantes Group contributions were received from Thomas Amossé, Ota De Leonardis, Corine Eyraud, Martine Mespoulet, Robert Salais, Ousmane Sidibé, and Laurent Thévenot, which gave room to invite further contributors: Peter Miller (London School of Economics and Political Science), Boris Samuel (Sciences-Po Paris and Fellow at the IAS-Nantes in 2014–2015) and Uwe Vormbusch (FernUniversität Hagen, Germany), who participated also in the meetings held at the IAS-Nantes and ENS Cachan.

This book is the result of a lengthy process of discussions, several rounds of reviews and revisions, copyediting and translation. We warmly thank the contributors to this volume for their unfailing patience, commitment, reactivity, and support, which have made our work on this book a truly collective undertaking. We also thank the editors and editorial managers of the Executive Politics and Governance series at Palgrave Macmillan for their excellent service and assistance.

Andrea Mennicken gratefully acknowledges the financial support provided by the Economic and Social Research Council (Grant Ref:

ES/N018869/1) under the Open Research Area Scheme (Project Title: QUAD—Quantification, Administrative Capacity, and Democracy), co-funded by the Agence Nationale de la Recherche (ANR, France), Deutsche Forschungsgemeinschaft (DFG, Germany), Economic and Social Research Council (ESRC, UK), and the Nederlands Organisatie voor Wetenschappelijk Onderzoek (NWO, Netherlands).

Last not the least, we especially thank the Wissenschaftskolleg zu Berlin and the IAS-Nantes, where the seeds for this undertaking were planted, for their confidence in our capability to handle this project and drive it to its end.

Andrea Mennicken
Robert Salais

CONTENTS

Notes on Contributors

Thomas Amossé is a Public Statistician and Researcher in sociology at the Conservatoire National des Arts et Métiers (CNAM). His research interests revolve around three areas: the socio-economics of work and employment, the sociology of social class, and the socio-history of statistics. He has published several articles, for instance: 'Professions au féminin: Représentation statistique, construction sociale', in *Travail, genre et sociétés*, 2004; 'Hommes et femmes en ménage statistique: Une valse à trois temps', in *Travail, genre et sociétés*, 2011 (with Gaël De Peretti); 'La nomenclature socio-professionnelle: une histoire revisitée', in *Annales, Histoire, Sciences Sociales*, 2013; 'The Centre d'Etudes de l'Emploi (1970–2015): Statistics—on the Cusp of Social Sciences and the State', in *Historical Social Research*, 2016.

Ota De Leonardis is a retired Professor of Sociology at the Department of Sociology and Social Research, University of Milano-Bicocca. She is the President of the advisory board of the Nantes Institute for Advanced Study. She was a visiting scholar at UBA and FLACSO (Buenos Aires), EHESS (Paris), Université de Montréal, Université de Sherbrooke (Quebec), and the Institute for Advanced Study Nantes. She was involved in several national and European research projects, as well as in international research networks promoted by the Battelle Foundation, WHO, and Rockefeller Foundation. Her research focuses on institutions theory and methods, on the management and instrumentation of the public sector, particularly in the social policies field, and on the transformations of the public sphere, citizenship, and democracy. She is the editor with Negrelli and Salais of *Democracy and Capabilities for Voice. Welfare, Work and Public Deliberation in Europe*, Brussels: Peter Lang, 2012, and recently of *Covid-19. Tour du monde*, Paris: Editions Manucius, forthcoming.

Emmanuel Didier is a Full Professor at the Centre Maurice Halbwachs at École Normale Supérieure, Paris, and a member of the Centre for the Study of Invention and Social Process at Goldsmiths, University of London. He is a founding member of EpiDaPo (Epigenetics, Data, Politics), initially a joint research unit of CNRS and UCLA. He taught at the University of Chicago and at UCLA and now teaches at Ecole Normale Supérieure and Ecole Nationale de la Statistique et de l'Administration Economique, both in Paris. He works on the sociology of quantification, statistics, and big data. He wrote several books, including *En quoi consiste l'Amérique, les statistiques le New Deal et la démocratie* (La découverte, 2009), *Benchmarking, l'Etat sous pression statistique* (La Découverte 2013, with Isabelle Bruno), and he edited with Isabelle Bruno and Julien Prévieux *Statactivisme, comment lutter avec des nombres* (La Découverte 2014). Recently, he published *America by the Numbers: Quantification, Democracy, and the Birth of National Statistics* (MIT Press, 2020). Didier was a Fellow of the Wissenschaftskolleg zu Berlin (2013–2014).

Corine Eyraud is Associate Professor at the Department of Sociology at Aix-Marseille University. She is also an Associate of the Maison française d'Oxford and of the Centre for Higher Education Futures (Aarhus University, Denmark). Her work resides at the intersections of public policy, economic sociology, critical management studies, and sociology of quantification. Eyraud was a visiting scholar at Oxford University (Wolfson College, St Antony's College, Department of Politics and International Relations), Aarhus University, and at the Chinese Academy of Social Sciences. She has been involved in several national and European research projects and has been a member of the Reflection Group on 'the Future of Official Statistics—Horizon 2030' organized by Eurostat and the European Political Strategy Centre of the European Commission. Her recent publications include 'Quantification Devices and Political or Social Philosophies. Thoughts inspired by the French State Accounting Reform', in *Historical Social Research*, 2016; 'Réflexions pour une sociologie de la quantification statistique et comptable', in *Entreprises et Histoire*, 2015;

and the books *L'entreprise publique chinoise* (L'Harmattan 1999), *Le capitalisme au cœur de l'État. Comptabilité privée et action publique* (Editions du Croquant, 2013).

Andrei Guter-Sandu is ESRC Postdoctoral Fellow at the Centre for Analysis of Risk and Regulation, London School of Economics and Political Science. He obtained his PhD in International Political Economy from City, University of London, and currently works on a project investigating how managerialist ideas and instruments of quantification change the manner in which public services are governed and controlled. His latest publication includes 'Theorising resilience: Russia's reaction to US and EU sanctions' (with Elizaveta Kuznetsova), in *East European Politics*, 2020.

Tong Lam is Associate Professor of History at the University of Toronto and Director of the Global Taiwan Program at the Asian Institute. His research is on the modern and contemporary history of China, with emphases on empire and nation, governmentality, knowledge production, as well as urban space and ruins. He is the author of *A Passion for Facts: Social Surveys and the Construction of the Chinese Nation-State, 1900–1949* (Berkeley: University of California Press, 2011) which analyses the profound consequences of the emergence of the technologies of the 'social fact' and social survey research in modern China. Lam's ongoing research also examines the prevalence of designer architectures, urban ruins, and derelict spaces in post-socialist China's spectacular and speculative development. As a visual artist, Lam uses photographic techniques to carry out ethnographic studies of contemporary China's hysterical transformation. At present, he is working on a photo essay book on industrial and post-industrial ruins and abandonment from around the world. See here also his book *Abandoned Futures: A Journey to the Posthuman World* (Carpet Bombing Culture, 2013). Lam was a Fellow of the Wissenschaftskolleg zu Berlin (2013–2014).

Andrea Mennicken is Associate Professor of Accounting at the London School of Economics and Political Science and Co-Director of the Centre for Analysis of Risk and Regulation (LSE). She is associate editor of *Accounting, Organizations and Society* and *Valuation Studies*. Her work has been published in *Accounting, Organizations and Society, Financial Accountability and Management, Foucault Studies, Organization Studies, Sociologie du travail, Actes de la recherche en sciences sociales* and different edited volumes. Recently, she has completed work on an international research project funded by the ESRC under the Open Research Area Programme for the Social Sciences, exploring the changing relationships between quantification, administrative capacity and democracy in health care, correctional services, and higher education. She was a Fellow of the Wissenschaftskolleg zu Berlin (2013–2014).

Martine Mespoulet is Professor emerita of Sociology at the University of Nantes and Researcher at the CNRS Centre of Sociology in Nantes (CENS). Her research focuses on two main fields: the use of quantification in public action and state and societies in Central and Eastern Europe. Her work has appeared in a wide range of sociology and history journals. Her books include *Construire le socialisme par les chiffres* (Ined, 2008), *L'anarchie bureaucratique. Statistique et pouvoir sous Staline* (La Découverte, 2003) with Alain Blum and *Statistique et révolution en Russie. Un compromis impossible (1880–1930)* (Presses Universitaires de Rennes, 2001).

Peter Miller is Professor of Management Accounting at the London School of Economics and Political Science, and an Associate of the Centre for Analysis of Risk and Regulation. He is an editor of *Accounting, Organizations and Society* and has published in a wide range of accounting, management, and sociology journals. He co-edited *The Foucault Effect* (1991), *Accounting as Social and Institutional Practice* (1994), and more recently *Accounting, Organizations, and Institutions* (2009). In 2008, he published (jointly with Nikolas Rose) *Governing the Present*.

Robert Salais is an Associated Researcher of the CNRS interdisciplinary laboratory 'Institutions and Historical Dynamics of the Economy and the Society' (IHDES). Working at the crossings of history and socio-economy, he is one of the founders of the economics of convention. He was a visiting researcher at the Institut Universitaire Européen, the Wissenschaftszentrum Berlin für Sozialforschung (WZB), the Centre Marc Bloch at Berlin (CMB), Warwick University, and the Norwegian Social Research (NOVA) at Oslo. He has published in a wide range of economical, historical, sociological and law reviews and edited books. He coordinated (EUROCAP, CAPRIGHT) or participated in several European research programmes on employment and social policies. Main books include *L'invention du chômage* (with Nicolas Baverez and Bénédicte Reynaud), Presses Universitaires de France, second edition 1999;

Worlds of Production (with Michael Storper), Harvard University Press, 1997 (French edition 1993); *Europe and the Politics of Capabilities* (editor with Robert Villeneuve), Cambridge University Press, 2005; *Le viol d'Europe. Enquête sur la disparition d'une idée*, Presses Universitaires de France, 2013; *Qualitätspolitiken und Konventionen* (editor with Marcel Streng und Jakob Vogel), Springer, 2019. He has received fellowships from the Wissenschaftskolleg zu Berlin (2005–2006) and the IAS-Nantes (2011–2012).

Boris Samuel is Researcher at the French National Research Institute for Sustainable Development (IRD), Centre d'études en sciences sociales sur les mondes africains, américains, et asiatiques (CESSMA), University of Paris. His research examines statistical and macroeconomic practices to provide an analysis of social relations and political trajectories. His current research projects deal with the new technologies of the social state in Morocco, with planning practices in Africa, and with the role of macroeconomists and statisticians in francophone West-African countries. He holds a PhD in political science from Sciences Po Paris (2013). Before joining the academy, he has worked for ten years as an expert in statistics, public finance and macroeconomics for international organizations and governments in Africa. He currently teaches at Paris Dauphine University and at the Faculty of Human Sciences and Education Sciences of Bamako. Boris Samuel is co-editor of *Politique Africaine* and member of the editorial board of *Sociétés contemporaines*. He was a Fellow of the IAS Nantes in 2014–2015.

Ousmane Oumarou Sidibé is Professor of Law at the University of Bamako in Mali. His research interests centre around employment law and social security law and the use of performance indicators in public policies in Africa. After graduating from the Ecole Nationale d'Administration of Mali in 1978, he earned a doctorate in Law (option Employment Law) in Bordeaux in 1983. He previously held several positions: Minister for Employment, Civil Service and Labor, managing director of the Ecole Nationale d'Administration of Mali, Commissioner for Institutional Reform, African Program Coordinator for the participation of workers. Along with his administrative and political activities, he continued his academic career teaching at the University of Bamako since 1992.

Laurent Thévenot is Directeur d'études at the École des hautes études en sciences sociales. Following his research on 'social coding' and 'investment in (conventional) forms' he initiated with Luc Boltanski what has become known as pragmatic sociology: *On Justification: Economies of Worth*, Princeton University Press, 2006 (French edition 1991). He also co-founded with economists the critical institutionalist current of the economics of convention. More recently in *L'action au pluriel. Sociologie des régimes d'engagement* (La Découverte, 2006), he has identified several constructions of commonality (the liberal grammar of interests, the grammar of personal affinities to common-places) and various modes of engaging which confer both consistency and dynamism on the community and the personality, while allowing to extend the convention theory's critical analysis of power and oppression. His empirical investigations have focused on governance by standards and objectives, politics of quantification, and have drawn on collaborative programmes comparing architectures of communities and criticism in Western Europe (*European Journal of Cultural and Political Sociology*, special issue on 'Politics of Engagement in an Age of Differing Voices' co-edited with Eeva Luhtakallio, 2018 5[1–2]), Russia (Revue d'Etudes Comparatives Est-Ouest, special issue on 'Critiquer et agir en Russie' co-edited with Françoise Daucé et Kathy Rousselet, 2017 48[3–4]) and the United States (*Rethinking Comparative Cultural Sociology: Repertoires of Evaluation in France and the United States*, co-edited with Michèle Lamont, Cambridge University Press, 2000).

Uwe Vormbusch is Professor for the Analysis of Contemporary Societies at the Institute of Sociology of the FernUniversität Hagen in Germany. He has been a visiting scholar at the London School of Economics and Political Science in 2009. His research areas are the sociology of quantification and valuation; economic and financial sociology; and general sociology. Vormbusch recently completed a research project on the quantified-self movement funded by the German Research Foundation (DFG). Recent publications include *Wirtschafts- und Finanzsoziologie. Eine kritische Einführung*, Springer VS, 2019; Kalkulative Formen der Selbstthematisierung und das epistemische Selbst, in: *Psychosozial*, vol. 41, issue 3/2018, pp. 16–35 (with Eryk Noji); *Die Herrschaft der Zahlen. Zur Kalkulation des Sozialen in der kapitalistischen Moderne*, Campus Verlag, 2012; *Soziologie der Finanzmärkte* (with Herbert Kalthoff), transcript, 2012.

LIST OF FIGURES

LIST OF TABLES

CHAPTER 1

The New Politics of Numbers: An Introduction

Andrea Mennicken and Robert Salais

The business of government is increasingly run with a calculator to hand. Both policymaking activities and administrative control are increasingly structured around calculations such as cost-benefit analyses, estimates of social impacts and financial returns, measurements of performance and risk, benchmarking, quantified impact assessments, ratings and rankings, all of which provide information in the form of a numerical representation. Through quantification, public services and policies have experienced a fundamental shifting from "government by democracy" towards "governance by numbers", with implications not just for our understanding of the nature of public administration itself, but also for wider

A. Mennicken (✉)
Department of Accounting and Centre for Analysis of Risk and Regulation, London School of Economics and Political Science, London, UK
e-mail: a.m.mennicken@lse.ac.uk

R. Salais
Institutions et dynamiques historiques de l'économie et de la Sociète (IDHES), École Normale Supérieure Paris-Saclay, Gif-sur-Yvette, France

1

debates about the nature of citizenship and democracy. This book scrutinizes the relationships between quantification, administrative capacity and democracy across different policy sectors and countries. In so doing, it seeks to offer unique cross-national and cross-sectoral insight into how managerialist ideas and instruments of quantification have been adopted and how they have come to matter.

More than thirty years ago, Alonso and Starr (1989) edited the by now classic "The Politics of Numbers", which was amongst the first books that scrutinized relations between quantification and democratic government in North America (but see also Cohen, 1982). Amongst other things, Alonso and Starr's collection of essays showed how government statistics had become vital to pursuing essential goals of a democratic polity, such as accountability and representation of diverse interests. The book also highlighted that a nation's number system creates new invisibilities (e.g. of minorities). It showed how and to what extent political judgements and bias are embedded in the statistical systems of the modern state, or as Rose (1991, p. 675) put it, "how the domain of numbers is politically composed and the domain of politics is made up numerically".

In parallel, also European scholars had begun to question the relation between numbers and democracy, and between government and numbers. In France, in particular Alain Desrosières and his colleagues, including Robert Salais and Laurent Thévenot, who contributed to this volume and at the time worked with Desrosières for the French National Institute of Statistics and Economic Studies (INSEE), interrogated the conventions and classifications underpinning the production, use, and consequences of statistics (e.g. Desrosières, 1987 [1983]; Desrosières, 1998 [1993]; Thévenot, 1979, 1981; Salais, 1986; for useful overviews see also Diaz-Bone & Didier, 2016; Diaz-Bone & Salais, 2011). In the UK, it was first and foremost Anthony Hopwood who triggered a critical-reflexive turn in the study of numbers, focusing on the multifaceted roles of accounting in representing and intervening in social and organizational life (see e.g. Hopwood, 1983; Hopwood & Miller, 1994; S. Burchell et al., 1980; but see also Miller, 1992; Miller & Rose, 1990). Further important early works on the production, history and influence of statistics were conducted by scholars with backgrounds in the history and philosophy of science (Daston, 1988; Gigerenzer et al., 1989; Hacking, 1990; Krüger et al., 1987; Porter, 1986).

Since then, particularly over the past fifteen years or so, there has been an increased interest in, and surge of, articles and books on governance by numbers, albeit in different fields (for a review see Mennicken & Espeland, 2019). In the field of public administration and public policy, especially the rise of New Public Management led to heightened attention to the roles of performance indicators in the governance of public services (see also Bruno et al., 2016; Bruno & Didier, 2013; Supiot 2015). Here, consideration has not only been given to problems of measurement (i.e. the limits of performance measures to capture what matters), but also gaming (Bevan & Hood, 2006; Strathern, 1997), reactivity (Espeland & Sauder, 2007) and the "audit explosion", characterized by the rise and expansion of formal systems of performance evaluation and assessment aimed at making elusive notions, such as quality, auditable (Power, 1997; Strathern, 2000).

Bevan and Hood (2006) showed how governance by targets changed the behaviour of individuals and organizations in the English National Health System (NHS). They coined the famous phrase "hitting the target and missing the point" (ibid., p. 521). Building on Goodhart's eponymous law that "any observed statistical regularity will tend to collapse once pressure is placed on it for control purposes" (Goodhart, 1984, p. 94, cited in Bevan & Hood, 2006, p. 521), they queried to what extent "governance by targets" subverts public service ethos, contributes to output distortions and a general narrowing of quality definitions. Also, Strathern showed for the case of the higher education sector in the UK that "when a measure becomes a target, it ceases to be a good measure" (Strathern, 1997, p. 308). More recently, Espeland and Sauder (2007) investigated reflexive interactions between people and measures by looking at the reactivity of US law school rankings. Amongst other things, they showed that these rankings contributed to a proliferation of gaming strategies, the redefinition of work and a redistribution of resources (Espeland & Sauder, 2016; but see also Bruno et al., 2016). Research has also drawn attention to new practices and strategies of gaming and manipulation that have emerged in the academic world over the past few years (Biagioli & Lippman, 2020a). These practices are different from the predictable gaming of academic performance indicators and may take the form of "massaging the definition what counts as a 'successful student' in metrics about schools' performance, or of what counts as a 'peer-reviewed' paper in faculty evaluation protocols" (Biagioli & Lippman, 2020b, p. 1).

Rankings, ratings and other governance indicators, such as the Human Development Index, Gender Inequality Index or Social Progress Index, rest on multiple levels of aggregation (Mennicken & Espeland, 2019, p. 232). They are seductive as they allow for easy comparison and ranking of countries and organizations, which can lead to oversimplification and homogenization if not grounded in qualitative, locally informed knowledge production (Davis et al., 2012; Merry, 2016; Rottenburg et al., 2015; Salais, 2006; Supiot, 2015; Thévenot, 2009). This and the rise and proliferation of "governance by numbers" makes it all the more important to understand how such numbers are produced, calculated and aggregated, and with what consequences.

The essays collected in this volume interrogate what has changed in the relation between numbers and democracy, and between government and numbers, since the publication of "The Politics of Numbers" (Alonso & Starr, 1989). What is "new" in the politics of numbers and our approach to their study?

First, we observe an unprecedented expansion, acceleration and intensification of quantification not only in political life but also in everyday life. To a large extent such an expansion and intensification of quantification has been aided by the rise of new computer technologies for (big) data collection and data processing. Such new digital technologies, including machine learning algorithms, have changed how public administrations deploy resources and make decisions. They promise new possibilities of governing, as the contribution of Lam in this volume on the rise of China's social credit system shows (Chapter 3). They also transform how we understand ourselves, what we attend to and consider important, as Vormbusch's study of the "quantified self" movement in Germany in this volume demonstrates (Chapter 4).

Second, we witness an increasing decline in the "trust in numbers" (Porter, 1995), an antipathy to government statistics, and a disillusionment and tiredness with New Public Management's "governance by targets". In the age of "post-truth" politics, many have come to believe that numbers are manipulated. As Davies (2017) argues, many well-recognized indicators have lost their legitimacy. He quotes a study from the US that discovered that 68% of Trump supporters distrusted the economic data published by the federal government. For the UK, Davies highlights "that a research project by Cambridge University and YouGov looking at conspiracy theories revealed that 55% of the population believes that the government 'is hiding the truth about the number of immigrants living here'" (Davies, 2017). Such distrust in numbers is further fuelled by the increasing distance between two objectives of public services and policies that service users have experienced over the past two decades: an increasing distance between attempts aimed at truly improving the situations of the people concerned, on the one hand, and the maximizing of quantitative performance, on the other hand. To counteract such developments, in the UK, an independent regulatory body was established in 2016, the Office for Statistics Regulation, to safeguard the "trustworthiness, quality and value of statistics" and to assure that "statistics serve the public good" acknowledging that statistics should meet the needs of a much wider range of users than public policy-makers and parliamentarians (see https://osr.statisticsauthority.gov.uk/publication/osr-vision, accessed 30 October 2020).

At the same time, and ironically, the presence of numbers in our lives seems to have never been greater. Such an increase in quantification can be the result of an "overproduction" of an administrative policy, a reaction (or even overreaction) to crisis, cognitive bias and uncertainty (Maor, 2018, 2019). As Miller points out in the Afterword to this book (Chapter 14), numbers have acquired an unassailable power, particularly in the wake of the Covid-19 pandemic. As he writes, "the phrase 'follow the science', and its numerical counterpart the 'R' number, has attained an ascendancy that none of us could have imagined only a few months ago". The R number, and associated statistics, such as the 14-day or 7-day notification rate of newly reported COVID-19 cases per 100,000 of the population, have been used as devices to regulate our lives, to make decisions about whether schools, shops, restaurants, and much else besides, should be open or closed, whether, and in what constellation, one can meet with others or not. Miller's contribution to this volume also shows that the R number is far from being a straightforward measure. It is not only incredibly difficult to calculate, it can also be potentially misleading or at least uninformative, because it does not necessarily tell you what is happening in your local area.

The essays collected in this volume query the rise and spread, as well as resistance to, and disappointment in, the tools of quantification that have come to govern our lives. In so doing, they do not so much ask what quantification *is*. Rather, they are interested in describing and analysing what quantification *does*. This volume is concerned with the tracking and unpacking of various practices of quantification and their manifold consequences in different contexts of public and private life. Such a close-up focus on quantification practice, or "quantification in action" to para-phrase Hopwood (1983), contributes to our understanding of the roles of numbers in public policy and public administration, and scholarship on quantification more generally, in at least four distinct ways.

First, it helps in developing a more nuanced understanding of the capacities and roles of the various calculative practices that have come to populate different domains. The contributions to this book highlight that quantification does much more than provoke gaming and reactivity. They offer valuable insights into the inner workings of (different) accountability regimes, their changing nature and the emergence of new regulatory spaces and practices.

Second, the contributions show how quantification is implicated in dreams and schemes of doing things differently, of creating new worlds and bettering society. In many respects, it appears now that the socialist countries have been pioneers in the political use of numbers. In the USSR, statistics played not only a key role in the operationalization of central planning; statistics were also inextricably linked to the articulation and specification of the "socialist dream" (see Mespoulet's contribu-tion, Chapter 2). Likewise, China's current social credit system has to be seen as part of a bigger, long-term commitment to creating "good" citizens via the controlling power of numbers (see Lam's contribution, Chapter 3). It is often such dreams and schemes, or programmatic ambi-tions, as Miller and Rose (1990) would put it, that animate the rise and spread of numbers and need to be attended to, rather than (unintended) behavioural effects, if we want to understand what keeps the machinery of quantification running despite its continuous failings.

Third, the book is concerned with providing deeper insight into how quantification travels. This book brings together works on governing by numbers by leading and emerging French, German, Italian, British and Anglo-American scholars. In so doing, the book makes not only French, Italian and German works on quantification accessible to an English-speaking audience. It also enhances understanding of the implication of quantification in different modes of regulating and governing. It gives insight into how different forms of quantification have been developed and deployed in the public administrations of varied countries (China, Mali, Guadeloupe, USSR, UK, France) as well as European institutions in different historical periods. The contributions collected here examine whether and how quantification—the shift from government by democ-racy to governance by numbers—has given rise to a shift in demands on administrative capacities of public administrations (e.g. expectations regarding analytical skills, regulatory capabilities, legal staff or financial resources). To what extent do tools of quantification advance the capacities of public administrations, regulatory agencies, and other organizations across sectors and states in terms of being able to monitor and steer?

Fourth, and finally, the book revisits the power of numbers, and the changing relationship between numbers and democracy. It engages two central programmatic strands of research in a critical dialogue with each other: namely Foucault inspired studies of governmentality (see here in particular the contributions by Miller, Guter-Sandu and Mennicken, Vormbusch, Lam and De Leonardis), which first and foremost flourished in the English-speaking world, and studies of state statistics and economic conventions, which have their origins in France, in particular the INSEE (Institut national de la statistique et des études économiques) (see here the contributions by Thévenot and Salais, two of the founding fathers of the economics of conventions approach, and the chapters by Samuel, Mespoulet, Amossé, Didier and Eyraud). Such a critical dialogue helps advance debates about the power of numbers and relations between quantification and democracy, a topic that in our view is in need of renewed discussion and theoretical reflection. For when quantification becomes a "technology of government", power may be understood, expressed and resisted differently, depending also on the specific characteristics of the instruments of quantification used and what these afford actors to do. For numbers are not only a device of rational rule, public administration and domination. As the contributions by Thévenot (Chapter 7), Samuel (Chapter 11) and Salais (Chapter 12) in this volume demonstrate, they can also aid social mobilization and empowerment, a theme that particularly more recently published works on quantification have often overlooked (but see the recent French works on statactivism published by Bruno et al., 2014).

While some recent literature, particularly social studies of finance, has devoted a lot of attention to the technological infrastructures of calculation, it has tended to neglect or downplay the roles that political ideas, programmes or myths play in articulating and mobilizing them. Foucault-inspired studies of the governmentality of quantification (Miller, 1992, 2001; Miller & O'Leary, 1987; Miller & Rose, 1990) emphasize that we need to attend to both instruments and ideas of calculation, and the interplay between them, as it is through that interplay that each dimension finds its conditions of operation. Similarly, also the economics of convention approach (Desrosières, 1985; Diaz-Bone, 2016; Diaz-Bone & Salais, 2011; Eymard-Duvernay, 1989; Salais, 2016; Salais & Thévenot, 1986; Thévenot, 2011) stresses the significance of ideas and conventions for the shaping of practices of quantification, and it assumes a plurality of possible ways in which numbers can come to govern. Yet, as Demortain (2019) points out, whereas governmentality studies have tended to approach quantification in terms of its (variable) political rationalities and disciplinary effects, the economics of convention approach has paid more attention to the collective mobilization capacities it offers (or not). We argue that it is important to understand processes and consequences of quantification from both angles. The contributions to this volume show that it is important to look at quantification as both a "technology of government, that reproduces a power structure" and as "a tool that can facilitate political action towards this structure, and its change" (Demortain, 2019, p. 974).

In our view, the time is ripe for such a joined-up, historically and contextually sensitive approach to get to grips with the multiplicity of quantification and "governing by numbers". This volume assembles contributions from different disciplines and contexts to deepen our understanding of "quantification in action". It is structured into three main parts. The first (Part I), explores *Quantification as Utopia*. Here, contributions scrutinize the implication of quantification in imaginaries of the future, "the ideal city" or "imagined community" as Anderson (1983) would say, that possesses highly desirable or nearly perfect qualities for its citizens. Statistics and other forms of quantification offer the promise of anchoring such ideals and imaginations in something tangible (Davies, 2017). But under what conditions and with what consequences?

The second part (Part II), revisits the roles of numbers in *Politics of Evidence*. Here, contributions probe the facticity of numbers and interrogate to what extent evidence-based quantification is in itself utopian. The third and final part (Part III, entitled *Voicing for Democracy*) scrutinizes changes in the relation between quantification and democracy. Nowadays democracy is less than ever a peaceful and established social activity. New relationships of power have emerged that dissimulate themselves under the cloak of technicalities which can lead to a weakening of democratic rights. Contributions in this part query the implication of quantification in the rise of what Crouch (2004) has termed "post-democratic society", a society "that continues to have and to use all the institutions of democracy, but in which they increasingly become a formal shell". Yet, we also ask what would be needed to disrupt this trend. Under what conditions can the exercise of political voice be made possible in and through quantification (see here in particular Samuel's and Salais' contributions, Chapter 11 and Chapter 12, but see also the contribution by Guter-Sandu and Mennicken, Chapter 10)?

In the following, we introduce each of these parts and the individual contributions that make them up in more detail to help readers navigate and comprehend the overall architecture of the book. We conclude our overview with a return to the question of where the economics of convention approach meets Foucault and Foucauldian studies of governmentality and what such a joining up brings to the furthering of debates about quantification and governing by numbers.

QUANTIFICATION AS UTOPIA

As Frank Manuel and Fritzie Manuel's (1979) witty and erudite book *Utopian Thought in the Western World* shows, imagining utopias has been a long-term characteristic of the western world since the Ancient Greeks. Yet, types, subjects, and political uses of utopia have changed over time and place. Whereas in the seventeenth century "utopia came to denote general programs and platforms for ideal society, codes and constitutions that dispensed with fictional apparatus altogether" (Manuel & Manuel, 1979, p. 2), more rationalist, systematic utopias appeared around the end of the eighteenth century. According to Manuel and Manuel (1979, p. 2), "the means of reaching utopia was transformed from an adventure story or a rite of passage to Elysium into a question of political action". The way of attaining the ideal city came to affect the very nature of the city itself (Manuel & Manuel, 1979, p. 3). This is still the case today. Utopias can reveal themselves as good or bad, but that is not our concern here. We are interested in the different ways utopian thought has influenced, and continues to influence, the building of the knowledge a society has of itself; concepts of the common good; the making of policies and their implementation; and the roles that come to be ascribed to citizens and their formatting.

Utopias could be said to set forth a horizon of expectations that is believed to open a new future viewed as enlarging and facilitating the actions of those who form it, ultimately aimed at making them more powerful. These can be scientists, philosophers, ideologues, managers, politicians, all people who more or less occupy the role of "conseillers du prince" in government matters. Policy-driven quantification often carries with it a utopian perspective. It is implicated in the promise and dream of creating an infrastructure that can facilitate the making of a new (better) order. In this respect, quantification is deeply ideational (Power, 2019; Miller & Rose, 1990); and it is often utopian thought that motivates its production and expansion.

Quantification is further sustained by beliefs in its own rationality. Simply put, the belief that often motivates in the political domain the recourse to quantification is the assumption that *more equals better*. Qualities are reduced to quantities, and difference is transformed into magnitude (Espeland & Stevens, 1998). As Espeland and Stevens (1998, p. 316) have highlighted, this transformation "allows people to quickly grasp, represent, and compare differences". It offers "ways of constructing proxies for uncertain and elusive qualities" (ibid.). For us the key issue here is not so much of a technical nature (i.e. how this transformation is achieved), but concerns the shift in cognition it entails. How can we be sure that the resulting loss of information, the moving away from in-depth and shared qualitative knowledge, will not affect the very way political decisions are taken, the way political objectives are formulated, motivated and legitimized? How is it conceivable not to worry about the claim of a universal toolbox that is able to emancipate itself from all material, socio-historically-built and nationally-rooted specificities whatever the domain in which we are acting, as if such particularities should have no meaning and impact?

For by their very nature utopias cannot be achieved. The political and ethical prudence (Raynaud & Rials, 1992) should be to recognize this impossibility of realization and to conceive of utopian thought in terms of a series of markers that guide, but not determine, an endless pragmatic progress towards something that can only be loosely referred to. The problem with any utopia begins when the prince's advisors and the prince himself try to implement it as it is formulated, take it literally, as a blueprint for action and to transform social reality and expectations so that they come to strictly obey to the precepts of the utopia in question. In that case, far from liberating, utopia comes to confine freedom and initiative in a straitjacket, eventually destroying them and utopia itself.

In this volume, Martine Mespoulet (Chapter 2) examines how statistics and national accounting came to be invested with dreams and schemes aimed at creating a new socialist society and a new human being. Statistics should help symbolically construct the (new) Soviet social and economic world, and reform collective spirit and behaviour. Mespoulet traces the tensions and twists and turns this involved. She shows how statistical debates were mingled with internal political rivalries and rifts. The new Bolshevik state had inherited an already well advanced statistical apparatus that had been developed in the Tsarist Empire in exchange with other western European countries. How could such an apparatus be adapted to the new political representations of society, where numbers were not only an information and decision support tool, but also an instrument of power designed to prove the soundness of state action?

The old guard of statisticians came to clash with the Bolsheviks. For the Bolsheviks, the production of figures was to be controlled by their immediate practical applicability, that is their usefulness in guiding actions and decisions. The autonomy of statistics as an independent science was thus questioned. The search for information became selective and had to satisfy political choices rather than statistical laws. As Mespoulet writes, "statistical law came to be perceived as contradicting the principle of political action, as it portrayed social processes as fatalistic [random, added] and the effects of political action on society as illusory". A new quantification language had to be invented which did not resort to statistical theory as a resource (such as the law of large numbers and randomization). Yet, although national accounting seemed to provide such a new language (the accounting for input and output), statistics were never abandoned, Mespoulet shows. Statisticians learned to adapt to the new world. They "became a kind of 'right hand' providing the economistic planners with statistics needed for elaborating the plan, for instance for building blocks of flats, schools or leisure facilities". Furthermore, a whole new area of statistics developed, based on economic forecasting and the construction of indexes. And especially from the 1950s onwards, Soviet statistics began to increasingly resemble those used in capitalist countries. As Mespoulet remarks, "the Soviet socialist system [...] had to demonstrate its superiority in the same economic and social fields as the capitalist countries, which meant that they tended to adopt the same tools, while re-interpreting their uses". What separated socialist statistics from their western capitalist counterparts was thus far from clear. Such separation had to be actively forged, time and again and was often rooted in similarity.

Also Tong Lam (Chapter 3) investigates the multifaceted roles of numbers and calculative expertise in state-building, albeit of a different kind. He turns our attention to China's new social credit system, a national system that is currently being developed under Xi Jinping's administration as part of the project of "The Chinese Dream" aimed at rejuvenating the Chinese nation. For long, China's political and intellectual elites have had the desire to create national citizens, motivated and enlightened enough to participate in China becoming, as Lam puts it, "a unified body politic to counter the encroachment of foreign countries". More than one century later, the dream of engineering new citizens is being drastically reformulated. Moving away from a paper-based bureaucratic surveillance system that rests on the maintenance of personal dossiers, Lam shows how the new social credit system seeks to track, evaluate and modify the financial, moral, social and political behaviour of citizens and companies with the help of new digital "evaluative infrastructures" (Kornberger et al., 2017).

Lam highlights that this new system seeks to modulate behaviour not via oppression, but gamification, individual responsibilization and incentivization. Drawing on Foucault's writings on governmentality, he points out that this system is "a technology of subjectivity and citizenship that seeks to calibrate and modify the behaviour of individuals and groups, compelling them to align themselves with the desired social and political order as defined by the state". Power is thus exercised indirectly through what Foucault referred to as "the conduct of conduct" (Foucault, 1991 [1979]). Responsibility and self-regulation are instilled through a credit-point based reward and punishment system that affects career and promotion prospects, possibilities of travel and mobility, and much else besides. The numerical reward system creates incentives not to simply obey, but to maximize one's score so as to enlarge one's field of possibilities. The new focus, Lam argues, "is no longer on ideological purity for political purposes but on trustworthiness as a basic condition for economic efficiency". In this respect, Chinese policy thinkers have come to share many of the assumptions of (Western) rational choice and game theorists. Lam cautions us to be careful not to demonize China as the foreign Other, as the social credit system looks "at once dystopian and strangely familiar". Despite the many differences that exist between China and Western market democracies, Lam emphasizes that the two sides converge significantly in how their corresponding surveillance infrastructures have produced a new governing paradigm that, according to Zuboff, has replaced "the engineering of souls with the engineering of behavior" (Zuboff, 2019, p. 376, quoted in Lam).

Lam's contribution prepares the ground for a number of questions on which other chapters in this volume seek to shed light. First, we should not forget that the effectiveness of the system relies on its ability to adequately capture the behaviour it seeks to modify, which is by no means a trivial (and often unattainable) task. Second, following the economics of convention (Diaz-Bone and Salais, 2011) social coordination does not necessarily obey to formal institutions, but is shaped by shared informal conventions of appropriate conduct. Such conventions are much more stable in the long run than the recourse to formal institutional rules. Hence, governing via credit scores might be much more difficult to achieve than envisaged, and we need to pay attention to the consequences of such uncertainty. Relatedly, we should also not underestimate people's ability to game the system's rules. Third, and finally, the chapter invites us to critically revisit and unpack what neoliberal governing entails. It reminds us of neoliberalism's multiple manifestations and the increased blurring of differences between East and West.

Uwe Vormbusch's chapter (Chapter 4) transposes us from dreams and schemes of national quantification to the Quantified Self (QS) movement. Born in California, the movement aims to obtain self-knowledge through self-tracking, using new digital mobile and wearable technologies. Self-tracking as such is not new (see e.g. the history of diaries and bathroom scales). Yet, digitization and automation equip quantifications of the self with new possibilities, allowing for ever more detailed and accelerated measurement that can be shared, compared and circulated on the web. Vormbusch explores how such quantifications bring forth new "taxonomies of the self" drawing special attention to the diversity of representational forms and moral conflicts contained in such taxonomies. Combining a Foucauldian approach (e.g. Miller, 1992) with studies of the economics of convention (Diaz-Bone & Salais, 2011; Boltanski & Thévenot, 2006 [1991]), which have highlighted the practical capabilities of individual actors enmeshed in conventions, he argues that self-quantification emerges as a contemporary "institution of the self" that does not displace but co-exists with other established "technologies of the self" (Foucault, 1988), such as religious confession and therapeutic and psychoanalytic approaches to identity and authenticity.

Following Foucault, Vormbusch considers power "as a productive network which runs through the whole social body, much more than as a negative instance whose function is repression" (Foucault, 1980, pp. 118). He stresses that self-measurement does not only lead to an intensification of surveillance, coercion and self-discipline. It also opens up new spaces for self-discovery and self-modulation. It is this tension between autonomy and subordination that he is interested in further exploring. Vormbusch shows that self-quantifiers are not necessarily using new technologies and data uncritically. He portrays them as capable and reflexive actors (Boltanski & Thévenot, 2006 [1991]) who are actively inventing and manipulating technology to explore who they are and could become. To be sure, such use of technology is not without problems and conflicts. Vormbusch shows how quantifying your self is as much about coping with new forms of cultural incertitude and precariousness as well as keeping oneself "fit" and adept to today's neoliberal "flexible capitalism" (Boltanski & Chiapello, 2007).

There is a striking proximity between the social utopias of the self-quantifiers and the Chinese Dream of forging the body into a new object of knowing and governing via digital tracking, although both are situated at opposing ends of the spectrum of neoliberal governing. In the former case, we are dealing with a network of supposedly autonomous individuals who are seemingly free to choose what to measure and how to adjust their behaviour accordingly; in the latter case, we are dealing with a highly centralized apparatus of authoritarian control. Yet, both resort to benchmarking as a key instrument of neoliberal, market-oriented governance (Bruno & Didier, 2013).

The dream of the self-quantifiers is to make visible the self, to make it comparable and manageable, which requires classification, scales of evaluation, the decontextualization of observations and the application of the same methodologies for all. How can one be sure not to lose essential information about the self? The self-quantifiers Vormbusch interviewed oscillate between confidence in numbers and worries, even despair, about the numbers' ability to grasp their (distinctive) selves. The metrics introduce an irreducible distance between forms of objectified external judgement and quests to learn more about ones' (unique) self. Such distance is also problematized in the contribution by Ota De Leonardis (Chapter 5).

De Leonardis examines measurements of inequality and their implication in dreams of "indifferent power". She shows how attempts aimed at quantifying poverty and inequality were accompanied by a semantic shift in which the meaning of inequality as a (historical) bond of dominationand subjection was being obscured. Inequality as defined by quantification, she argues, "tends to designate a distributive difference, a gap, a disparity: a distance, and no longer a tie". It has no longer a relational meaning that refers to power relations between unequal people, where a bond is recognized between the "high" and "low" that creates for both sides obligations and claims. She focuses in her analysis on the dispositif of the *threshold*, in particular poverty thresholds, and their effects on redefining welfare policies in Europe (above all Italy).

She highlights that "threshold" is not only a quantitative measure but also a spatial metaphor which shapes social spaces with divisions and separations. It is aimed at "governing at a distance" as Miller and Rose (1990) would say. De Leonardis distinguishes between space and territory (or place), a distinction also developed by Supiot (2008). Space refers to abstract geometries (distance, direction, size, shape, volume) detached from any material, living, cultural and historical contents. De Leonardis interrogates how a "spatialized" configuration of notions of inequality, helped by measurement and the definition of quantitative poverty thresholds, represents the negation of any relationship between privileged and deprived people, and especially one of domination. As she writes, when inequality is captured in a quantitative format, a poverty threshold, the "vertical configuration, which anchored inequality to burning issues of power, politics, and institutions, is being obfuscated". Questions of inequality are reduced to "a comparison between linear positions and intended merely as a matter of plus or minus, more or less, yes or no, according to a binary code". De Leonardis emphasizes that quantitative poverty thresholds render inequality relative, and no longer relational (see here also Townsend's definition of relative poverty from 1979, quoted in De Leonardis). Inequality loses its "absolute" that is, its societal and political, dimension, as Sen (1983) would put it. Such "flattening effects" of quantification come also to light in other contributions to this volume. Quantification is thus far more than an instrument to envisage and realize a new, "better" world. Quantification technologies, such as poverty measurements or social credit scores, reconstitute the very object they are asked to help create ("the ideal city" or "the ideal citizen"). As Espeland and Stevens (1998, p. 323) write, quantification "reconstructs relations of authority, creates new political entities, and establishes new interpretive frameworks".

Part II of this volume, to which we now turn, takes a closer look at the implication of quantification in "politics of evidence"—attempts aimed at undergirding policies with an "undebatable truth" produced via numbers. The creation of such truth is not only difficult and laborious. What counts as an undisputable fact, and the ways how such facts are produced, have changed over time, and therewith also the very meaning of facticity itself.

THE POLITICS OF EVIDENCE

Numbers have come to be integral to how democracy is justified and operationalized (Rose, 1991). To paraphrase Rose (1991, p. 684), to govern legitimately does no longer mean to govern at the mercy of opinion and prejudice, but to govern in the light of quantifiable facts (see also Porter, 1995). Although "evidence-based policy making" has a long history that spans several centuries, it gained renewed popularity—particularly in the Anglo-Saxon world—with the rise and spread of New Public Management. Evidence-based policy making is based on the assumption that public policy decisions should be grounded in rigorously established "objective evidence" (facts), rather than "common sense" or ideology. To become fact, knowledge should be detached from both the context of observation in which it was generated and contemporary theoretical controversies (that are relegated to the rank of ideologies) (Salais, 2016).

Daston (1992) has coined such objectivity as "aperspectival". Numbers are often seen as a crucial element in realizing an aspiration to "escape from perspective" (Daston, 1992) and to obtain a univocal, impersonal interpretation of the phenomena around which political decisions come to be framed (Samiolo, 2012). The objectivity of numbers, in turn, following Porter (1992), is rooted in standardization, a process whereby decisions are "linked to replicable calculative methodologies which are seen to transcend individual subjectivity and deemed universally applicable" (Samiolo, 2012, p. 383). Yet, how the objectivity of numbers is generated, stabilized or disrupted, is culturally contingent and context-specific (Fourcade, 2011; Samiolo, 2012).

Alain Desrosières' works (see e.g. Desrosières, 1998 [1993], 2008) as well as studies by the economics of convention (see e.g. Diaz-Bone, 2016; Salais 2016; Thévenot, 2001) have shown that the production of objective quantitative evidence relies on conventions of measurement, categories and classifications that are rooted in specific socio-historical processes and projects of nation-building. These conventions, categories and nomenclatures are historically and geographically contingent, as for instance studies of changes in the measurement and categorization of employment and unemployment or the professional occupations have demonstrated (Boltanski, 1987 [1982]; Desrosières & Thévenot, 1988; Salais et al., 1986).

The Baconian dream of eliminating variation due to the specificity of context and different theoretical foundations of debates—when transposed to the social—is impossible to achieve. Evidence-based quantification in itself is a utopia one might say. The contributions collected in this part of the volume trace different aims and modes of intervention that have come to be attached to quantification-based politics of evidence over time and in different settings. Following the economics of convention, they highlight, amongst other things, that it is important to accept a plurality of positions and perspectives as valid in the search for evidence. Amartya Sen referred in this context to "positional objectivity" (that he distinguished from subjectivity) arguing that "the idea of objectivity requires explicit acceptance and extensive use of variability of observations with the position of the observer" (Sen, 1990, p. 114).

A key element of evidence-based politics are state statistics. In this volume, Thomas Amossé (Chapter 6) traces the history of France's public statistical infrastructure aimed at knowing its population. The chapter distinguishes between three different models of quantification that came to sustain the relationship between the French State and its statistical citizens since 1950. Although these models appeared successively, they co-exist today. These are the "representative household survey" (which emerged in the 1950s), the "biographical investigation" (which was developed in the early 1980s) and "the matched panel" (which emerged at the end of the 1990s). According to Amossé, "these models articulate, each in a specific way, social science theories, statistical methodologies, and public action conceptions", and they correspond to three different types of "statistical being" (homo statisticus)—subject, person and individual—modelled on Supiot's exploration of the development of the legal subject (Supiot, 2007 [2005]).

Imported from the US in the 1950s, the representative household survey sought to provide a snapshot of society. The surveys were aimed at representing the entire population and became one of the most important sources of social and demographic statistics. Their main variables (such as age, gender, income and socio-professional categories) were primarily of an administrative nature and assumed to be unanimously and uniformly understood. Amossé shows that, today, such surveys have not disappeared, but they have become supplanted with other statistical tools. Biographical investigations, born out of social scientific reflection and critique in the 1980s, gave less primacy to standardized, institutional categories and questions of statistical representativeness. They focused on the tracking of a respondent's individual trajectory (birth, move, promotion). Grounded in interpretive sociology, they were based on the assumption that the individual bears multiple identities which must be explored. Both models (household surveys and biographical investigations) respected the respondents as owners of a practical experience deemed valuable for the formulation and implementation of public policies. In contrast, the third form of statistical infrastructure (the matched panel) that Amossé explores is of a very different nature.

It responds to a micro-causalist agenda, where the statistical infrastructure is not focused on individuals as persons or members of the collective entities to which they belong (e.g. a household). Rather, the "statistical beings" making up matched panels can be seen as "dividuals" (Appadurai, 2016; Moor & Lury, 2018) resulting from an assembling of matched data, multiple data marks and imprints. As Amossé writes, matched panels are based on "libraries of information which, for each respondent, store a great number of fragmentary and heterogeneous pieces of a puzzle, which are put back together once they have been re-aggregated to represent the targeted individuals". In this form of statistical infrastructure, the metrological objective is preeminent. The collected variables rarely, or only secondarily, refer to practical experience or institutional categories. Statistical data are "purified" and abstracted from context and convention.

The theme of abstraction is also of importance in the politics of evidence examined by Laurent Thévenot (Chapter 7). Thévenot examines sustainable palm oil certification as a new mode of global governing that operates away from states. Voluntary governing schemes, such as that of the Roundtable for Sustainable Palm Oil (RSPO), have gained legitimacy through the implementation of multi-stakeholder governance. Thévenot interrogates what such governance arrangements entail. He argues that "governing by standards" has shifted the political debate about power, legitimacy and the common good to questions of measurable certifiable characteristics of products and services that are to be chosen by "autonomous opting individuals". What kind of alternative to the rule of law do certification standards offer? How do actors, including local smallholders in developing economies, cope with the standards' reliance on measurable objectives? What does this format of governance, and the liberal grammar underpinning it, do to voices of concern? These are some of the central questions Thévenot explores.

Building on his earlier research on the politics of statistics (Thévenot, 1990), economics of convention (Eymard-Duvernay et al., 2006), investments in form (Thévenot, 1984), orders of worth (Boltanski & Thévenot, 2006 [1991]) and the plurality of regimes of engagement (Thévenot, 2006, 2015), Thévenot studies governing by standards "in action". This demands close-up fieldwork, including following the most vulnerable actors, from their daily life in remote rural areas to their participation in the "open spaces" of the public roundtables that are a key element of the certification process.

As Thévenot writes, each year the general assembly of the "Roundtable on Sustainable Palm Oil" (RSPO) meets in a "convention" which votes for resolutions and changes to the standard (of sustainable palm oil). Thévenot describes the arrangements and procedures designed to allow for wide participation in deliberations over the standard. These procedures urge participants to formulate their voice in a particular manner—as "engagement in a plan" and "in a format of individual choice between optional plans". Discussions are organized in an open space akin to a marketplace, where participants shop for information and ideas. Scrutinizing the liberal grammar of the "open space technology", Thévenot shows how it transforms (and limits) participants' (in this case a smallholder and representative from a local community) ability to engage in critique. The "open space technology" does not fit "grammars of practice which support pluralist constructions of commonality and difference". Thévenot shows that to be truly heard and understood requires the training of rhetoric abilities, the learning and use of the right language, and concepts specific to the object.

The registration of a complaint with the RSPO (here a complaint that the rules of the certification standard were not being properly implemented by the corporation) demands its "right formatting" in accordance with RSPO's standards. "Local familiar" and "customary formats" of evidence have to be translated into evidence legible by the organization (in this case the RSPO)—quantifiable and auditable markers, which in turn limit the very possibility to articulate dissent. For such transformation deprives the most vulnerable (local smallholders, communities and NGOs) of their primary resource: practical knowledge.

Such "dogmas of universal competition" (Supiot, 2009) and standardized "best practice" are also present in the case of implementation of development aid in Mali, which Ousmane Ousmarou Sidibé studies in this volume (Chapter 8). When writing his contribution, Sidibé was a senior civil servant of the Malian government. Hence, he speaks not only as a researcher, but also as a participant.

In a quest to make development aid more efficient and effective—to reduce waste and transaction costs and enhance accountability—governmental and non-governmental organizations involved in the provision and use of development signed on 2 March 2005 the Paris Declaration on Aid Effectiveness which, amongst other things, commits donors and recipients of aid to "management by results", a shift from "project aid" to "budget support", increased coordination and mutual accountability. Performance indicators have become a cornerstone in the implementation of the declaration.

Sidibé explores how "management by indicators" contributed to the undermining of public policies in Mali focusing on the cases of education and public health. Amongst other things, he highlights that a national indicator, by its very nature, does not reflect the conditions and results of a public policy in distinct geographic areas. In Mali, there are large disparities between regions. Such regional differences are not captured by the national indicators that average local results. Likewise, the public health indicator stating the percentage of the population that lives within five kilometres of a functional healthcare centre (56% in 2013) does not say much about the actual accessibility of healthcare. Accessibility to healthcare is shaped by many more factors than geographical distance, such as availability of appropriate medications, staffing levels, and staff's expertise. Another problem of performance-based management to which Sidibé draws attention consists in the difficulty to reach consensus on targets. As Sidibé states, "the aims of public policy are multiple, and sometimes contradictory, so reaching consensus can be extremely laborious".

How can under such circumstances a new social contract be devised, which sets reciprocal obligations for families, communities, local authorities, civil society and the national government, Sidibé asks. According to him, "performance indicators have become a formidable tool in the hands of a few international experts, who insidiously impose far-reaching public policy choices on states, in the absence of any real debate among citizens". Paradoxically, the states with the least institutional capacity to handle the weighty apparatus of "management by indicators", are often those on whom donors impose the most stringent terms of conditionality, because these states lack the capacity to negotiate. This does not mean that management by indicators should be abandoned altogether. But Sidibé warns against its blind, mechanical application. As he concludes:

"An indicator should be just that, literally furnishing an indication of the quality of governance, but not in itself a full appraisal of quality. From this point of view, indicators can send a warning on the state of public policy, and foster discussion, without issuing a judgement with no appeal, and are much less a guilty sentence".

Also Corine Eyraud examines a case of "management by indicators" (Chapter 9). She turns our attention to the development of governmental performance indicators for French universities. As a participant observer, she followed the numbers from their birth through their detailed construction to their concrete uses. The development of these indicators began in France in 2001 with the "*Loi organique relative aux lois de finance*" (LOLF) that took effect in 2006. This law sought to bring New Public Management (NPM) ideas to France. One of its main objectives was to make the government and the public services accountable to parliament for the results of their actions. The law further sought to give more power to parliament over budgetary policies and choices. Lastly, it sought to reform the allocation of resources to administrations and public services. Yet, from the three projected uses of "management by indicators" only the third one was actually realized.

Eyraud shows that MPs did not pay much attention to numerically based performance reports in budgetary debates. First, the LOLF did not link performance to budgetary funding. Second, the MPs did not find the indicators particularly meaningful in relation to public policy making. They did not participate in their construction and found the indicators too technical and too detached from their concerns. Yet, a few years later, the performance indicators (e.g. number of publishing academics, number of PhD degrees delivered, undergraduate success rate, number of master degrees delivered, the rated research quality of a research centre) became an essential element in a new system of university resource allocation, called "Sympa". From 2009 onwards, a university's performance became financially rewarded, while non-performance was financially punished. This new way of funding introduced competition amongst universities and lead to various forms of "reactivity" (Espeland & Sauder, 2007), including gaming and the redirection of financial and human resources from teaching towards research. Yet, at the same time, Eyraud also describes how this system became contested, especially by academics, and eventually abandoned under Hollande's government in 2012.

Eyraud highlights that the various policy reforms and implied quantification instruments did not form a very integrated assemblage. The changes that the performance measures were supposed to help bring about were at the same time significant and very limited. Similar to Sidibé, also Eyraud does not seek to argue against quantification per se. But, she cautions, citing Supiot (2012), that one should not confuse measurement with assessment. Furthermore, the setting of indicators should not become "a way of surreptitiously encapsulating values and hierarchies". Rather, she pleads for the opening up of quantification, the setting and use of performance indicators, to public debate and scrutiny, which brings us to the third and final part of this volume, Part III, devoted to exploring the changing relation between numbers and democracy.

Voicing for Democracy

In democratic political systems, the law guarantees citizens the freedom of opinion and expression, including the right to demonstration and protest (see here also Articles 19, 20 and 21 of the 1948 Universal Declaration of Human Rights of the United Nations). How does the rise and spread of "governance by numbers" (Miller, 2001; Supiot, 2015) affect individual and collective capacities to exercise voice? Quantification may be motivated by a democratizing ambition, the desire to hold to account, and to counteract despotism and arbitrariness (Alonso & Starr, 1989; Cohen, 1982; Kurunmäki et al., 2016; Porter, 1995). Yet, numbers also change how we perceive things, how decisions are framed and how concerns become articulated. Numbers promise a "de-politicization" of politics. But as Rose (1991, p. 676) remarks, "numbers do not merely inscribe a pre-existing reality. They constitute it". And with that new conduits of power are brought into being (see also Miller & Rose, 1990). "Numbers delineate fictive spaces for the operation of government" (Rose, 1991, p. 676). In so doing, they generate new forms of visibility and invisibility. They create new actors and new relations which might thwart political engagement and participation.

Contributions assembled in this part interrogate possibilities for the incorporation of voice in numbers. To paraphrase Morgan (2010), they examine the ways in which quantification can (or cannot) express citizens' experience about political, economic, and social arrangements that affect them. They also scrutinize the relationship between quantification and public debate (Mennicken & Espeland, 2019, pp. 232–233), and investigate the role that numbers play in generating and framing public discussions and deliberations, for example about public goods and services, such as public security, fiscal policies or employment. Finally, they query prospects for a pluralization of quantification (ibid.). They examine how numbers can come to be unsettled, challenged and changed through the development of alternative measures, "counter-quantifications", as the literature on "statactivism" has highlighted (Bruno et al., 2014).

Andrei Guter-Sandu and Andrea Menpicken (Chapter 10) use the case of prison privatization in England and Wales to scrutinize what it means to "economize the social" through numbers. Guter-Sandu and Mennicken show how ratings and rankings of organizational entities, such as hospitals, universities or prisons, are often closely linked to aspirations aimed at facilitating competition and the establishment of (quasi) markets in the public services. Further, they highlight that particularly over the past fifteen years or so, we have witnessed an increase in quantified social impact assessments seeking to make the value of public sector work knowable in financial terms not only for evaluation purposes but also to attract (private) investment. Guter-Sandu and Mennicken analyse this multiplicity of quantification and its implication in different processes of economizing distinguishing between activities aimed at curtailing, marketizing and financializing.

Tracing the history of quantification and performance measurement in the HM Prison Service of England and Wales, they attend, first, to the introduction of prison performance metrics in the 1990s which, amongst other things, was stimulated by prison privatization and concerns with the accountability of prison governors. They examine then the evolution of these metrics, including their translation into aggregated prison ratings from the early 2000s onwards. Underlying the introduction of the prison ratings, the authors argue, was a belief in the power of market incentives and the aspiration to govern through competition. But the chapter also shows that we need to be careful not to equate quantification with economization. It describes how measures seeking to capture the quality of prison life from a prisoner's perspective were introduced to "moralize" the prison ratings (Liebling, 2004). Of course, we need to be cautious when labelling this as democratization. Nonetheless, the performance measures served as an important platform for debate about prison values and reform, not least because of the public attention and criticism they attracted.

Yet, these "moralized" measures were, in turn, destabilized in the wake of the government's austerity policies. Concerns with the measurement and management of costs came to the fore which overruled the Prison Service's "balanced" performance measurement system. Ironically, contracted-out private prisons were largely spared from these cuts, due to the inflexible nature of the 15- or 25-year contracts under which they were operating. Thus, they were largely shielded from economization in the form of budgetary savings requests. This does not mean that private prisons are not economized, but different mechanisms of economization (and quantification) are at work here, which the authors disentangle examining experimentations with Social Impact Bonds (SIBs). These mechanisms of economizing relate first and foremost to processes of financialization and a logic of (financial) "capitalization" (Muniesa, 2016; Muniesa et al., 2017) where prisons and prisoners come to be viewed as (financial) assets, as vehicles for the generation of future returns. Guter-Sandu and Mennicken conclude by cautioning us to be mindful of the multiple ways in which quantification and financial concerns come to be interlinked—reinforced, mitigated or undermined.

Subsequent chapters (see in particular Chapters 11 and 12) explore in what ways and under what conditions quantification can be turned against programmes of marketization, financialization and austerity. They examine instances when quantification is subjected to scrutiny, debate and critique; when forms of disruption are sought that go beyond "gaming the numbers"; when numbers become (re)attached to dreams and schemes of doing things differently.

Boris Samuel's analysis (Chapter 11) focuses on the struggle against high living costs in Guadeloupe in 2009. He investigates a case of "stat-activism" (Bruno et al., 2014), where calculation and figures came to be used as a "weapon" by a social and civic movement, led by local trade unions, to fight what they identified as "pwofitasyon". "Pwofitasyon" [in French: profitation] is a Creole term that denotes the capturing of undue profit resulting from the existence of excessive sales prices, particularly in relation to basic consumer goods, such as food. Samuel shows how the movement, called LKP [l'alliance contre la profitation], engaged in the calculation and collection of data to revaluate purchasing power and challenge the setting and regulation of prices by the administrative authorities. Samuel assesses to what extent "the statactivistic momentum" of the LKP and other non-state actors was capable of shifting generally accepted price measurement methods and, more generally, debates on price.

His analysis shows that the struggles with numbers lead to unequal outcomes, in which technical but also relational resources determined the balance of power between the actors. The alternative calculations (counter-quantifications) put forward by the LKP were effective in generating a public debate on "pwofitasyon", yet, at the same time the measurements came to be deemed as "too simplistic" to be considered legitimate by the public authorities. Instead, a statistically more sophisticated analysis of price differences between Guadeloupe and mainland France undertaken by the INSEE (France's Institute of Statistics and Economic Studies) became an important reference point in the negotiations. But here the reported price differences rested on averages, and the reported average gap of approximately 15% made it impossible to expose the abusive pricing of individual goods, for example chocolate powder or yoghurt, where price differences could be as large as 100% and more.

The common and rigorous statistical methods used by the INSEE were thus not socially acceptable and led to further controversy and social unrest. Yet, the report formed the basis for the development of a new strategy of communication on price differentials. This strategy represented prices no longer on the basis of average values, but focused on the representation of particular extreme values which, in the end, came to be considered a fairer representation of the inequitable situation lived by many Guadeloupians. This strategy was supported by a report of the Competition Authority which identified several violations of competition law in the large-scale retail and import sectors. This report, Samuel writes, "turned each price, as experienced by the consumers in their everyday life, into an indication of abuse and injustice, deserving to be discussed and publicly denounced".

Samuel's study helps gain a deeper understanding of the multiplicity of price level measurement and the shifting legitimacies associated with it. It also shows that the increased complexity and sophistication of statistical techniques may reduce possibilities for public scrutiny and the utilization of quantification as an emancipatory device. Further, Samuel's analysis raises important questions about what should count as "right and fair" statistical proof in a democratic debate, and in what way statistical data are, or should be, able to capture lived experience, a topic which also Robert Salais' contribution addresses (Chapter 12), to which we turn next.

Salais investigates the development and transformation of a social and statistical category, namely that of unemployment, in three major European countries (France, Germany, Great Britain) and at European level (Salais, 2007). In his analysis, Salais differentiates between statistics and governance-driven quantification. He shows that both rely on quantification, but differ with regards to the political status of social and cognitive conventions. As he writes: "The purpose of statistics is *to build general knowledge* 'extracted' from the plurality and variety of *social conventions* people use in daily life to understand their world [...]. The purpose of governance-driven quantification is to find ways *to rationally transform social conventions* toward some pre-given political objective, judged by the centre as optimal". According to Salais (but see also Desrosières, 1998 [1993]), in statistics, the establishment of facts logically precedes the design and implementation of public policies. In governance-driven quantification, the "fabrication" of facts is internal to public policies and driven by the search of maximizing these policies' quantitative performance. Put differently, the ultimate outcome (or aim) of "governance by numbers" is to modify the conventions upon which people rely as landmarks to identify what can be their legitimate claims.

Rather than being rooted in collective deliberation, the informational basis of governance-driven quantification is predetermined and imposed by "the Centre"; it incorporates norms without discussion and directs the decisional process towards prefixed political outcomes. These norms are mostly incorporated into technicalities (e.g. definitions of operational categories). All this creates a move towards what Salais terms an "a-democratic regime" of quantification. According to Salais, "a-democracy" is a political regime that maintains the formal procedures of democracy, but impedes any effective participation of citizens and other actors who could speak on their behalf (see here also Crouch's definition of post-democracy, 2004).

Salais shows how European employment policies came to be based on such governance-driven quantification and "a-democracy". European authorities substituted the search of Keynesian full employment for the maximizing of the rate of employment as their main target. In this new regime, employment took on a very different meaning, encompassing any job, regardless of wage, working conditions, duration, or type of labour contract. Salais goes on to explore ways by which this turn could be counter-acted and social justice expectations be brought back into quantification processes so that another understanding of the collective issue to deal with can be developed. For him, producing and interpreting data should thus be seen as a collective undertaking. Informants (e.g. people who are asked to respond to a survey on employment) should be considered as active interpreters bringing with them valuable practical experience and knowledge that should be used in defining the issue at hand (e.g. helping to arrive at a collectively shared understanding of what should or should not count as employment).

Quantification, in this context, has to become legitimate in terms of both fairness and correctness of the data it helps produce. Drawing on Amartya Sen's concept of "the informational basis of judgement in justice" (IBJJ) (Sen, 1990; but see also Salais, 2019), Salais argues that a major aim of quantification should be the generation of an informational basis which has not only been produced with the help of rigorous methodologies, but satisfies also accepted requirements of justice. This informational basis is thus no longer merely objectified "aperspectival" evidence reflecting reality; it is also elaborated by (lay) people through the prism of their own feelings on what is or is not justice and injustice, taking into account relevant features of their "factual territory". Such an understanding of quantification opens up the possibility of a plurality of "data-makings" for the same situation. It also does not oppose qualitative and quantitative methodologies and, more importantly, has as its foundation effective freedom.

Also the chapter by Emmanuel Didier (Chapter 13) problematizes the divide between qualitative and quantitative methodologies. Examining the history of what is now called "qualitative sociology" in the US, Didier asks how we can account for the political production of the border of qualitative enclaves which exclude quantities. The founding fathers of sociology never chose between quantification and non-quantification. So, how did the conceptual pair "qualitative vs. quantitative" come to settle within sociology?

Didier pays attention to both the epistemic and political forces that participated in the production of the border between the qualitative and quantitative in sociology. His analysis begins with a discussion of Herbert Blumer's *Symbolic Interactionism*, who located the social primarily in situations of interaction. Deeply influenced by American pragmatism, he examined how a member's action is guided and formed by a process of interpretation. Blumer was against surveys (as e.g. conducted by Paul Lazarsfeld at the time), but he was not against quantification per se. Next to symbolic interactionism, ethnomethodology, originally developed by

Garfinkel in the mid-1950s, elaborated another criticism of statistics. As Didier writes, ethnomethodology is aimed at developing "a generalized social system built solely from the analysis of experience structures". But also ethnomethodological criticisms of quantification did not oppose all and every quantification. Indeed, Garfinkel was probably the first to take a sociological interest in examining the production of statistics, as early as 1956, arguing that the process through which quantitative rates are produced should be conceived as a socially organized activity.

Later, qualitative sociologists, such as Becker and Horowitz, labelled as "the radicals" by Didier, became more defensive when trying to fight the wave of quantitative scholarship that washed over their discipline. Glaser and Strauss's publication of *The Discovery of Grounded Theory* consolidated the divide between the categories of "qualitative" and "quantitative" further. Today, Didier shows, qualitative sociology has become a category where two sets of "good examples" of published papers he examined are in opposition. On the one hand, there are those which belong to a "Lazarsfeldian" cluster where qualitative and quantitative are in a hierarchy and the former is serving the latter. On the other hand, there are papers pertaining to an "interpretative" definition of the qualitative which seeks to set itself apart from "quantitativist" uses of qualitative information. Didier concludes by cautioning us not to forget that such a divide had been produced over time, and that many of the classical qualitative sociologists had criticized a certain method of quantification (surveys and opinion polls), not the general use of quantities. It is neither possible (nor desirable) to completely wipe out quantities from an epistemic system. Rather, we ought to scrutinize their production.

In this respect, both ethnomethodology and symbolic interactionism bear strong proximities with the economics of convention, an approach which was developed in France in the 1980s, amongst other things, by statisticians who at the time were working for the INSEE (France's National Institute of Statistics and Economic Studies) (see also the special issue of *La Revue économique*, March 1989 [edited by Dupuy et al., 1989]). Now the approach has taken hold more widely in sociology, socioeconomics and history (Desrosières, 2011; Diaz-Bone, 2018 [2015]; Salais et al., 2019). The developers of the economics of conventions approach tried to converge American pragmatist sociology with their own lived experience as state statisticians. This unique positioning enabled them to move beyond "positivism" and "structuralism", beyond a sterile division between qualitative and quantitative approaches, which also Didier argues against.

In this context, it is also worthwhile to recall Alain Desrosières' notion of the "convention of equivalence", which has largely escaped North-American sociology, including more recent (qualitative) studies of quantification. Conventions of equivalence (commensurate scales as Espeland & Stevens, 1998, would put it) allow the rise to generality, from the singular (by essence qualitative) to the general (based on numbers). They ensure continuity from the qualitative to the quantitative, instead of separation, opposition or domination. Such a perspective considerably enlarges the scope and ambition for the study of social life, including practices of quantification.

Many of the contributions assembled in this volume draw on the economics of convention in their study of quantification. Others, as we have highlighted earlier, examine governing by numbers with reference to Foucault, in particular Foucault's works on governmentality. This volume seeks to engage these two strands of research in a critical dialogue. Where does the economics of convention approach meet Foucault? And what does such a joining up bring to the furthering of debates about quantification and the politics of numbers?

QUANTIFICATION: WHERE THE ECONOMICS OF CONVENTION APPROACH MEETS FOUCAULT

Neither the economics of convention (EC) nor Foucauldian studies of quantification represent coherent and unitary research programmes. As Diaz-Bone (2019, p. 311) recalls, "EC was projected not as a coherent paradigm, but as a scientific movement, organized around some core concepts and methodological positions. EC has developed the concept of conventions as logics of coordination". Also Foucauldian studies of quantification do not form a "school". To borrow Colin Gordon's (1991) apt way of characterizing governmentality research three decades ago, studies in this area amount to a "zone of research", rather than a "fully formed product" (Mennicken & Miller, 2012, p. 18).

Both the EC and Foucauldian studies of quantification are concerned with relations between "quantified objectivity", as Thévenot writes in this volume (Chapter 7), "and modes of governing that make the world calculable". Both seek to examine the various ways in which the administering or governing of lives in a wide range of settings has been made thinkable and practicable through quantification. Peter Miller, who laid the foundations for governmentality studies of quantification with Nikolas Rose in the early 1990s in the UK (Miller & Rose, 1990; Rose & Miller, 1992; but see also Hopwood, 1992), highlighted that accounting numbers have a distinctive capacity for "acting on the actions of others", one that goes far beyond the abstract injunctions of economic theory (Miller, 1992; but see also Mennicken & Miller, 2012 for a review of Foucauldian studies in accounting). Also Alain Desrosières, who initiated a sociologically reflexive turn in the study of statistics in France and was involved in the development of the economics of convention approach, was interested in examining questions related to the governing by numbers. He associated different modes of government with different modes of quantification, distinguishing, amongst other things, between five forms of state and associated (different) modes of quantification, the neoliberal state being one amongst others (Desrosières, 2003) (but see also Diaz-Bone & Didier, 2016, pp. 14–15).

Nonetheless, for long, both approaches did not take much note of each other. Curiously enough, in France, the founders of the economics of convention were aware of Foucault's work on nomenclatures (Foucault, 1966), but they did not take cognizance of Foucault's works on governmentality (Diaz-Bone & Didier, 2016, pp. 14–15). One reason for this might be the fact that these works became only widely available in French in 2004 (see Foucault, 2004), whereas Foucault's lecture on governmentality delivered at the Collège de France was translated into English as early as 1979 by Colin Gordon (Foucault, 1979), a British philosopher who had attended Foucault's lectures at the time. Some years later, Gordon co-edited *The Foucault Effect* with Graham Burchell and Peter Miller (G. Burchell et al., 1991), which contained further English translations of Foucault's works and became a cornerstone in Anglo-Saxon governmentality studies.

For the EC, statistics and other forms of calculative practice have to be related to the foundational conventions of data production (measurement) and data interpretation, modes of justification and orders of worth, to be fully understood (see Thévenot in this volume, Chapter 7, but see also Diaz-Bone, 2019). In contrast, Foucauldian analyses of quantification draw attention to the inherently political character of calculation. They highlight that accounting and other numbers, through their ability to produce certain forms of visibility and transparency, both create and constrain subjectivity (Miller, 1992; Miller & O'Leary, 1987). This creates distinctive possibilities for intervention while potentially displacing others (see also Mennicken & Miller, 2012, p. 7). Yet, despite differences in theoretical heritage, in our view, and as the contributions to this volume demonstrate, the EC and Foucauldian approaches converge in at least three important respects.

First, both are concerned with the study of practices of quantification. Instead of concentrating on the interests of politicians, scientists or bureaucrats, both seek to investigate and unpack different stages in "the statistical [or accounting, added] production chain" as Thévenot puts it in this volume, from data collection, classification, codification, to the processing of information and its effects on the "making up" (Hacking, 2002) of people and entities. As Diaz-Bone (2019, p. 309) writes, "both have an anti-substantialist ontology: properties, qualities and valuations of people, objects and actions are results of practices". Such a focus on practices of quantification helps to develop a more nuanced understanding of the different capacities and roles that numbers have come to assume in political life. It helps generate insight into the inner workings of accountability regimes, their changing nature, and the emergence of new regulatory spaces and practices.

Second, both approaches are interested in examining the variable ways in which the capacities and attributes of subjects are constituted, shaped and changed, through quantification. Both do not "posit a universal form for the human subject" (Mennicken & Miller, 2014, p. 15). See here, for instance, Vormbusch's study of the emergence of new taxonomies of the self in this volume (Chapter 4) or Thévenot's unfolding of different regimes of engagement with standardization and quantification (Chapter 7).

Third, as Thévenot highlights (Chapter 7), both seek to move beyond "the state" and "the neoliberal" as an all-encompassing notion (see here also the contributions by Mespoulet, Lam, De Leonardis, Guter-Sandu and Mennicken, and Salais in this volume). Both are interested in developing what Raffnsøe et al. (2016) have termed a "dispositional analytic" of quantification (see also Diaz-Bone, 2019). Both draw attention to the implication of quantification in what Foucault termed "dispositif" or "apparatus", "a thoroughly heterogeneous ensemble consisting of discourses, institutions, architectural forms, regulatory decisions, law, administrative measures, scientific statements, philosophical, moral and philanthropic proportions – in short: the said as much as the unsaid" (Foucault, 1977, p. 299, quoted in Raffnsøe et al., 2016, p. 278).

In so doing, both approaches share a relational and processual understanding of power and the state. Power is not a property—something that can be possessed. For what defines a relationship of power, according to Foucault, "is that it is a mode of action that does not act directly on others. Instead, it acts upon their actions, whether an existing action or one that may arise in the future. And, the 'other' over whom power is exercised remains resolutely a person who acts, who is faced with a whole field of possible actions and reactions" (Foucault, 2001 [1982], pp. 341–342, cited in Mennicken & Miller 2014, p. 15).

However, one concept that is absent from Foucauldian studies of quantification which is central to the EC is the notion of "convention". As Thévenot points out in this volume, in the economics of convention, activities of "in-forming, trans-forming and formatting through invested conventional forms are central operations, because they sustain coordination power under uncertainty" (see also Boltanski & Thévenot, 2006 [1991]; Thévenot, 1984). Comparing France and USA, Storper & Salais (1997) focus on the plurality of conventions of the state on what people agree on the reciprocal roles and actions of citizens and state institutions with regards to defining, quantifying and contributing to common goods (see also Salais, 2015; Salais & Storper, 1993). The economics of convention approach is interested in the pluralism of (different) modes of evaluation (and quantification) constituted by conventions, orders of worth, modes of coordination, worlds of production and regimes of engagement (Boltanski & Thévenot, 2006 [1991]; Eymard-Duvernay, 1986, 1989; Salais, 2006; Salais et al., 1986; Salais & Storper, 1993; Storper & Salais, 1997; Thévenot, 2001, 2007).

Here, authors seek to break down policies and politics into a variety of modes of coordination (e.g. market, industrial, civic) or worlds of production (e.g. interpersonal, market, industrial, immaterial) (see the chapters by Thévenot, Salais and Vormbusch). This allows the EC to account more systematically for differing voices and evaluative orientations, conflict and contestation. It also makes it possible to draw more explicitly attention to the agency of participants and "their capacities of critique", as Vormbusch points out (see Chapter 4). Put differently, the EC makes us query to what extent people "have a grip" (Bessy & Chateauraynaud, 2014 [1995]) on the specificity of the situation, are able to challenge and change its definition, and the way in which numbers work.

Yet, in both approaches, human freedom is always present. How can we then conceive the dynamics of socio-historical processes? How can we draw renewed attention to the diverse arts of government, resistance and freedom—to the conflicts and compromises that are the engine of history? Foucault deeply acknowledged the contradictory and dialectic (Grant, 2010) nature of historical processes and was always concerned "with the multiple and dispersed surfaces of emergence of disparate and often humble practices" (Miller & Napier, 1993, p. 633). Nonetheless, as Raffnsøe et al. (2019, p. 162) highlight, a preoccupation of scholarship with the image of "discipline as subjugation" led some to reinstate the very dualism between power and freedom that Foucault's notion of power sought to overcome (see also Foucault, 2010, 2011) (for a critique see Jameson, 1984).

The economics of convention approach seeks to overcome such dualism by explicitly recognizing that coordination in real situations is structured by a plurality of conventions (Diaz-Bone, 2018). In so doing, it makes recourse to actors' competencies to master and recognize the plurality of conventions; to their ability to use different grammars for interpreting the situation of coordination they face. Yet, under what conditions and circumstances can such competence be assumed and assured? This, in our view, is something that warrants further exploration, also in relation to the production and use of numbers in economic and political life.

We ought to acknowledge that "real situations" are often characterized by inequality and asymmetrical power relations. Inequalities exist, for example, with regards to access to public life; the knowledge and expertise participants possess; economic resources; control over conditions of work (and life more generally); recognition and respect for cultural differences; age; gender; class; race; and much else besides. Such inequalities shape the capacity to exercise voice, including the capacity to exercise voice in the production and use of numbers. Amartya Sen has developed the idea that society should promote equality in the space of *capabilities* (see e.g. Sen, 1992). We propose that both Foucauldian studies of quantification and the economics of convention would benefit from making use of Sen's concept of *capability* (see also De Leonardis et al., 2012).

Sen acknowledges that persons are not similarly situated in their capability to convert resources brought by the situation to freedom of choice. Sen's notion of capability captures both competency and effective freedom to act—the ability to choose the life one wants to live from amongst a wide variety of valued functionings to which one has effective access (Sen, 1992, 1993). According to Sen, capabilities are not only a function of fixed personal traits and divisible resources, but also social relations, organizational and institutional environments, and the practical configuration of situations, in short, what Foucault termed "dispositif" (see also Salais' chapter in this volume and Salais, 2011). Struggling against inequalities in capabilities to understand numbers and their production should be at the heart of concerns with effective democracy, even more so today, where numbers have taken an unprecedented rise in the governing of our lives.

In conclusion, we think it is time for a renewal of the study of the politics of numbers. The two strands of scholarship brought together in this volume move quantification scholarship beyond the tired dichotomies between autonomy and discipline, compliance and resistance, power and freedom by moving our attention to the processes of quantifying and calculating, and the possibilities for action these open up or foreclose. In so doing, they do not only provide in-depth empirical insight into the multi-varied nature of governing by numbers and its consequences in different sites. They also help more generally to rethink the study of the politics of numbers, acknowledging plurality, contingency and "the differential structuring of freedom, performative and indirect agency" (Raffnsøe et al., 2019, p. 155).

References

Alonso, W., & Starr, P. (Eds.). (1989). *The politics of numbers: Population of the United States in the 1980s*. Russell Sage Foundation.

Anderson, B. (1983). *Imagined communities: Reflections on the origin and spread of nationalism*. Verso.

Appadurai, A. (2016). *Banking on words: The failure of language in the age of derivative finance*. University of Chicago Press.

Bessy, C., & Chateauraynaud, F. (2014 [1995]). *Experts et faussaires. Pour une sociologie de la perception*. Editions Petra.

Bevan, G., & Hood, C. (2006). What's measured is what matters: Targets and gaming in the English public health care system. *Public Administration, 84*(3), 517–538.

Biagioli, M., & Lippman, A. (Eds.). (2020a). *Gaming the metrics: Misconduct and manipulation in academic research*. MIT Press.

Biagioli, M., & Lippman, A. (2020b). Introduction: Metrics and the new ecologies of academic misconduct. In M. Biagioli, & A. Lippman (Eds.), *Gaming the metrics: Misconduct and manipulation in academic research* (pp. 1–22). MIT Press.

Boltanski, L. (1987 [1982]). *The making of a class: Cadres in French society.* Cambridge University Press.

Boltanski, L., & Chiapello, E. (2007). *The new spirit of capitalism.* Verso (French edition, 1999).

Boltanski, L., & Thévenot, L. (2006 [1991]). *On justification: Economies of worth.* Princeton University Press (French edition, 1991).

Bruno, I., & Didier, E. (2013). *Benchmarking: L'Etat sous pression statistique.* Zones.

Bruno, I., Didier, E., & Prévieux, J. (Eds.). (2014). *Statactivisme: Comment lutter avec les nombres.* Zones.

Bruno, I., Jany-Catrice, F., & Touchelay, B. (Eds.). (2016). *The social sciences of quantification: From politics of large numbers to target-driven policies.* Springer.

Burchell, G., Gordon, C., & Miller, P. (Eds.). (1991). *The Foucault effect: Studies in governmentality.* University of Chicago Press.

Burchell, S., Clubb, C., Hopwood, A. G., Hughes, J. S., & Nahapiet, J. (1980). The roles of accounting in organizations and society. *Accounting, Organizations and Society, 5*(1), 5–27.

Cohen, P. C. (1982). *A calculating people: The spread of numeracy in early America.* University of Chicago Press.

Crouch, C. (2004). *Post-democracy.* Polity Press.

Daston, L. (1988). *Classical probability in the enlightenment.* Princeton University Press.

Daston, L. (1992). Objectivity and the escape from perspective. *Social Studies of Science, 22*(4), 597–618.

Davies, W. (2017, January 19). How statistics lost their power—And why we should fear what comes next. *The Guardian.* https://www.theguardian.com/politics/2017/jan/2019/crisis-of-statistics-big-data-democracy.

Davis, K. E., Fisher, A., Kingsbury, B., & Merry, S. E. (Eds.). (2012). *Governance by indicators: Global power through classification and rankings.* Oxford University Press.

De Leonardis, O., Negrelli, S., & Salais, R. (Eds.). (2012). *Democracy and capabilities for voice: Welfare, work and public deliberation in Europe.* Peter Lang.

Demortain, D. (2019). The politics of calculation: Towards a sociology of quantification in governance. *Revue d'anthropologie des connaissances, 13*(4), 973–990.

Desrosières, A. (1985). Histoire de formes: Statistiques et sciences sociales avant 1940. *Revue française de sociologie, 26*(2), 277–310.

Desrosières, A. (1987 [1983]). Des métiers aux classifications conventionnelles: l'évolution des nomenclatures professionelles depuis un siècle. In J. Affichard (Ed.), *Pour une histoire statistique* (pp. 35–56). INSEE & Economica.

Desrosières, A. (1998 [1993]). *The politics of large numbers: A history of statistical reasoning*. Harvard University Press.

Desrosières, A. (2003). Managing the economy. In T. M. Porter & D. Ross (Eds.), *The Cambridge history of science* (pp. 553–564). Cambridge University Press.

Desrosières, A. (2008). *Pour une sociologie historique de la quantification*. Presses de l'Ecole des Mines de Paris.

Desrosières, A. (2011). The economics of convention and statistics: The paradox of origins. *Historical Social Research, 36*(4), 64–81.

Desrosières, A., & Thévenot, L. (1988). *Les catégories socioprofessionnelles*. La Découverte.

Diaz-Bone, R. (2016). Convention theory, classification and quantification. *Historical Social Research, 41*(2), 48–71.

Diaz-Bone, R. (2018). Economics of convention and its perspective on knowledge and institutions. In J. Glückler, R. Suddaby, & R. Lenz (Eds.), *Knowledge and institutions: Knowledge and space* (Vol. 13, pp. 69–88). Springer.

Diaz-Bone, R. (2018 [2015]). *Die "Economie des conventions". Grundagen und Entwicklungen der neuen französischen Wirtschaftssoziologie*. Springer VS.

Diaz-Bone, R. (2019). Economics of convention meets Foucault. *Historical Social Research, 44*(1), 308–334.

Diaz-Bone, R., & Didier, E. (2016). Introduction: The sociology of quantification—Perspectives on an emerging field in the social sciences. *Historical Social Research, 41*(2), 7–26.

Diaz-Bone, R., & Salais, R. (2011). Conventions and institutions from a historical perspective. *Historical Social Research, 36*(4), 5–247.

Dupuy, J.-P., Eymard-Duvernay, F., Favereau, O., Salais, R., & Thévenot, L. (1989). L'économie des conventions. *Revue économique, 40*(2), 141–406.

Espeland, W. N., & Sauder, M. (2007). Rankings and reactivity: How public measures recreate social worlds. *American Journal of Sociology, 113*(1), 1–40.

Espeland, W. N., & Sauder, M. (2016). *Engines of anxiety: Academic rankings, reputation, and accountability*. Russell Sage Foundation.

Espeland, W. N., & Stevens, M. L. (1998). Commensuration as a social process. *Annual Review of Sociology, 24*, 313–343.

Eymard-Duvernay, F. (1986). La qualification des produits. In R. Salais, & L. Thévenot (Eds.), *Le travail. Marché, règles, conventions* (pp. 239–247). INSEE-Economica.

Eymard-Duvernay, F. (1989). Conventions de qualité et pluralité des formes de coordination. *Revue économique, 40*(2), 329–359.

Eymard-Duvernay, F., Favereau, O., Orléan, A., Salais, R., & Thévenot, L. (2006). Valeurs, coordination et rationalité : Trois thèmes mis en relation par l'économie des conventions. In F. Eymard-Duvernay (Ed.), *L'économie des conventions, méthodes et résultats* (pp. 23–44). La Découverte.

Foucault, M. (1966). *Les mots et les choses*. Editions Gallimard.

Foucault, M. (1979). Governmentality. *Ideology and Consciousness, 7*, 5–26.

Foucault, M. (1980). Truth and power. In C. Gordon (Ed.), *Power/knowledge: Selected interviews and other writings, 1972–1977*. Pantheon Books.

Foucault, M. (1988). Technologies of the self. In L. H. Martin, H. Gutman, & P. H. Hutton (Eds.), *Technologies of the self*. Tavistock.

Foucault, M. (1991 [1979]). Governmentality. In G. Burchell, C. Gordon, & P. Miller (Eds.), *The Foucault effect: Studies in governmentality* (pp. 87–104). Harvester Wheatsheaf.

Foucault, M. (2004). *Sécurité, territoire, population*. Seuil et Gallimard.

Foucault, M. (2010). *The government of self and others: Lectures at the Collège de France, 1982–1983*. Palgrave Macmillan.

Foucault, M. (2011). *The courage of truth: The government of self and others. Lectures at the Collège de France, 1983–1984*. Palgrave Macmillan.

Fourcade, M. (2011). Cents and sensibility: Economic valuation and the nature of "nature". *American Journal of Sociology, 116*(6), 1721–1777.

Gigerenzer, G., Swijtink, Z., Porter, T. M., Daston, L., Beatty, J., & Krüger, L. (1989). *The empire of chance*. Cambridge University Press.

Gordon, C. (1991). Governmental rationality: An introduction. In G. Burchell, C. Gordon, & P. Miller (Eds.), *The Foucault effect: Studies in governmentality* (pp. 1–51). The University of Chicago Press.

Grant, J. (2010). Foucault and the logic of dialectics. *Contemporary Political Theory, 9*(2), 220–238.

Hacking, I. (1990). *The taming of chance*. Cambridge University Press.

Hacking, I. (2002). Making up people. In I. Hacking (Ed.), *Historical ontology* (pp. 99–114). Harvard University Press.

Hopwood, A. G. (1983). On trying to study accounting in the contexts in which it operates. *Accounting, Organizations and Society, 8*(2/3), 287–305.

Hopwood, A. G. (1992). Accounting calculation and the shifting sphere of the economic. *European Accounting Review, 1*(1), 125–143.

Hopwood, A. G., & Miller, P. (Eds.). (1994). *Accounting as social and institutional practice*. Cambridge University Press.

Jameson, J. (1984). Postmodernism, or the cultural logic of late capitalism. *New Left Review, 146*(September), 53–64.

Kornberger, M., Pflueger, D., & Mouritsen, J. (2017). Evaluative infrastructures: Accounting for platform organization. *Accounting, Organizations and Society, 60*, 79–95.

Krüger, L., Daston, L., & Heidelberger, M. (Eds.). (1987). *The probabilistic revolution, vol. 1: Ideas in history*. MIT Press.

Kurunmäki, L., Mennicken, A., & Miller, P. (2016). Quantifying, economising, and marketising: Democratising the social sphere? *Sociologie du travail, 58*, 390–402.

Liebling, A. (2004). *Prisons and their moral performance: A study of values, quality, and prison life*. Oxford University Press.

Manuel, F. E., & Manuel, F. P. (1979). *Utopian thought in the western world*. Harvard University Press.

Maor, M. (2018). Rhetoric and doctrines of policy over- and underreactions in times of crisis. *Policy and Politics, 46*(1), 47–63.

Maor, M. (2019). Overreaction and bubbles in politics and policy. In A. Mintz, & L. Terris (Eds.), *The Oxford handbook on behavioural political science*. Oxford Handbooks Online. Oxford University Press.

Mennicken, A., & Espeland, W. N. (2019). What's new with numbers? Sociological approaches to the study of quantification. *Annual Review of Sociology, 45*, 223–245.

Mennicken, A., & Miller, P. (2012). Accounting, territorialization and power. *Foucault Studies, 13*, 4–24.

Mennicken, A., & Miller, P. (2014). Foucault and the administering of lives. In P. S. Adler, P. du Gay, G. Morgan, & M. I. Reed (Eds.), *The Oxford handbook of sociology, social theory, and organization studies: Contemporary currents* (pp. 11–38). Oxford University Press.

Merry, S. E. (2016). *The seductions of quantification: Measuring human rights, gender violence, and sex trafficking*. University of Chicago Press.

Miller, P. (1992). Accounting and objectivity: The invention of calculating selves and calculable spaces. *Annals of Scholarship, 9*(1–2), 61–86.

Miller, P. (2001). Governing by numbers: Why calculative practices matter. *Social Research, 68*(2), 379–396.

Miller, P., & Napier, C. (1993). Genealogies of calculation. *Accounting, Organizations and Society, 18*(7–8), 631–647.

Miller, P., & O'Leary, T. (1987). Accounting and the construction of the governable person. *Accounting, Organizations and Society, 12*(3), 235–265.

Miller, P., & Rose, N. (1990). Governing economic life. *Economy and Society, 19*(1), 1–31.

Moor, L., & Lury, C. (2018). Price and the person: Markets, discrimination, and personhood. *Journal of Cultural Economy, 11*(6), 501–513.

Morgan, M. S. (2010). 'Voice' and the facts and observations of experience. In W. J. Gonzalez (Ed.), *New methodological perspectives on observation and experimentation in science* (pp. 51–69). Netbiblo.

Muniesa, F. (2016). Setting the habit of capitalization: The pedagogy of earning power at the Harvard Business School, 1920–1940. *Historical Social Research, 41*(2), 196–217.

Muniesa, F., Doganova, L., Ortiz, H., Pina-Stranger, A., Paterson, F., Bourgoin, A., et al. (2017). *Capitalization: A cultural guide.* Ecole des Mines.

Porter, T. M. (1986). *The rise of statistical thinking, 1820–1900.* Princeton University Press.

Porter, T. M. (1992). Objectivity as standardization: The rhetoric of impersonality in measurement, statistics, and cost-benefit analyses. *Annals of Scholarship, 9*(1–2), 19–59.

Porter, T. M. (1995). *Trust in numbers: The pursuit of objectivity in science and public life.* Princeton University Press.

Power, M. (1997). *The audit society: Rituals of verification.* Oxford University Press.

Power, M. (2019). Infrastructures of traceability. In M. Kornberger, G. C. Bowker, J. Elyachar, A. Mennicken, P. Miller, J. R. Nucho & N. Pollock (Eds.), *Thinking infrastructures. Research in the sociology of organizations* (Vol. 62, pp. 115–130). Emerald.

Raffnsøe, S., Gudmand-Høyer, M. T., & Thaning, M. S. (2016). Foucault's dispositive: The perspicacity of dispositive analytics in organizational research. *Organization, 23*(2), 272–298.

Raffnsøe, S., Mennicken, A., & Miller, P. (2019). The Foucault effect in organization studies. *Organization Studies, 40*(2), 155–182.

Raynaud, P., & Rials, S. (Eds.). (1992). *Une prudence moderne?* Presses Universitaires de France.

Rose, N. (1991). Governing by numbers: Figuring out democracy. *Accounting, Organizations and Society, 16*(7), 673–692.

Rose, N., & Miller, P. (1992). Political power beyond the state: Problematics of government. *British Journal of Sociology, 43*(2), 172–205.

Rottenburg, R., Merry, S. E., Park, S.-J., & Mugler, J. (Eds.). (2015). *The world of indicators: The making of governmental knowledge through quantification.* Cambridge University Press.

Salais, R. (1986). L'émergence de la catégorie moderne de chômeur : les années 1930. In R. Salais, N. Baverez, & B. Reynaud (Eds.), *L'invention du chômage. Histoire et transformations d'une catégorie en France depuis les années 1980* (pp. 77–123). PUF.

Salais, R. (2006). Reforming the European Social Model and the politics of indicators. From the unemployment rate to the employment rate in the European Employment Strategy. In M. Jepsen, & A. Serrano (Eds.), *Unwrapping the European Social Model* (pp. 189–212). The Policy Press.

Salais, R. (2007). Europe and the deconstruction of the category of unemployment. *Archiv Für Sozialgeschichte, 47*, 371–401.

Salais, R. (2011). *Convention, sens commun et action publique*. Paper presented at the Colloque Cerisy September 2009 "Conventions, l'intersubjectif et le normatif". https://halshs.archives-ouvertes.fr/halshs-00577036.

Salais, R. (2015). Revisiter la question de l'État à propos de la crise de l'Europe: État extérieur, absent, situé. *Revue française de socio-économie, special issue* (2), 245–262.

Salais, R. (2016). Quantification and objectivity: From statistical conventions to social conventions. *Historical Social Research, 41*(2), 118–134.

Salais, R. (2019). Freedom in work and the capability approach: Towards a politics of freedoms for labour? In B. Langille (Ed.), *The capability approach to labour law* (pp. 311–331). Oxford University Press.

Salais, R., Baverez, N., & Reynaud, B. (1986). *L'invention du chômage*. PUF.

Salais, R., & Storper, M. (1993). *Les mondes de production*. Ed. de l'EHESS.

Salais, R., Streng, M., & Vogel, J. (2019). Einleitung: Qualitätspolitik und Konventionen. In R. Salais, M. Streng, & J. Vogel (Eds.), *Qualitätspolitik und Konventionen. Die Qualität der Produkte in historischer Perspektive (18.–20. Jahrhundert)*. (pp. 9–61). Springer.

Salais, R., & Thévenot, L. (Eds.). (1986). *Le travail. Marché, règles, conventions*. INSEE-Economica.

Samiolo, R. (2012). Commensuration and styles of reasoning: Venice, cost-benefit and the defense of place. *Accounting, Organizations and Society, 37*(6), 382–402.

Sen, A. (1983). Poor, relatively speaking. *Oxford Economic Papers, 35*(2), 153–169.

Sen, A. (1990). Justice: Means versus freedoms. *Philosophy and Public Affairs, 19*(2), 111–121.

Sen, A. (1992). *Inequality reexamined*. Harvard University Press.

Sen, A. (1993). Capability and well-being. In M. Nussbaum & A. Sen (Eds.), *The quality of life* (pp. 30–53). Oxford University Press.

Storper, M., & Salais, R. (1997). *Worlds of production: The action frameworks of the economy*. Harvard University Press.

Strathern, M. (1997). 'Improving ratings': Audit in the British University system. *European Review, 5*(3), 305–321.

Strathern, M. (Ed.). (2000). *Audit cultures: Anthropological studies in audit, ethics and the academy*. Routledge.

Supiot, A. (2007 [2005]). *Homo Juridicus: On the anthropological function of the law*. Verso.

Supiot, A. (2008, November). L'inscription territoriale des lois. *Esprit*, 151–170.

Supiot, A. (2009, February). Justice sociale et libéralisation du commerce international. *Droit social* (2), 131–141.

Supiot, A. (2012). *The spirit of Philadelphia: Social justice vs. the total market*. Verso.

Supiot, A. (2015). *La gouvernance par les nombres*. Fayard (English edition, Verso 2016).

Thévenot, L. (1979). Une jeunesse difficile : les fonctions sociales du flou et de la rigueur dans les classements. *Actes de la recherche en sciences sociales* (26–27), 3–18.

Thévenot, L. (1981). Les catégories socioprofessionnelles et leur repérage dans les enquêtes. *Archives et documents 38*. Paris: Institut National de la Statistique et des Etudes Economiques.

Thévenot, L. (1984). Rules and implements: Investment in forms. *Social Science Information, 23*(1), 1–45.

Thévenot, L. (1990). La politique des statistiques : les origines sociales des enquêtes de mobilité sociale. *Annales ESC, 45*(6), 1275–1300.

Thévenot, L. (2001). Organized complexity: Conventions of coordination and the composition of economic arrangements. *European Journal of Social Theory, 4*(4), 405–425.

Thévenot, L. (2006). *L'action au pluriel. Sociologie des régimes d'engagement*. La Découverte.

Thévenot, L. (2007). The plurality of cognitive formats and engagements: Moving between the familiar and the public. *European Journal of Social Theory, 10*(3), 413–427.

Thévenot, L. (2009). Governing life by standards: A view from engagements. *Social Studies of Science, 39*(5), 793–813.

Thévenot, L. (2011). Conventions for measuring and questioning policies: The case of 50 years of policies evaluations through a statistical survey. *Historical Social Research, 36*(4), 192–217.

Thévenot, L. (2015). Making commonality in the plural, on the basis of binding engagements. In P. Dumouchel & R. Gotoh (Eds.), *Social bonds as freedom: Revising the dichotomy of the universal and the particular* (pp. 82–108). Berghahn.

Zuboff, S. (2019). *The age of surveillance capitalism: The fight for a human future at the new frontier of power*. Public Affairs.

PART I

Quantification as Utopia

CHAPTER 2

Creating a Socialist Society and Quantification in the USSR

Martine Mespoulet

When the Bolshevik government came to power in October 1917, it based the legitimacy of its action on the scientific nature of its decisions, figures being one of the core elements thereof. Its leaders asserted the ambition of building a state in which science would form the basis for political decisions, for the well-being of everyone and fundamentally for the fulfilment of the communist plan to create a new society and new human being. From this perspective, statistics—an information and decision support tool—was also an instrument of power designed to prove the soundness of state action. Statistics would help symbolically construct the Soviet social and economic world. That notwithstanding, can one discern specific forms of production of figures in such a state, that is to say, as Alain Desrosières has conceptualized, a particular relationship between concepts, methods, technical instruments, statistical institutions and representations of the social and economic world, that could be construed as characterizing the nature of such a state (Desrosières, 1985)?

M. Mespoulet (✉)
University of Nantes, Nantes, France
e-mail: martine.mespoulet@univ-nantes.fr

© The Author(s) 2022
A. Mennicken and R. Salais (eds.), *The New Politics of Numbers*,
Executive Politics and Governance,
https://doi.org/10.1007/978-3-030-78201-6_2

In particular, did certain tools take on a specific form in the USSR, in a context of construction of a socialist society underpinned by centralized and planned management of the economy? If that is the case, how were they adapted to new political representations of society, the economy and the role of science in such a state? This newly developed form of producing figures serving a new type of state and a new political plan included the creation of a new central government statistics department and an effort to formulate a new theory of statistics and design new methods and new tools. Various attempts to that effect gave rise to debates and tensions between leaders and statisticians, and between the statisticians themselves. After trying to characterize some of them, we will endeavour to underscore the specific nature of state statistics in the Soviet Union.

INVENTING A NEW FORM OF STATISTICS
FOR A NEW MODEL OF SOCIETY

A New State Statistics Administration

The Bolshevik state's new Central Statistics Administration, the TsSU (Tsentral'noe Statisticheskoe Upravlenie), was set up on 25 July 1918. One can consider that it was the product of two plans tending towards change, but plans in which the role of statistics was defined differently. For the Bolshevik leaders, the production of figures had to play a key information role for developing the plan and for managing the economy and society. But for the TsSU's statisticians, most of whom were formerly employed by the statistical offices of the *zemstva*, local self-governed authorities of the Tsarist state provinces founded in 1864, the challenge consisted above all in creating the statistical institutions and tools of a modern state, in line with the recommendations of the international statistics congresses of the nineteenth century (Mespoulet, 2001; Mespoulet & Blum, 2003; on statistical internationalism in the late nineteenth century, see Brian, 1989).

So notwithstanding the political message conveyed by the Bolshevik government, the TsSU was organized on the model of the statistical agencies of late 19th-century European states, structured around a series of departments reflecting the main divisions of statistics of the time, and moreover constructed on the institutional and methodological bases of regional statistics as practised by the *zemstva*. The texts regulating its foundation and missions adhered to the spirit of the debates and resolutions of the international statistics congresses of the nineteenth century, as asserted by its director, Pavel I. Popov, speaking at the national congress of Russian statisticians held between 8 and 16 June 1918 (Popov, 1918). Encouraged by the experience of the statisticians in the regional offices of the *zemstva*, the Bolshevik state statistics administration was also resolutely organized in line with statistical internationalism, its scientific positions and organizational principles (on this point, see Brian, 1989).

This continuity with the pre-revolutionary period, relying also on continuity of the staff, was simply the visible part of other forms of inheritance in the representation of statistical work, methods and observational tools, and in the maintenance of certain administrative practices behind the apparent institutional changes (see Mespoulet, 2001, chapters 5 and 7). This strong continuity of individuals and practices coexisting with a disruptive political message was at the root of a great deal of tension between statisticians and political leaders from the early 1920s (see Blum & Mespoulet, 2003).

For the Bolsheviks, planned management, centralization and accounting had to be the bedrock of the organization of production in the future socialist state, and the statistics administration had to serve this political plan. While in line with the new plan model formulated by Lenin and the Bolsheviks prior to 1917, it should be noted that this conception of a centralized and planned economy was strengthened by the experiences of the First World War (see Holquist, 2002) and the civil war between 1918 and 1921 (Sapir, 1997), which provided a breeding ground for formulating and experimenting with a state-controlled economy (Stanziani, 1998), in particular to resolve supply chain issues.

In such a context, it is hardly surprising that the statisticians in charge of setting up the TsSU subscribed to this goal, in particular those who had played a part in the management of public and economic affairs during the war, in the Tsarist Ministry of Agriculture, or within the framework of the All Russian Union of towns and of the *zemstva*, or in the Provisional Government. However, such an endorsement of the Bolshevik plan did not preclude tensions and clashes between leaders and statisticians about the very definition of the role, methods and tools of state statistics. What form of quantification should be adopted to construct a socialist economic and social model tending towards communism? The first challenge consisted in how to define the connection between accounting and statistics.

A Complicated Demarcation Between Accounting and Statistics

The expression "socialist accounting" was used by Lenin as early as 1917 to describe the form of quantification aimed at providing figures to construct a socialist economy and society. In his speech to the Petrograd Soviet, Lenin said: "Socialism is accounting. If you want to record each piece of iron and fabric in the books of account, then that is socialism" (Lenin, 1955, vol. 26). Before the Revolution, he had already stressed the key role of accounting in his book entitled *The State and Revolution:* "Accounting and control, these are the chief things necessary for the organizing and correct functioning of the first phase of Communist society" (Lenin, 1955, vol. 21).

Initially, and for reasons of efficiency, the production of figures was to be controlled by their immediate practical applicability, that is their usefulness in guiding actions and decisions. This conception tended to strengthen the comparison of statistics with accounting, regarding it as a set of tools rather than a science. This demarcation of the role of statistics clearly comes across in the letter Lenin sent to the TsSU's director on 16 August 1921: "For practical work, we need to have figures, and the TsSU should have them before anyone else. But we will defer the verification of the accuracy of the figures, the estimated percentage errors, etc., to a later period" (Lenin, quoted in Iastremskii & Khotimskii, 1936, pp. 14–15).

Lenin does not provide a precise distinction between accounting and statistics. Rather, he differentiates between statistics as a "bourgeois science" inherited from the nineteenth century, and a newly emerging field of (socialist) statistical practice, accounting, which he sees as a practice of factual recording and counting, which does not involve any analysis or interpretation of reality, contrary to what, in his view, (bourgeois) statisticians did when producing their data. For Lenin, it was accounting which was needed for the construction of a communist economy and society. In his eyes, this construction project was an urgent undertaking and the "academic" efforts of (bourgeois) statisticians a waste of valuable time, standing in the way of proceeding with it as quickly as possible.

Economic and political urgency justified paying less attention to the use of statistical theory to verify the accuracy of figures, which led statisticians to restrain their scientific ambitions. The scope of statistics had to relate above all to the need to plan economic activity, which demanded a state statistics administration totally dedicated to this task:

> The central administration of statistics must not be an "academic" and "independent" body, which it currently is, for 9/10[ths] following old bourgeois habits, but one for constructing socialism, for verifying and controlling the accounts of what the socialist state needs to know now, today. (Lenin, 1955, vol. 28)

These Lenin quotes are not provided merely for form's sake. For it was in the name of his own vision of the role of statistical surveys in the management of Soviet economy and society that he regularly intervened in the TsSU's affairs until his death in January 1924 (see Kotz & Seneta, 1990). Then, at the 13th Congress of the Party in May 1924, Stalin in turn linked statistics back to accounting, while remaining very vague about the nature of the connection between them:

> No construction work, no work for the state and no planning is imaginable without correct accounting. But accounting is inconceivable without statistics. Accounting without statistics will not make a single step forward.[1]

The link between statistics and accounting was clearly reaffirmed in the early 1930s and symbolized by the TsSU being taken over by the State Planning Commission *Gosplan* in 1930. In the reference manual of statistics published that year under the direction of B. I. Iastremski and V. I. Khotimski (1931),[2] the authors reiterated the tasks Lenin had assigned to the TsSU in the introduction: statistics had to be associated with accounting, even in the study of social phenomena. In such a conception, the attention paid to the tools was essential. To better understand the decisions made for their development and use, we need to recall the definition of the scientific nature of statistics given in the 1930s, a key period in the formation of the Soviet system under Stalin.

In fact, just like scientists or professional specialists in other fields, the statisticians had to justify the status of statistics as a science in the early 1920s (on the status of science in the USSR, see Graham, 1993; Krementsov, 1997). They strove to develop an analytical framework for social and economic phenomena adapted to the plan to construct a socialist economy and society, setting themselves apart from the writings of European statisticians, described as "bourgeois" by the Soviet political leaders.

In this task of revision and reinterpretation, the Soviet statisticians were faced with a profound contradiction. Though rooted in an intellectual tradition inherited from the nineteenth century, which regarded statistics as a science on a par with chemistry or mathematics, during various conflicts they were forced to justify the socialist nature of their work and its conformity with the political plan to construct a new state, and its economy and its society. But for these statisticians, who associated figures with objectivity and scientific truth, all science sought to establish was a truth that could not be imposed by political leaders (on the notion of objectivity, see Porter, 1995). The TsSU's statisticians were subjected to various forms of political pressure that they nonetheless had to come to terms with, at times under duress, and that drove them to work out a theory in conformity with the Bolshevik political message about the construction of a socialist economy and society. How much of their theoretical constructs were based on compromise? In the initial stages, they embarked on a task of theoretical deconstruction.

A Task of Theoretical Deconstruction

In reality, it was rather a hollow definition of Soviet state statistics that was produced, as opposed to the statistics qualified as "bourgeois" to describe the theories developed in the nineteenth century. The manual published by B. S. Iastremskii and V. I. Khotimskii in the early 1930s begins with a denunciation of the latter:

> Like any science, statistics is a science of class. The whole system of statistics in capitalist countries is constructed to serve the interests of the ruling classes. Bourgeois statistics does not just serve to significantly reduce income tax of different capitalist groups, it also hides the actual amounts of military spending. Furthermore, bourgeois statistics gives a false picture of the situation of the capitalist economy by embellishing it and attesting to the absence of antagonisms. (Iastremskii & Khotimskii, 1931, p. 6)

The production of statistics in capitalist countries is presented as an undertaking of falsification serving a non-egalitarian state. Accordingly, its fundamentals cannot be used in a state that plans to achieve equality between all human beings. On that basis, the authors of the manual are led to refute certain theoretical contributions of the pre-revolutionary period by associating them with the fundamentals of statistics as used by the bourgeoisie of capitalist countries:

> Bourgeois statistics as practised is based on the bourgeois theory of statistics, which is organically connected to the whole system of bourgeois political economy and philosophy. The authors and theorists of the bourgeois science of statistics (Süssmilch, Quetelet, Lexis, Bortkiewicz, Pearson, Mittchell, Bowley, Moore, Chuprov, the fascists Pareto and Gini, and others) give arguments, with the aid of statistical constructs, extolling the "unshakeable" and "eternal" nature of the capitalist system. (Iastremskii & Khotimskii, 1931, p. 6)

The authors discredit the statistical reasoning developed in the nineteenth century (on statistics in the nineteenth century, see Porter, 1986), accusing it of encouraging the status quo and of curbing economic and social progress by basing natural and unchanging laws on the "regular stability of statistical figures":

> In the first half of the 18th century, pastor Süssmilch spoke of "the divine order" that manifested itself in the regular stability of statistical figures. In the 19th century, Quetelet, who admittedly said so without referring to God, spoke of a natural order that found expression through statistical figures, of an "average man" having a specific number of crimes, good deeds, etc. (Iastremskii & Khotimskii, 1931, p. 6)

This accusation (nineteenth-century statistics being a curb on progress) led to the plan to construct a system of "socialist accounting" serving a political programme of figures being used by the people:

> Lenin on many occasions said that accounting would become the business of the masses only after the overthrow of capitalism. But such accounting is impossible without statistics, which is also essential for the good of the masses. "In capitalist society, statistics was the exclusive reserve of 'people of the state' or narrow-minded specialists; we have to bring it to the masses, popularize it so that workers can gradually learn to understand themselves and see how and how much they have to work, how and how much they can rest..." (Lenin, 1955, vol. 21). The essential nature of disseminating

socialist accounting among the people requires maximum simplification of its technique. (Iastremskii & Khotimskii, 1931, p. 13)

The production of figures in a socialist state thus becomes a tool for the people, one that the people must appropriate, which requires the contents of statistics to be reduced primarily to technical mechanisms, assumed to be accessible to the greatest number. However, the link between accounting and statistics is not broken, even though it is not clarified. Such a conception gave rise to numerous discussions between political leaders and statisticians and among the statisticians themselves.

Debates and Tensions Surrounding Statistical Theory

Scientific debates were mingled with internal political rivalries (see Maksimova, 1996). These discussions were all the more heated when statisticians trained in Marxism in the new Soviet higher education institutions, such as the Plekhanov Institute for the National Economy[3] or the Institute of Red Professors,[4] started being recruited by the TsSU. The rifts between statisticians keen to maintain theoretical mathematical statistics and those who preferred more descriptive statistics based on surveys were reinterpreted during their controversies. The interpretation of the mean and of the law of large numbers, as well as the use of probabilities in statistics, were in particular the subject of heated debate.

Tension Surrounding the Mean and the Law of Large Numbers

As already mentioned earlier, the interpretation of the mean by Quetelet was accused of leading to moral fatalism, which negated the freedom of individualwill, and facilitated a form of social fatalism whereby society depended on superior laws that one could not combat (on Quetelet and the mean, see Desrosières, 1988; Porter, 1986, ch. 2 and 4; Thévenot, 1994). The fact is that these conceptions were in contradiction with the Bolsheviks' Promethean transformation plan.

The law of large numbers, which was one of the essential factors legitimizing statistics as a science serving the measurement of social facts, was condemned for the same reasons from the early 1920s. The Marxist theorists emphasized the fact that observed trends were not inevitable, and that such a concept led to putting the state's action into perspective. In their view, this statistical law contradicted the principle of political action, as it portrayed social processes as fatalistic and the effects of political action on society as illusory. Moreover, it helped to justify the idea of stability of capitalism and thus its unchanging dimension.

On the other hand, for the TsSU statisticians, statistics could not exist without the law of large numbers. Consequently, they strove to present a use of the said law that did not challenge its theoretical foundation while at the same time taking into account the attacks on it. In 1936, Vladimir N. Starovskii, head of the Central Statistics Administration,[5] proposed a wording full of stylistic acrobatics in a work published that year (Starovskii, 1936). True to Marxist thinking, he did not question the existence of laws that transcend individuals when having to deal with large populations, but he challenged the ineluctable nature of phenomena that appeared to him to be suggested behind the statistical regularities and the lack of references to history. He presented as essential the reference to the historicity of these laws, which he presented as specific to a given social and political system. That being the case, the state could transform the said laws and replace them with others. Human and political action was no longer at risk of being powerless. But a socialist use of the law of large numbers and the mean still had to be defined.

In fact, as they failed to establish the theoretical foundations of a form of statistics that could be described as socialist, the statisticians in charge of this task of reinterpretation endeavoured first of all to clarify what statistics should not be. Their reasoning consisted in calling into question the purportedly erroneous interpretations inherited from the pre-revolutionary past rather than in elaborating a new set of concepts adapted to a socialist economy and society. On the substance, it was therefore more a political debate on the use of the tools than disagreements on their theoretical foundations. Beyond the views and reasoning of convenience, the statisticians were forced to think about the interpretation of the tools they used and the creation or use of other tools better suited to the context of the moment. The tension around the concept of randomness and the use of probabilities is one such example.

Tensions Around the Shift to the Random Model

In 1917, Russia had already acquired extensive experience in the field of sample surveys. The first tests of surveys on a "part of the whole" probably took place in the 1870s. The period from 1885 to 1917 was marked by numerous efforts to perfect sampling methods, and by serious thinking by the Russian statisticians in this respect (Mespoulet, 2002). From 1895, they had the same discussions between themselves as their colleagues in other European countries about the shift from an exhaustive count to a partial survey, then about the shift from a sample constructed in a reasoned manner, based on the use of type, to a random sample based on calculated probabilities (on the history of sampling methods and sampling surveys, see Desrosières, 2002; Gigerenzer et al., 1989; Stigler, 1986). At the European level, the two questions of representativeness and confidence raised by a survey of "a part of the whole" were at the heart of a debate that started with Kiær's first communication at the Congress of the International Institute of Statistics (IIS) in 1895 and ended in 1925 at the IIS Congress in Rome, where both sampling methods—reasoned selection, also called purposive selection, and random selection—were accepted after heated discussions about their respective merits (on the debates during the period 1895–1925, see Desrosières, 1998 [1993], ch. 7; Kruskal & Mosteller, 1980).

The exuberant production of statistics in the USSR in the early 1920s created particularly favourable conditions for the rapid spread of sample surveys in the country (Mespoulet, 2002). Intensive use of such surveys from 1919 onwards stimulated thinking about sampling methods. After 1925, such thinking indeed developed much more as a logical extension to the questions raised by the already long-standing practice of Russian statisticians in this field than in relation to the questions discussed at the Congress of Rome, even though a summary on this subject was drawn up by the TsSU.

However, the shift to the random model took particular forms from the 1930s, in a country where the political leaders propounded a representation of society in which the collective prevailed over the individual, and in which an average collective behaviour was preferred to individual—and therefore diverse and dispersed—forms of expression. Moreover, the Bolsheviks' purposeful plan to construct a new state and their Promethean representation of human action on the environment gave a new dimension to the debates about the adoption of random sampling: what place could chance have in a planned world, in which there was precisely no place for uncertainty, and in which the unpredictable nature of individual behaviour and the variability of individual cases could not be factored in?

Planning concerns influenced the thinking about statistical representation and uncertainty, and consequently the forms of recourse to the calculation of probabilities. The treatment of statistical dispersion and of the mean, on the one hand, and the construction of sampling methods for sample surveys, on the other hand, depended on the status of the individual in society. The way of considering the question of confidence in the method for constructing samples influenced the forms of the shift to random sampling in the USSR.

Random selection, the product of the law of randomness, could only be considered within the bounds of a prior breakdown of reality controlled by human reason, in accordance with the Marxist concept of the primacy of human action over the economy and society. Chance could only operate if confined within a framework demarcated by human reasoning and compliant with political choices and directives. From that stemmed the basic Soviet statistical sampling method, developed in the 1930s, subsequently modified in the early 1950s. Whether under the name "random stratified sample" (*raionirovannyi sluchainyi otbor*) or "typical stratification" (*tipicheskoe raionirovanie*), the method for creating a sample based on a combination between the demarcation of "typical groups", firstly, and random selection, secondly, addressed the need to perform a prior breakdown of social reality into classified categories deemed relevant for analysing class structure or different types of agricultural or industrial production structures. This way of combining use of type and a specific interpretation of randomness characterized Soviet statistics from the 1930s to the early 1990s.

Mechanical selection was preferred to probabilistic random selection in the strict sense. The former was considered to provide more accurate results than random sampling, in the strictly random sense of the term, inasmuch as it guaranteed regular distribution of the sample's units over the entire set. In actual fact what was called mechanical selection was a systematic selection (1/5 or 1/10 for example) without replacement. This preference for mechanical selection was also a constant of Soviet statistics until 1991.

This form of constructing samples combines the two symbolic characteristics of the preference for typical stratified sampling: attachment to territory as the basis for division into typical groups, which dates back to the association between territory and type of the nineteenth century, and the principle of an a priori breakdown of observed reality into classifications. Random selection could only occur subsequently. From the 1930s, the classifications developed served to form "typical groups" based on an economic breakdown of reality into branches of production. In the process, they resulted in excluding a part of the population of the samples used in budget surveys for instance, casting out to an unknown world all those who were not counted in the productive sphere, pensioners among others. They thus departed from the principle of representativeness in relation to the population as a whole. This way of focusing statistics on groups or entities that mattered economically, socially and politically in the eyes of the leaders can be explained by the priority given to targets. This appears no less in contradiction with the desire to know everything, which is often ascribed to the Soviet leaders. In this case, the search for information appears to have been selective, satisfying political choices rather than statistical laws.

At the end of the 1960s, A. Boiarskii clearly addressed the problems raised by this sampling method:

> A sample formed on the basis of production is obviously a sample of individuals who work, and in that case the rest of the population is only included in the field of view inasmuch as it has a connection with those who work. However much we then adjust and correct the results, this sample will never replace a sample of another kind of population. With this sampling process, observation is still based on the habitual recourse to administrations and companies. But from data concerning all companies one cannot obtain an appropriate reflection of the life of all the population, all socialist society in its entirety. (Boiarskii, 1968, p. 16)

This tension between a realistic conception of statistics, which produces figures "reflecting" reality and was defended by the statisticians, and a representation of the figures, treated as one of the instruments for constructing a socialist reality and preferred by the political leaders, is regarded as one of the main components of Soviet statistics. In such a situation, the statisticians had to make efforts to adapt to the requirements of the moment; this took various forms according to the period and the circumstances. A. Boiarskii's remark implies moreover that the adjustments and corrections of results may in certain cases have resulted from this effort aimed at better matching the data with reality, despite the choices of methods or techniques that resulted in ignoring or masking certain aspects or dimensions thereof.

What Form of Statistics for Constructing a New Order?

The Soviet state was founded on a system of representations of management of the country predominated by the image of a stable universe, in which a planned economy was based on an organizational method designed to lead to an increase in general living standards and to an egalitarian society. Figures occupied a central position, being a tool for information and action, a tool for evaluating results and an instrument for proving the rightfulness of the measures taken. Central to such a system of administration of the economy and society, quantification was a basis for legitimizing the state and power. Accounting seemed to be the new language that the Soviet Union needed to construct a system of quantification serving the construction of socialism in the economy and society. But that also needed new categories of classification to be constructed to characterize the composition of society and to redefine the criteria on which confidence in the data produced by state statistics was based.

The Relationship Between Statistics and Accounting

In 1932 Valerian V. Osinskii, then head of the Central Directorate of Accounting for the National Economy (TsUNKhU),[6] defined the role of statistics in relation to accounting:

> During the transition towards socialism, accounting must encompass all spheres of economic and social life, and must penetrate all its links, even the smallest.
>
> [...] As the remains of capitalism were being swept from the economy and the consciousness of people, *statistics* was increasingly superseded by national accounting itself. The former TsSU did not become the TsUNKhU by accident. This change in name did not happen by accident, it characterizes the transformed orientation of activity of the system as a whole. (Osinskii, 1932)

V. V. Osinskii regarded statistics as a relic of capitalism. Henceforth serving the plan, statistics had to be transformed into accounting to draw up the national accounts. This unified form of accounting, from initial recognition to centralized processing, had to ensure uniform auditing and processing both of demographic and social phenomena and of economic activity, and guarantee continuity between companies' accounts and the national accounts. This idea of a unified accounting system was not exclusive to the USSR. It is found with variants, after the Second World War, in European countries that set up their own system of national accounting (see Studenski, 1958; Vanoli, 2002). In the USSR, it was intensified by bureaucratic management at different levels, from the basic department of a company to the central planning bureau of the *Gosplan* (Kornai, 1996, pp. 141–164).

Was accounting destined to replace statistics? The answer to this question remained ambivalent until the end of the 1930s, reflecting the difficulty in inventing a new quantification language centred primarily on accounting devices without resorting to statistical theory as a resource. So Stalin's formulation, at the 16th Party Congress in 1930, of the idea that "there can be no accounting without statistics" then gave rise to much prevarication and many changes in attitude in the definition of the connection between statistics and planning through accounting.

One such example is provided by the fate of the theory of the decline of statistics formulated by Osinskii in 1932 and subsequently denounced at his trial in 1937. The idea that statistics had to be transformed into accounting to serve the preparation of the national accounts was put to the test. This prompts one to put into perspective the formulations made during the 1930s about the conception of the scientific role of statistics, as the use of ideas for political ends was standard practice at the time. However, even if the role of statistics was once again officially recognized at the end of the 1930s, it was not at all well-defined. In reality, it was only in 1948 that statistics was fully restored as a field in its own right, when the TsSU regained institutional independence from the *Gosplan*, the figures of which it was tasked to audit. Its rehabilitation was completed in 1960 when Starovskii, then director of the TsSU, gave himself up to a form of self-criticism on the occasion of the publication of his work on the history of statistics in the USSR:

> In economic literature, all sorts of leftist 'theories' were formulated, for instance the 'theory' of the decline of money under socialism or the 'theory' of the decline of statistics under socialism (the transformation of statistics into accounting, which appeared for the first time in the articles of academic V.V. Osinskii) [...]. The author of this article, in his works, in the period 1932-1935, also asserted this 'theory'. Life has clearly demonstrated the pointless and erroneous nature of these 'theories', which distracted the attention of scientific managers from current questions surrounding statistical theory. (Starovskii, 1960, p. 16)

Nevertheless, the asserted integration of statistics into accounting had far-reaching consequences on the forms of the use of statistics and the tools used. Firstly, the period 1930–1955 was marked by the virtual disappearance of social statistics. The scientific and universalist ambition of nineteenth century statisticians, defended by the early heads of the TsSU, was supplanted by an immediate use of figures for accounting purposes to provide the *Gosplan* with the data needed to work out and evaluate five-year plans. The social figures produced between 1930 and 1955 were mainly demographic in nature. Statisticians became a kind of "right hand" providing the economistic planners with statistics needed for elaborating the plan, for instance for building blocks of flats, schools or leisure facilities.

On the other hand, a whole new area of statistics developed, based on economic forecasting and the construction of indexes. The statisticians elaborated new approaches directed at constructing a national accounting system, the reflections of which are already evident in the manuals published in the 1930s (see in particular Iastremskii & Khotimskii, 1936). The chapters dealing with "relative magnitudes" and indexes occupied a more important place in them, helping steer practices towards a purely accounting use of statistical techniques.

When the Methodological Council of the Central Directorate of Accounting examined the training programme on corporate accounting techniques and statistical theory in 1934, it stressed the need to limit statistical theory to what it described as the purpose of statistics: categorization, the calculations of averages, indexes and the sampling method.[7] The construction of indexes, for instance, was based on ratios aimed to compare two quantities, for example industrial production to agricultural production. The indexes were built for planning purposes and for the evaluation of the plan objectives. However, not by any means unbiased, categorization brought into play considerations that were much more than merely technical, namely social and political issues, in particular when it was a matter of constructing other tools, such as classifications of population censuses.

Categorization of the Population and Censuses

The work of constructing occupational classifications used for censuses combined the desire of the statisticians to reflect reality as closely as possible and the classification models officially accepted by the Party.

The tension between a "realistic" conception of statistics, which produces figures 'reflecting' reality defended by the statisticians, and an understanding that treated statistics as one of the instruments for constructing a socialist reality, which was preferred by the political leaders, can be regarded as a key feature of Soviet statistics.

Each census involved a confrontation between the desired construction of a socialist reality and forms of resistance to transformation of society or the economy, and it was important for the results to provide a snapshot of the volume and structure of the population. The census, which had to confirm the successful construction of a socialist society, was the central link in the statistical tools for constructing reality. From the mid-1920s, its production was subjected to increasingly frequent attempts at intrusion by the Party at various different data collection, processing and publication stages.

While as in any country the elaboration of classification categories for the individuals and phenomena under study was the subject of negotiation between various administrations or institutions, it raised particular issues in the USSR, where the administration of society was based on a classification of the population into categories defined by the Party. The TsSU's statisticians had to find the adaptations needed to match the bill of materials with, among other things, the structure of the social classes adopted by the Party or the official list of nationalities (on the classification of nationalities in the censuses, see Cadiot, 2007; Hirsch, 1997). This work of adjustment, which varied according to period, the political views and priorities of the moment, was a source of tensions between statisticians and political leaders and calls to order from the latter to the former. The example of the definition of occupational categories in the 1920s and 1930s clarifies certain forms of adjustment. In the 1926 census, the category "occupation" replaced "profession", which had been used by choice in the 1920 census, but which created difficulties for constituting the main social groups defined in the theoretical Marxist social analysis model(Mespoulet, 2008). Subsequently, behind a simplified classification into major social groups, called "classes" at certain periods, there remained a very detailed classification of "occupations" that reflected social diversity.

The purges of statisticians following the 1937 census halted the TsSU statisticians' efforts to reflect such diversity through censuses (on the Great Purges in the Soviet state statistical administration, see Blum & Mespoulet, 2003). From the 1939 census onwards, a stable image of social structure dominated through a classification of social groups that remained broadly unchanged until 1989. This virtual stability of the classification, which can be an advantage for comparisons over time, nonetheless resulted in obscuring some of the diversity of the social situations of individuals from the 1960s onwards. Generally speaking, the model of stable social aggregates, based on major groups, overlooked changes affecting small groups of individuals, regarding them as residual or marginal phenomena. In the process, it left blind spots with regard to the actual state of Soviet society and failed to spotlight its underlying transformations that broke out after the perestroika.

What room was left for the social in this model where the collective prevailed over the individual? In reality, social phenomena were studied mainly when they could be reduced to economic management. Furthermore, reflecting on how social matters were taken into consideration raises another question, namely the confidence one has in the data on society collected by the statisticians and in what the respondents say.

Confidence in the Data and the Status of the Statistician

As already mentioned, the question of how much confidence we can have in statistics explains the reservations one can have about using a random selection in the sample surveys method practiced in the USSR. This question is also expressed in a number of ways in connection with the collected data and the processing of individuals' responses to the surveys. How much confidence can one have in what respondents say, and thus in the data based on their responses? Beyond that, this point also raised the question of confidence in the work carried out by the statistician.

From the end of the nineteenth century, surveys were considered a social situation in their own right, in which the effects of interaction between interviewer and respondent played their part (Mespoulet, 2008). However, whereas this thinking had helped to consolidate the role of the professional statistician in conducting field surveys up to the early 1920, in particular in questioning individuals, it went in quite another direction from the early 1930s. The effects induced by the leeway respondents had in responding made them suspect in the eyes of the authorities, and the Party started intervening in the organization of data collection operations, more particularly by controlling recruitment and the work of local census takers.

In areas other than state statistics, in the early 1960s, the local Soviets and trade unions appointed "civic controllers and engineers" to audit certain stages of statistics work at a local level (see Anisimov, 1968; Sbeglov, 1967). In the late 1970s, an article in *Vestnik statistiki*, the TsSU's journal, pointed out: "The state's statistics bodies, with the active help of the Party, Soviet organizations and trade unions, do their work by recruiting civic activists on a large scale to audit the accounts, administrative registers and the authenticity of the data in the registers" (Vestnik statistiki, 1978, p. 67). This auditing of statistics by the citizens, in reality by social activists, presupposes a careful choice of those hired to do the work, who were also known as "civic inspectors of state statistics": "They were selected and confirmed nationwide by branch, according to worker qualification, and also according to the possibility of combining an inspector's *obshchestvennaia rabota* with his main job" (Vestnik statistiki, 1978, p. 67). Who were these inspectors, also called *obshchestvenniki,* civic activists?

> In principle, the most experienced specialists in the various branches of the national economy were recommended as civic inspectors of state statistics, and the nature of each district is also taken into account. These inspectors are accountants, economists, engineers in factories, administrations, organizations and kolkhozes (collective farms), workers in the plan and financial bodies. Applications are put forward with the agreement of the managers of the companies and organizations where the applicants work. They are examined with the leaders of the Party and the trade unions in the workplace, and only after that are they confirmed by the local executive committee of the district or town, the soviets of the people's deputies. (Vestnik statistiki, 1978, p. 67)

What precise tasks were expected of these inspectors? First and foremost, their function entitled them to demand that the managers of factories, work sites, government departments, organizations and kolkhozes disclose their accounting records and work registers for inspection purposes. This consisted in auditing accounts and figures in factories and other workplaces, by examining the original documents. On that basis and with the aid of the managers of these establishments, the inspectors could take measures in situ aimed at eliminating the defects brought to light, and also ask for the collectives of workers subject to this inspection process to be convened in order to discuss the results thereof.

In 1978, there were nearly 30,000 inspectors covering most of the districts of the Soviet republics, half of whom were in the Federal Republic of Russia (Vestnik statistiki, 1978, p. 68). Most of them had a secondary school or higher level of educational attainment, and many of them were members of the Party or candidates for Party membership.

Whether it be a control of the data provided by factories or administration registers, by censuses or budget surveys, this inclusion of data collection operations in the sphere of civic activities demonstrates a special way of treating interaction between interviewers and respondents in an authoritarian state such as the USSR. The question of the degree of confidence one can have in the information provided by the respondents, a subject of discussion for Russian statisticians from the 1880s onwards, remained an important factor in Soviet statistics, but the way it was resolved took on a different form. Before 1917, Russian statisticians shared the idea that it was up to the interviewer himself to behave in such a way as to gain the trust of the respondents and thereby elicit more accurate and comprehensive responses from them. After 1917, the Party had a quite different view of the way of handling trust in relations with respondents.

In the censuses of 1919 and 1920, against a backdrop of grain requisition campaigns, the Bolsheviks suspected the peasants of holding back information, especially about their harvests and stocks of grain production. So they were considered guilty *in principle*. This suspicion of the leaders vis-à-vis the peasants was a constant in the 1920s and 1930s. So, an assessment of the confidence one would have in the information provided by the respondents was not left solely to the statisticians, suspected *in principle* of possibly colluding with the respondents. The degree of confidence one could have in the collected information therefore had to be handled differently, both by the Party and its social organizations. The control of the statistics would be done by new forms of social and political control. The census preparation and collection operations in the various districts were controlled by Party bodies and social organizations until 1989.

The intrusion of the Party and the social organizations in the conduct of field survey operations, both for censuses and for surveys of family budgets, socialized this moment of statistical work while at the same time divesting part of the statistician's professional competence.

In that light, Lenin's statements regarding the need to "popularize" statistics, in the sense of making it accessible to the people, take on a particular relevance. The ambition of the Soviet regime was to create a "new human being". At the centre of this project, all aspects of social life were deemed susceptible to control by the mass organizations representing the people. Who better than the collectors of the people's data to control the statements of citizens subject to survey? The verification of the reliability of the gathered information compared with reality took the form of social control. Such inclusion of certain statistical operations in the sphere of civic activity restricted the statistician even more to a primarily technical role, cutting him off from the very source of the information he had to process.

What Statistical Tools for a New Order?

Up to the 1950s, the management tool aspect of statistics was regarded as the main one in the apparatus of state statistics and was closely linked to planning. However, surveys of family budgets, which were precisely at the point where statistics and planning intersected, offer a good example of a technical mechanism that in certain cases could also be used as a tool for acquiring knowledge about society, which Boiarskii, then Director of the TsSU Research Institute, himself acknowledged in 1968 when, following new directives from the 1967 Party Congress aimed at developing the social sciences, the TsSU had to "intensify the statistical study of social phenomena" (Boiarskii, 1968). For Boiarskii, surveys of family budgets were the preferred tool for assessing what he included in the subjects of study of social statistics: problems at work, wages, consumption, services, housing, everyday life. Apart from that, health, the composition and movements of the population and the family structure were in his view in the realm of demography, which also encompassed social statistics. In fact, the demarcation of the latter was dictated by an imperative: all these questions had to be studied as "a reflection of the economy's effects on all other phenomena of social life" (Boiarskii, 1968, p. 14).

In reality, from the 1950s, and even more so from the 1960s, many of the tools and mechanisms used in Soviet social statistics resembled those used in capitalist countries (Elisseeva, 2003). The difference was not so much in the type of tools used but rather in the way they were used, which Boiarskii himself pointed out, for instance in connection with the methods of sample surveys conducted by Soviet sociologists in the late 1960s:

We cannot compare our surveys with those of bourgeois sociologists without due consideration, on the basis of specific criteria. The main difference lies in the fact that their programme (both of observation and of processing) is constructed in such a way that by expressing itself in a manner full of imagery one can't see the wood for the trees. Their specific nature does not stem from the fact that one proceeds by conducting sample surveys or by conducting a survey, but rather from the fact that the whole construction of these surveys is not designed to bring to light what matters most: the role of relations of production that, all things considered, in reality determine all these processes. (Boiarskii, 1968, p. 17)

Much more so than the tools as such, and the statistical theory underpinning them, the difference resided in the approach and the model for interpreting the reality in the framework of which they were used:

Our sample surveys are constructed on quite different bases and principles – the principle of the historical materialism of sociology, in our Soviet, Marxist conception of the word. It is precisely in that, and not in the technical question, that resides the dividing line between the surveys of sociologists in socialist countries and those of bourgeois sociologists. The very forms of the work may be similar. (Boiarskii, 1968, p. 17)

National accounts, demographic censuses, budget surveys, all these tools were used for a similar purpose in different European countries after the Second World War, whether to measure production levels or living standards or to analyse trade circuits for the purpose of macroeconomic regulation and increasing material well-being in a country. In the case of the USSR, victory against the United States to claim the title of superpower also depended on success in the economic competition. The Soviet socialist system thus had to demonstrate its superiority in the same economic and social fields as the capitalist countries, which meant that they tended to adopt the same tools, while re-interpreting their uses.

Conclusion

Quantification can be analysed as a way of legitimizing a fragile and young power, if one considers the USSR of the 1920s and 1930s, but it can be said that this was still true in the 1950s and 1960s, in the period of confrontation with the capitalist states. Adam Tooze has shown how statistics have been used to boost the legitimacy of the German state under the Weimar Republic (Tooze, 2001). In his view, statistics offered the Weimar Republic a new, attractive and credible language for government. Theodore Porter for his part has shown how the use of quantitative language also goes hand in hand with a transformation in the bases of authority of the expert, in this instance the statistician (Porter, 1995). To what extent can one speak of socialist statistics in the case of the USSR?

At first sight, what is striking is the difficulty that political leaders and managers of the TsSU had in defining a new conceptual framework specific to state statistics, both in the 1920s and 1930s (the period of formation of the Soviet system) and subsequently. Marking a break with the pre-revolutionary economic management systems, planning had to have a quantitative language and new tools, radically different from those of statistics nourished by 19th century scientific representations and the bourgeoisie.

Accounting seemed to provide this new language that Soviet Russia needed to construct a quantification system serving the construction of communism. In reality, statistics was never abandoned, even if its existence was hotly contested in the period of formation of the Soviet system. The expression "socialist statistics" was used to refer to a different way of producing figures from the way that capitalist countries produced them. However, although we can speak of socialist statistics, it is not so much to refer to a specific theory, concepts and tools, but rather to a set of uses of them, which, depending on the period, were reinterpreted and adjusted to the objectives set by the political plan of the leaders of the Soviet state.

Against this backdrop, the Soviet statistician was caught between two poles: heroes of science, or engineers of figures in the service of the authorities. A way of resolving this dichotomy consisted in the state restricting the role of the statistician to a primarily technical role in a Directorate of Statistics or a computational centre, without any expectation of a scientific role, which was entrusted to various research institutes (on the Soviet organizational system of science after the Second World War, see Graham, 1993; Krementsov, 1997). This transformation of the status of the statistician is inseparable from a conception of quantification adapted to a planning model, but also a form of state where power was exercised in an authoritarian manner and figures were considered a propaganda tool.

The authority of the expert statistician was no longer based on his independence vis-à-vis the political leaders. On the contrary, the components of his professionalism were defined by the latter. The expert was kept at arm's length from the authorities, but at their disposal. The part of his work that was closest to the population, such as data collection for censuses or household budget surveys, was until the 1970s placed under the political control of the Party, being included in the sphere of civic activities.

NOTES

1. Source: *XIII s"ezd VKP (b), Stenograficheskii ochet* (XIIIth Congress of the Communist Bolshevik Party of Russia (b), Shorthand report), p. 130.
2. A new copy of this handbook was edited each year from 1930 to 1936. One of its editors, Boris S. Iastremskii (1877–1962) was a mathematician specialized in the field of probability calculation. From 1918 to 1933 he headed the Department of Statistical Methodology of the TsSU. He became a member of the Party in 1931. The other editor, Valentin I. Khotimskii (1892–1939), was a former student of Aleksandr A. Chuprov and was specialized in the field of the probabilities. Although he was close to the Party he never became a member. After teaching mathematics from 1924 to 1927 at the Plekhanov National Institute for the National Economy, he became a researcher in the mathematical section of the Communist Academy. He directed it until 1932 and then was appointed at the head of the controlling sector of the Gosplan for Russia. In 1935 he was appointed at the direction of the Statistics Department for Population and Health of the Directorate Accounting for the National Economy (TsUNKhU), that was in charge of the organization of the 1937 and 1939 demographic censuses. The TsUNKhU had replaced the TsSU in 1930. V. I. Khotimskii was arrested in 1937 during the Great Purges. Most statisticians taking part in the organization of the 1937 census were arrested and put in prison. V. I. Khotimskii died in 1939.
3. The Plekhanov Institute for the National Economy was re-formed in 1924 in Moscow. It had a Department of Statistics.
4. The Institute of Red Professors was created in 1921 in Moscow. It was placed under the authority of the Central Committee of the Party.
5. Vladimir N. Starovskii (1905–1975), was recruited as a statistician to the TsSU in 1925. He became a member of the Party in 1939 and was appointed to the direction of the 1939 demographic census after Stalin had rendered void the results of the 1937 census. After being Deputy Director of the TsUNKhU in 1939 and 1940, he was appointed to the direction of the new TsSU when it was re-formed in October 1940. He remained in this function until he died in 1975. From March 1941 he was also vice president of Gosplan.
6. The TsUNKhU replaced the TsSU in 1930. Valerian V. Osinskii (1887–1938) was a member of the Bolshevik Party since 1907. Immediately after October 1917 he was director of the State Central Bank and president of

The Supreme Council for the National Economy. After 1918 he fulfilled various leading functions. In 1926 he was appointed head of the TsSU to give a new political direction to the Central Statistics Administration. He was replaced in this function in 1928. A few years later he was appointed head of the new TsUNkhU. He was arrested in 1937 during the Great Purges and executed in 1938.
7. See RGAE, f. 1562, op. 1, d. 749, ll. 138–154.

REFERENCES

Anisimov, S. (1968). S pomoshchiu obshchestvennykh inspektorov (With the help of civic inspectors). *Vestnik statistiki, 49*(5), 73–74.

Blum, A., & Mespoulet, M. (2003). *L'anarchie bureaucratique. Statistique et pouvoir sous Staline.* La Découverte.

Boiarskii, A. I. (1968). Issledovanie sotsial'nykh yavleniy i gosudarstvennaia statistika (The study of social phenomena and state statistics). *Vestnik statistiki, 49*(3), 14–21.

Brian, E. (1989). Statistique administrative et internationalisme statistique pendant la seconde moitié du XIXe siècle. *Histoire & Mesure, 4*(3–4), 201–224.

Cadiot, J. (2007). *Le laboratoire impérial. Russie-Urss 1870–1940.* CNRS Editions.

Desrosières, A. (1985). Histoire de formes: statistiques et sciences sociales avant 1940. *Revue française de Sociologie, 26*(2), 277–310.

Desrosières, A. (1988). Masses, individus, moyennes: la statistique sociale au XIXe siècle. *Hermès, 2*(2), 41–66.

Desrosières, A. (1998 [1993]). *The politics of large numbers: A history of statistical reasoning.* Harvard University Press.

Desrosières, A. (2002). Three studies on the history of sampling surveys: Norway, Russia-USSR, United States. *Science in Context, 15*(3), 377–383.

Elisseeva, I. I. (2003). *Sotsial'naia statistika* (Social statistics). Finansy i statistika.

Gigerenzer, G., Swijtink, Z., Porter, T. M., Daston, L., Beatty, J., & Krüger, L. (1989). *The empire of chance.* Cambridge University Press.

Graham, L. (1993). *Science in Russia and the Soviet Union: A short history.* Cambridge University Press.

Hirsch, F. (1997). The Soviet Union as a work in progress: Ethnographers and the category nationality in the 1926, 1937 and 1939 censuses. *Slavic Review, 56*(2), 251–278.

Holquist, P. (2002). *Making war, forging revolution: Russia's Don Territory during total war and revolution, 1914–21.* Harvard University Press.

Iastremskii, B. S., & Khotimskii, V. I. (1931). *Teoria matematicheskoi statistiki* (Theory of mathematical statistics). Sotsekgiz.

Iastremskii, B. S., & Khotimskii, V. I. (1936). *Statistika. Osnovy obshchei teorii* (Statistics: The founding of a general theory). Soyuzorguchet.

Kornai, J. (1996). *Le système socialiste. L'économie politique du communisme.* Presses Universitaires de Grenoble.

Kotz, S., & Seneta, E. (1990). Lenin as a statistician: A non-Soviet view. *Journal of the Royal Statistical Society, 153*(A), 73–94.

Krementsov, N. (1997). *Stalinist science.* Princeton University Press.

Kruskal, W., & Mosteller, F. (1980). Representative sampling IV: The history of the concept in statistics, 1895–1939. *International Statistical Review, 48*(2), 169–195.

Lenin, V. I. (1955). *Pol'noe sobranie sochinenii* (Complete works), 55 volumes (5th ed.). Gosudarstvennoe Izdatel'stvo politicheskoi literatury.

Maksimova, V. N. (1996). Iz vospominanii (20-30e gody) (Extracts of my memories,1920–1930s). *Voprosy statistiki, 77*(10), 78-87.

Mespoulet, M. (2001). *Statistique et révolution. Un compromis impossible (1880–1930).* Presses Universitaires de Rennes.

Mespoulet, M. (2002). From typical areas to random sampling: Sampling methods in Russia from 1875 to 1930. *Science in Context, 15*(3), 411–425.

Mespoulet, M. (2008). *Construire le socialime par les chiffres. Enquêtes et recensements en URSS de 1917 à 1991.* Ed. de l'Ined.

Mespoulet, M., & Blum, A. (2003). Le passé au service du présent. L'administration statistique de l'Etat soviétique entre 1918 et 1930». *Cahiers du Monde russe, 44*(2–3), 343–368.

Osinskii, V. V. (1932). *Chto znachit uchet* (What accounting means). Soyuzorguchet.

Popov, P. I. (1918). Organizatsiya gosudarstvennoe statistiki (The organization of the State Statistics Administration). In RGAE (Ed.), *f. 1562, op.1, d. 28.*

Porter, T. M. (1986). *The rise of statistical thinking, 1820–1900.* Princeton University Press.

Porter, T. M. (1995). *Trust in numbers: The pursuit of objectivity in science and public life.* Princeton University Press.

Sapir, J. (1997). La guerre civile et l'économie de guerre, origines du système soviétique. *Cahiers Du Monde Russe, 38*(1–2), 9–28.

Sbeglov, N. (1967). Nashi dobrovol'nye pomoshchniki (Our voluntarily inlisted assistants). *Vestnik statistiki, 48*(11), 61–64.

Stanziani, A. (1998). *L'économie en révolution. Le cas russe, 1880–1930.* Albin Michel.

Starovskii, V. N. (1936). *Azbuka statistiki* (The ABC of statistics). Soyuzorguchet.

Starovskii, V. N. (1960). Sovetskaia statistitcheskaia nauka i praktika (Soviet statistical science and practice). In V. N. Starovskii (Ed.), *Istoria sovetskoi gosudarstvennoi statistiki. Sbornik statei* (The history of Soviet State Statistics. Collection of articles). Gosstatizdat.

Stigler, S. M. (1986). *The history of statistics: The measurement of uncertainty before 1900.* Harvard University Press.

Studenski, P. (1958). *The income of nations.* New York University Press.

Thévenot, L. (1994). Statistique et politique. La normalité du collectif. *Politix, 25,* 5–20.

Tooze, A. (2001). *Statistics and the German state, 1900–1945: The making of modern economic knowledge.* Cambridge University Press.

Vanoli, A. (2002). *Une histoire de la comptabilité nationale.* La Découverte.

Vestnik statistiki. (1978). Uchet i gosudarstvennaia ochetnost' – pod kontrol' obshchestvennosti (Statistics and national accounting under the control of the activity of the mass organisations). *Vestnik statistiki, 59*(12), 67–70.

CHAPTER 3

The People's Algorithms: Social Credits and the Rise of China's Big (Br)other

Tong Lam

In 2013, just a few months after Chairman Xi Jingping came into power in China, the government declared that the country was entering a new era and launched a project called "the Chinese Dream" (*Zhongguo meng*). In order to realize this dream of "national rejuvenation", the Central Committee of the Communist Party issued a set of guidelines aiming at the cultivation of what it referred to as "core socialist values" that it divided into three respective categories: national goals (prosperity, democracy, civility and harmony), social goals (freedom, equality, justice and the rule of law), and individual values (patriotism, dedication, integrity and friendship) (Gow, 2017). The making of state-defined "civilized" (*wenming*) and "high quality" (*gao sushi*) political subjects for the newly enounced social and economic order, in short, is an integral part of the so-called Chinese Dream.

Almost immediately, propaganda slogans began popping up everywhere, from giant LED billboards on main avenues to the pages of

T. Lam (✉)
University of Toronto, Toronto, Canada
e-mail: tong.lam@utoronto.ca

A. Mennicken and R. Salais (eds.), *The New Politics of Numbers,*
Executive Politics and Governance,
https://doi.org/10.1007/978-3-030-78201-6_3

school textbooks. Predictably, Western observers immediately contrasted the Chinese Dream with the American Dream, pointing out that while the Chinese version might have borrowed a concept from the United States, it focuses on collectivism rather than individualism (Kai, 2014). Implicit in their argument is not only just a criticism of authoritarianism's and nationalism's suppression of individual aspirations but also an embrace of a certain belief in the ability of the autonomous individual to exercise reason that is thought to be fundamental to the operations of the market and democracy. China, in their estimate, needs to unleash that potential if it is to fully align itself with the Western economic and political order and become a true global leader.

This narrative deploys a certain moral filter to make sense of the world and human behaviour and has a long genealogy traceable to the Enlightenment. And in spite of the recent financial crisis and repeated electoral catastrophes in liberal democracies on both sides of the Atlantic, the foundation of that moral conviction has remained largely unshaken. The recent critique of fake news, alternative facts and misinformation (all of which have been made possible by the deployment of politically motivated computer algorithms and analytics) is heavily grounded in the insistence on the importance of real facts and the human capacity to reason.[1] Yet in the face of popularist political division and even violence fed by data-driven technology and the feedback loop in our post-truth world, it has become clear that the notion of self-determination and self-governance of the autonomous individual is being called into question (Rahwan, 2018).

This essay equally questions the adequacy of this view of human capacity, especially in the contemporary digital landscape, by examining the recent introduction of the social credit (*shehui xinyong*) rating system in China. As a new technology of governance, the social credit system is intended to track and calculate the social credit scores of every Chinese citizen and organization based on their activities and performance. Significantly, this government-mandated big data and surveillance project involves more than just the mining and processing of data by the state and corporations; it also seeks to compel individuals and groups to regulate themselves tirelessly based on the social and ethical order sanctioned by the state. Under such practices of state-led neoliberalism, the government uses social engineering interventions—the idea of the social credit in this case—to promote the ideas of marketization, social harmony, innovation, entrepreneurship, the rule of law, and other state-defined "core socialist values". By subjugating the everyday to neoliberal logics and normalizing its citizens through self-regulation, postsocialist China is moving away from the older socialist system of surveillance. In doing so, it has also given up on the dream of creating enlightened and critical citizens once cherished by Chinese intellectuals and revolutionary vanguards a century ago. Instead, it edges towards a posthuman world where citizens are fast becoming calculable and mouldable data subjects.[2]

Of course, Chinese citizens are not alone in their subjugation to sophisticated digital surveillance, as it has been demonstrated by Shoshana Zuboff's (2019) study of surveillance capitalism in market democracies. Yet the ubiquitous and conspicuous way in which the everyday activities of the individual are being tracked and regulated by a single party-state with an explicit agenda of behaviour modification is unprecedented. Moreover, whereas surveillance capitalism focuses primarily on capital accumulation, the Chinese social credit system also includes a political and ideological dimension that cannot be subsumed completely under the logic of capital. As such, Big Other—the "instrumentarian power" enabled by the vast surveillance infrastructure for herding and moulding society as suggested by Zuboff (2019)—is even bigger in China. Given the resilience and strengthening of the authoritarian party-state, one can even refer it to as Big (Br)other.[3]

THE EARLIER CHINESE DREAM

The desire to reform the thought of the individual is at least partially rooted in China's looming existential crisis at the beginning of the twentieth century. When the multi-ethnic Qing empire led by the Manchu ethnic group (1644–1912) repeatedly suffered major military defeats and setbacks in political and institutional reforms in its final decades, many Han Chinese intellectuals came to believe that the failure of the empire was due to its inability to create a unified body politic to counter the encroachment of foreign powers. The prominent intellectual Liang Qichao (1873–1929), for example, argued that the Chinese nation emerging out of the crumbling empire was in dire need of an organic society. According to him, the prerequisite to forming such a society was to create national citizens who were motivated and enlightened. Those who led the top-down revolution that ultimately toppled the dynasty also shared this view. Sun Yatsen, the revolutionary leader and "father of the republic", also famously castigated the disorganized and disunited state of the Chinese nation (Lam, 2011, p. 9). Underlying this line of reasoning was a fundamental shift in the political logics of the state from that of the Manchu-led dynastic empire to the Chinese nation. Political legitimacy, similarly, was now derived from the people rather than from the imperial lineage and divine sources.

When the new republic disintegrated soon after its establishment, most intellectuals blamed the top-down approach to political change. They further emphasized that if China wanted to institute a modern political order, creating a functional society with politically awakened citizens was critical. Thus, during the 1910s and 1920s, many Chinese intellectuals spoke of the need to create a "new culture" based on science and democracy. They vernacularized language for the masses and carried an education campaign to the countryside with the hope of turning the nation's mass population into new citizens, making this period a sort of Chinese enlightenment (Lam, 2011, pp. 38–45; Schwarcz, 1986). Yet, owing to the political imperative of the time, the idea of turning individuals into enlightened citizens quickly gave way to the idea of producing a people who would adhere to the newly declared social and political order that was seen as vital to the survival of the nation. Being politically aware, in this new context, was to acknowledge the priority of the collective over the individual.[4]

Immediately after the Second World War, unsurprisingly, officials and academics of the Nationalist government also began to contemplate how to put the population under surveillance as part of the national reconstruction project. Nevertheless, it was only after the founding of the People's Republic in 1949 that the dream of engineering the new citizen on a large scale became possible. Among other things, a system of household registration was put in place, subjugating individuals, workplaces, schools, neighbourhoods and so forth, to a new administrative order legible to the surveillance state. While such a system was no doubt partially drawn from practices used in the Soviet Union and the Eastern Bloc, scholars have also noted that the Chinese population was put under surveillance in the imperial era (Lu & Perry, 1997). The social surveillance system in twentieth-century China can thus be seen as a case of the modern bureaucratic state appropriating both native and foreign ideas for its state-building needs. This essay takes up one aspect of this vast surveillance network, the personal file or dossier system (*renshi dangan* or *geren dangan*), as it offers a meaningful departure point for understanding the significance of the new social credit system in the era of big data.

PERSONAL DOSSIERS

In many ways, the specific idea of putting the behaviour and thought of the individual under constant surveillance followed directly from the way in which party cadres were managed within both the Communist Party and the Chinese Nationalist Party (Huang, 2002). After the founding of the People's Republic in 1949, ideology was seen as key to the Communist Party's consolidation of its control of the government and the country. In 1956, the Party issued a set of guidelines regarding the management of the personal files of its cadres (Huang, 2002). Soon, the system was expanded to cover all urban residents. At a time when Communists were struggling to bring the country under their firm control amid heightening Cold War anxiety, one main purpose of the more elaborated surveillance system was to identify and eradicate the so-called class enemies and foreign spies. Thus, seeking more than to just discipline the docile bodies of the people in order to prepare them to mobilize for war and economic production, the state now also strove to monitor and reform their minds in order to secure the revolution. In other words, as China transformed into a "dossier society", it departed further from the aspirations of creating the free-thinking new citizens the intellectuals of the turn of the twentieth century had hoped for.

Generally, the personal dossier for urban residents is created when a child enters the school system and tracks his or her character, attitudes, performance and social relationships. Although Chinese citizens have no direct access to these files themselves, these dossiers literally follow them throughout their lives, leaving no temporal and spatial gaps. During the socialist era when a large segment of the Chinese society was organized into work units (*danwei*), the local unit was responsible for the updating and storage of the dossiers. In schools, for example, student dossiers were kept up-to-date by teachers. Likewise, in workplaces, individuals were evaluated periodically by supervisors and peers. To a certain extent, the Chinese socialist dossier system was similar but not identical to its counterparts in the Eastern Bloc. For instance, in East Germany, unlike in China, information about targeted individuals was collected by recruited informants and secret state agents, and those records were centrally managed by the Ministry of State Security commonly known as the Stasi.[5]

The Chinese dossier system was a central pillar in the social surveillance system of the party-state, as it allowed the state to monitor the moral character, work ethic, ideological leanings and social relationships of its urban citizens, workers, students, not to mention its own cadres.[6] Furthermore, the content of these dossiers was often an important factor in determining the individual's eligibility for opportunities and benefits such as transferring to a better school, promotion, better housing or admission to the party.[7]

In short, even if the tracking of the individual through the dossier was only part of the larger surveillance infrastructure, it was an important one.[8] And the idea of having a dossier trailing the life of a citizen like a shadow, deciding his or her individual fate based on past behaviours and attitudes, certainly invokes the menacing imagery of Big Brother. Still, this sort of imagery may have overlooked the nuances, failures and contradictions of the system in practice. Not only were most rural citizens or the so-called peasants not subjected to the dossier system, but calling in personal favours, exacting revenge and seeking leniency were conceivably always part of the game for those who were. In the film *The Lives of Others (Das Leben der Anderen)* that depicts the surveillance programme in former East Germany, for instance, the Stasi agent assigned to monitor a subversive writer ends up empathizing with his subject and eventually refuses to properly report his illegal activities. In postsocialist China, ideological control is more relaxed and so the ability of the system to keep track of individual citizens' thoughts and behaviours has probably become even less effective.[9]

Indeed, the end of the socialist era in 1978 and the subsequent introduction of a mixed economy have produced new challenges for the dossier system. Since the 1980s, a growing portion of the population has not been employed by traditional work units, such as government or state enterprises. The non-government workforce has become even bigger since the 1990s due to intense privatization. In order to address the changing social and economic order, talent exchange centres (*rencai jiaoliu zhongxin*) with field offices in cities all over China were created. Among their many functions, these government-run centres are responsible for keeping files on urban residents who do not work for state-assigned work units. Under this new system, urban residents outside of the state employment system, along with their employers, such as private or foreign corporations, are required to make sure that their files are properly maintained by the relevant local field offices.

When these field offices first opened in the early- to mid-1980s, they only served a relatively small number of workers who were in high demand—normally experts or workers with foreign language skills—who worked for foreign companies or were part of Sino-foreign joint ventures. However, because more and more workers are no longer working for the government or state enterprises, these talent exchange offices have evolved into general employment centres for the public. Meanwhile, the dossiers maintained by these offices have begun to function as a kind of resume for school and employment and even as evidence when it comes to individuals' entitlements to social insurance and social security benefits (Wang, 2011, p. 27).

How does a surveillance programme that was initially designed to enable political and ideological control interact with the country's emerging new social and economic order? This is a central question that Chinese officials and policy thinkers have been grappling with (Edin, 2003). As relocation, job changes, business closures and restructuring have become common occurrences, so too has the misplacement and loss of dossiers, filing errors and other management mishaps. Since such occurrences have direct impacts on the livelihoods of affected individuals, disputes over the accuracy of the information in the dossiers have been on the rise. Policy thinkers are unsure whether they should classify these as labour disputes, administrative mishaps or civic disputes, as each of these categories has different legal ramifications (Wang, 2011, pp. 27–28). The stakes are certainly high, since any mishandling of these cases could contribute to social discontent and political instability.

Social Credit

Although the rise of the social credit system is not directly linked to the erosion of the original function of the personal dossier system, it does represent the latest attempt to create new citizens by the state. In fact, the idea of placing the moral character of each citizen under surveillance jibes with neoliberalism. The new focus, however, is no longer on ideological purity for political purposes but on trustworthiness as a basic condition for economic efficiency, because trustworthiness is thought to be vital to minimizing economic risks and facilitating transactions. Chinese policy thinkers share the belief of advocates of rational choice and game theory that economic development proceeds apace with the level of social trust (Liu, 2016, pp. 30–39; Zak & Knack, 2001). In this context, trust is more than an emotional or psychological issue; it is also an important economic variable. The key question then is how to turn trust into social capital, and how to turn social capital into quantifiable and calculable social credit. And it is in the light of this imperative of converting trustworthiness into creditworthiness that the constant surveillance of the moral character of the individual is thought to be highly relevant and even critical in establishing and maintaining the neoliberal social and economic order (Zhongguo Guowuyuan, 2014, 2015).

The social credit idea first began to circulate around 2000. Prior to that time, this concept was only mentioned rarely, even though the experiment with marketization had accelerated in the 1990s. Since 2000, however, thousands of articles mentioning this concept have appeared in magazines and academic journals, mostly in finance-related fields but also in governance.[10] Nevertheless, it was not until 2014 that a detailed outline of the new system, called the "Planning Outline for the Construction of a Social Credit System", first came to light (Zhongguo Guowuyuan, 2014). Jointly released by the Central Committee of the Chinese Communist Party (the highest administrative body of the Party) and the Chinese State Council (*Zhongguo Guowuyuan*) (the highest administrative body of the central government), the document reveals a central initiative of the government's ongoing effort to "strengthen and innovate social management" (Zhongguo Guowuyuan, 2014). As part of the proposed thirteenth five-year plan (2016–2020), the planning outline stipulates in no ambiguous terms that credit is the foundation of all market operations and that a market economy is essentially a credit economy. Moreover, it further argues that the social credit system is vital to the functioning of the socialist market economy and to social governance. In so doing, it lays out the rationale for radically economizing and financializing the social world in an unprecedented way.

Needless to say, the idea of using quantifiable data to rate the creditworthiness of an individual or an organization is neither new nor unique in China or elsewhere. Yet unlike in countries with well-established credit infrastructures, Chinese credit rating agencies often have difficulty tracking rural residents, migrant workers and students. Moreover, while the cash economy is increasingly replaced by app-based direct transfer and payment platforms such as WeChat and Alipay (Liu, 2016, p. 169), these methods of payment do not contribute to establishing credit in the conventional sense. Therefore, as the planning outline points out, the existing credit rating system in China is sporadic and fragmented at best. From the perspective of the government, the system misses the opportunity to piece together different databases and the different platforms consumers use to construct a fuller picture of individual citizens, organizations and society at large.

With this in mind, the newly introduced social credit system is designed to deliver an aggregated, albeit not necessarily total, information system that emphasizes uniformity, consistency, comprehensiveness, accuracy, efficiency and up-to-dateness. It tracks the creditworthiness of citizens, enterprises, institutions and even government agencies using a uniform framework. In order to facilitate the implementation of the new system, the government has also started to introduce new laws, regulations and standards for the social credit system that were intended to be fully implemented by 2020 (Zhongguo Guowuyuan, 2014, 2015). Just months after the outline was published, the development of a system of national unified social credit codes was named a top government priority. These codes, not to be confused with credit ratings themselves, are standardized identification numbers assigned to all citizens and organizations. Such a nationwide system of standardized credit codes is to facilitate the sharing and exchanging of credit information among governmental agencies, enterprises and social organizations.

Significantly, even if the installation of a social credit system for tracking individuals meticulously and constantly may sound like an Orwellian nightmare, the rationale of the system is generally not articulated in negative and repressive terms. More often than not, it highlights the importance of generating incentives to reward good behaviour. The language of the planning outline echoes that of the media's and Chinese social scientists in their frequent comments about the lack of morality in Chinese society, revealing anxiety about how a lack of morality and trust will harm the market economy and social stability (W. Zhang & Ke, 2003). The goal of the social credit system is, according to the planning outline, to "build mechanisms to incentivize the keeping of trust and to punish the breaking of trust" (Zhongguo Guowuyuan, 2014). An ideal social credit environment, in other words, will "encourage people to be sincere, keep trust, promote morality and uphold courtesy". Therefore, in line with previous attempts to cultivate moral and "civilized" citizens for the nation, social credit is meant to promote civic virtue and patriotism in order to foster a "harmonic society" (*hexie shehui*), which has been a state slogan for the past decade.

A central concept underlying the emerging incentive structure is the so-called natural person (*ziran ren*) that appears multiple times in the document (Zhongguo Guowuyuan, 2014). While the concept is certainly linked to the rights discourse of the Enlightenment, its immediate context is actually game theory in economics, which is paradoxically predicated on a dark vision of humankind, namely, that the individual is nothing but calculating, distrustful, suspicious and so forth (Brown, 2015). Still, despite this negative view, in modern economics, the natural person is nonetheless adorned as a "rational" being who makes "self-interested" decisions based on incentives. This conception of the human being informs public choice theory, which uses economic theories to address social and political problems. Under this logic, the purpose of governing is to provide incentives to reward individuals who behave in ways consistent with the objectives of the state and punish those who don't.

Economists have long argued that a sound credit system promotes the smooth functioning of the market (Bartels, 1964). In a way, the Chinese social credit system takes the idea of credit rating to a new level. By evaluating and establishing the creditworthiness of all citizens, businesses and organizations, the system is trying to make individuals and groups accountable for their actions by subjugating their behaviours to calculable economic and financial logic. In so doing, the government hopes to rein in the perceived growing culture of fraud, selfishness and callousness that are regarded as the most prominent problems of the postsocialist era. The social credit system is regarded as a way to help safeguard social order and build "social sincerity" and a "sincerity culture" (Zhongguo Guowuyuan, 2014). Once the social credit system and the national social credit codes are in place, the official newspaper *China Daily* predicts, it will "let credit weigh in for malfeasances and lawbreaking" (Shehui xinyong daima [Social Credit Codes], 2015).

In addition to emphasizing the construction of credit for the "natural person", the planning outline also discusses the importance of including businesses, institutions and government agencies in the same social credit system. Just like individual citizens and consumers, businesses, social organizations and government agencies must also be evaluated by people and other organizations in order gain respect and credibility. As if the invisible hand of the market will magically solve all problems, the planning outline specifies that the construction of the credit infrastructure will help to strengthen healthcare services, lead to better hygiene and birth control, deliver safer food, reinforce scientific and technological development, generate stronger environmental protection and bring improvements in many other sectors. Above all, it will help to construct and maintain social and political stability. This is like a form of credit fundamentalism, similar to free market fundamentalism, that believes an omniscient credit system will save China from social discontent, instability and other perils.

It is also important to point out that as much as the government is trying to use the social credit system to instil the so-called socialist core values, such as a "harmonic society", "Chinese virtue" and "socialism with Chinese characteristics", it is not promoting a brand of Chinese exceptionalism. Among the keywords, such as "sincerity", "trustworthy", "amity" and "patriotism", that can be found throughout the planning document, there are also explicit references to the desire to integrate the so-called Chinese "socialist market economy" with the global market economy. For instance, it maintains that a positive credit infrastructure will promote corporate responsibility, a productive and efficient workforce, and a transparent and accountable government, all of which, it maintains, are crucial for China's global competitiveness (Zhongguo Guowuyuan, 2014). After all, at the most fundamental level, the logic of economizing society through quantification is to break down and replace the old order with a self-proclaimed universal order that can be rendered in numerical and deeply statistical terms (Asad, 1994).

THE TOTAL INFORMATION SYSTEM

The dream of establishing a total information system, of amassing data and acting on this data, is not without precedent. In 1965, for example, a group of US social scientists and statisticians proposed establishing a national data centre in order to facilitate the storage, sharing and processing of large datasets owned by the government for use in carrying out research, designing social programmes and making policy decisions. However, the proposal was not adopted as it was vigorously opposed by the public and the US Congress precisely on account of the fear that this would lead to the infringement of privacy and the creation of an Orwellian dossier society (Kraus, 2013). Similarly, in the mid-1950s, some anthropologists and psychologists came up with the idea of a "database of dreams", where everyday human dreams, life stories and wandering thoughts could be stored and then made available for analysis (Lemov, 2015). In the end, the idea of totality in all these proposed and imaginary projects is more like a fantasy, and that fantasy has long been replaced by the more effective idea of networked information, which is explicitly manifested in the design of the internet that emerged during the Cold War. Driven by the fear of a nuclear apocalypse, architects of the system emphasized not just the importance of constant and real-time communications that the information network made possible, but also the necessity of its decentralization so that the entire system could not be incapacitated by a single strike (Naughton, 2016).

In his analysis of the decentralizing nature of networked surveillance, Roger Clarke (1988) has characterized such practice as "dataveillance". Writing long before the rise of the social media, he argues that the kind of surveillance based on the mining of data linked by networked information technologies is far more powerful than the Orwellian totalitarian state, since the monitoring and analysis of the data trails take place constantly in linked and automated local processes. By now, obviously, the phenomena observed by Clarke has already saturated our everyday life. In fact, it is not an exaggeration to say that tech giants today, such as Amazon, Google and Facebook, know many of us better than we ourselves and that they are in some respect more powerful than the government.

The traditional sense of surveillance or the ubiquitous Big Brother trope is therefore no longer adequate to describe the digital landscape of the twenty-first century. In his discussion of the rise of the "expository society", Bernard Harcourt (2015) argues that it is not just that our physical and online activities are being tracked constantly. It is also that we have become very eager to share our information in exchange for convenience, security and social belonging. To put this in lay terms, we are constantly posting and liking on social media in order to be liked and stay relevant. Similarly, we give out our most private information from secret login questions to biometric data in exchange for security. In this brave new world, we need to check in with the surveillance machine incessantly, and we have to constantly turn ourselves into spectacles for others to consume. In our desire to exhibit ourselves, we are like the incarcerated subject in Jeremy Bentham's classic panopticon who wants to be seen rather than just watched (Harcourt, 2015; Horne & Maly, 2014, pp. 110–142). In essence, we are both watching Big Brother and wanting to be watched.

If big data in the neoliberal age has altered our sociality in fundamental ways, it has equally transformed our practice of knowledge. In particular, the implication of the total information system is far more consequential than that of the total archive. Whereas "archive fever", as Jacques Derrida (1998) puts it, is driven by the desire to collect and hoard in anticipation of the future in a vague sense, the modus operandi of surveillance capitalism is not simply to collect but also to calculate, analyse and act on those data in or near real time for capital accumulation, which, among other things, has the effect of creating infinite behavioural feedback loops. We may know nothing about computer algorithms and learning machines that we are helping to train, but they know us. Moreover, they guide and shape us in the process.

The Chinese dream of creating a standardized and aggregated, if not total, information system of its population is no different in this regard. Arguably, this dream is readily shared by both the government and tech conglomerates even if their interests are different. For tech giants, this is surveillance capitalism par excellence as mining data of the everyday will allow them to reach a much larger segment of the population that is not covered by traditional credit rating organizations (Chai, 2015). For the one-party security state, the potential access to these otherwise dispersed and unconnected databases provides a new capacity to govern that has been unthinkable until now.

Indeed, even at this moment, the degree of Chinese internet companies' penetration into the everyday is already more pronounced than that of their non-Chinese counterparts. The messaging and payment app WeChat is the ultimate example that offers a glimpse into the future that is now. Introduced in 2011 by Tencent, China's largest internet company, WeChat developed the first cross-platform instant messaging service. It has since evolved into an app that functions as a clearinghouse for a wide range of online activities, including shopping, travel, banking, messaging and much more. In a way, it is like the combination of WhatsApp, Facebook, Google, Amazon, eBay, Expedia, Uber and a dozen of other commonly used platforms in a single app. In 2016, WeChat alone had at least 700 million subscribers, over 90% of which were in China (*The New York Times*, 2016). These days, as China becomes increasingly cashless, urban and even rural citizens cannot conduct most of their daily activities smoothly without using the app. In theory at least, the information collected by Tencent along with Alibaba, Baidu and other major online platforms together can provide a detailed picture of their users, including their movement, finances, reading habits, health conditions, social networks and so forth.

Critics of surveillance capitalism are deeply concerned about the erosion of freedom, democracy and privacy that has resulted from big corporations' amassing of data of their users with little transparency and government supervision. In China, however, criticism of this sort is generally muted because heavy-handed state surveillance has always been the norm, and that the boundaries between the private and public domains have always been blurry. In short, even though there is no evidence yet that the government is planning to incorporate commercial databases for its social credit calculation, there is no doubt the potential is tremendous and tempting. Furthermore, despite their occasional reluctance, most Chinese citizens, especially Han Chinese who are educated, urban based and affluent, are supportive of the initiative because they believe that social credits will bring them security, convenience and prosperity (Kostka, 2019; Lee, 2019). All in all, while the social credit system is not a total information system, it is a system that seeks to deepen the reconceptualization the human and the everyday in the hegemonic economic and financial order.

THE FUTURE NOW

In late 2016, more than two years after the Chinese State Council published its planning outline on the implementation of the social credit system, the British science fiction anthology series *Black Mirror* premiered an episode called "Nosedive", which tells the story of a woman who was not allowed to board a plane due to her recently reduced social credit score. Somewhat predictably, the spiralling narrative ends with tragedy.[11] Although there is no indication that the dystopic science fiction was inspired by China's emerging new reality, and that the two cases have some crucial differences, the parallels are still uncanny.

Many media reports have noted the resemblance between China's social credit system and *Black Mirror*'s dystopia. Unsurprisingly, the undertone of some of these observations is built on a long history of viewing China as an exotic and fearsome Other. In response to this renewed Cold War rhetoric, some critics (including some of those who had initially contributed to the sensational reporting mentioned above) have started to offer new "corrective" views, emphasizing that the official intention of the system is to guide morality, promote trust and facilitate law enforcement. In short, they contend that this is just a Chinese version of data-governance, and therefore the hysteria about the coming dystopia is unwarranted (Develle, 2019; Matsakis, 2019). Some have further pointed to the fact that the implemented system so far is only local, fragmentary and partially digital (Horsley, 2018). Lost in the back and forth between the persistent Sinophobia and the insistence on evaluating China's situation in its own terms, however, is a recognition of the growing convergence between postsocialist China and market democracies in spite of their many differences. All nuances and differences aside, the logic of financialization, capital and the security state are actually the shared underpinning of today's expanding surveillance infrastructures in various nations.

That the future as fantasized by a British science fiction should so closely resemble the emerging everyday reality in contemporary China is therefore astonishing and yet unsurprising. After all, the "unimaginable" is able to appear in the science fiction precisely because it is imaginable and even desired in certain contexts. Already in early 2015, just months after the State Council had issued its comprehensive guidelines for constructing the social credit system by 2020, the financial wing of the tech giant Alibaba introduced the beta version of its own personal credit rating system, Sesame Credit (*Zhima Credit*). In 2020, Tencent has also launched its own credit scoring system based on WeChat transactions, even though the system so far is more like a loyalty reward programme (Hu & Guo, 2020). Meanwhile, with at least four hundred million users across the various platforms maintained by its subsidiaries, the Sesame Credit programme has been quick to collect participants' information such as personal identity, credit history, contractual reliability, behaviours and social relationships. Based on this information, participating users are assigned with social credit scores that are visible to others (Shu, 2014). Some users even see the advantages of displaying high social credit scores in their dating profiles. Users with high credit scores are also offered perks, such as faster loan approvals and faster check-in at some airports (Hatton, 2015; Kostka, 2019).[12] In short, the social credit platforms introduced by tech giants have been gamified with rewards that are designed to modify behaviour.

These commercial social credit platforms are not related to the system implemented by the government, however (Daum, 2017). And it remains unclear whether or how commercial social credit platforms are linked to the larger surveillance state (Ahmed, 2019). But even if commercial platforms such as Sesame Credit and Tencent Credit remain unconnected to the government's system and even if user participation remains voluntary, the story for the government-run social credit system is entirely different. Soon after the planning outline had been announced, authorities at every level started to develop and implement their corresponding social credit infrastructure. For instance, in Chongqing municipality, local districts drew up blacklists of individuals and organizations whose conduct they regarded as "seriously untrustworthy", lists that they intended to share with all other government agencies at least within the municipality (Cqnews.net, 2017). Similarly, major transportation services, such as China Rail and many Chinese airlines, have reportedly created their own blacklists, leaving millions of individuals no longer eligible to use some of their services (Chin & Wong, 2016; He, 2019). Social credit scores have crept into many other aspects of life—for instance, people who switch jobs too often as well as people who do not visit their elderly parents often enough have lower scores (P. Wood, 2018; Zhang, 2019). Although local implementation of the social credit programme has been rather uneven, one wonders when and how far the central government will further standardize and centralize the social credit system at the national level.[13]

For now at least, unlike the *Black Mirror* story, national social credit scores do not exist, and there is also no indication that social credit scores will become viewable by the general public like those gamified social credits run by tech conglomerates. Yet, ultimately, a "loyalty programme" run by the state, especially an authoritarian state, will certainly lead to rewards (and punishments) that are far more consequential. Moreover, as decentralized practices of dataveillance, social credit programmes managed by local governments and tech giants have together substantially economized and financialized Chinese society by making citizens and consumers credit conscious, as well as turning them into mouldable data subjects.

BIGGER THAN BIG OTHER

The social credit system is a technology of subjectivity and citizenship that seeks to calibrate and modify the behaviour of individuals and groups, compelling them to align themselves with the state-sanctioned social, economic and political order. At one level, by using reward and punishment to instil responsibility and self-regulation, the government is exercising its power through what Foucault refers to as "the conduct of conduct" (Gordon, 1991, p. 48). As Foucault argues, "to 'conduct' is at the same time to 'lead' others (according to mechanisms of coercion which are, to varying degrees, strict) and a way of behaving within a more or less open field of possibilities. The exercise of power consists in guiding the possibility of conduct and putting in order the possible outcome" (Foucault, 1982, p. 789). No wonder so many Chinese citizens, especially middle-class Han Chinese who have benefited tremendously from China's uneven but rapid economic growth, are willing to accept or even embrace the idea of social credit as a way to ensure their economic prosperity. For them, high social credit scores are their tokens to become "civilized" and "high quality" citizens as defined by the state (Tomba, 2009).

Moreover, the Chinese social credit system is part and parcel of a state-led neoliberal model of development and governance. After all, neoliberalism is never just a set of laissez-faire practices. Behind the facade of the free market is always a political and legal structure created and guaranteed by state power.[14] In China, that very market ecology is maintained by a strong party-state that prioritizes economic growth and political stability. The emerging social credit system that seeks to economize and financialize the social world is therefore a political instrument as much as an economic one. In particular, using governing algorithms, predictive analytics, big data profiling and so on, the system meticulously tracks, archives, calculates and moulds the activities of all citizens and organizations. If the dossier society of the socialist era saw China moving away from its earlier dream of cultivating critical and enlightened citizens, the mandatory social credit infrastructure in the postsocialist era takes it even further away from that dream by producing calculating individuals who are nothing but normalized and optimized for the state-defined order.

Politics and security are therefore equally central to China's social credit system. By design or not, the social credit infrastructure has been unfolding together with an array of mass surveillance technologies with profound political and security ramifications. Driven by the imperatives of one-party rule, domestic stability, geopolitical ambitions, nationalism and capital accumulation, the party-state has introduced unprecedented technological measures to manage its population (BBC, 2019). Such technologies include an all-encompassing CCTV network with growing facial and gesture recognition capabilities, the collection of genetic and biometric information especially in the ethnic minorities areas, and the monitoring of online activities, as well as other forms of mass surveillance and censorship (Churchill & Delaney, 2019; Leibold, 2020). Granted that many of these practices can also be found in liberal democracies, as in the controversial cases of dataveillance linked to the National Security Agency (NSA), Cambridge Analytica and Palantir that have come to light in recent years, the totality of them and the aggressive way through which they have been weaponized in China is still far more menacing (Burke, 2020; Cadwalladr & Graham-Harrison, 2018; Steinberger, 2020).

With social credit systems of various kinds operating at all levels, China's Big (Br)other is indisputably more overt, ubiquitous and powerful. Even at this initial stage, what makes these social credit systems particularly ominous is that the practice has already amplified the existing systematic state violence against vulnerable individuals such as the poor, non-Han minorities and political dissidents by subjecting them to additional scrutiny and discrimination. As such, "dispossession by surveillance" as described by Zuboff (2019) has taken on yet another layer of meaning.

Nonetheless, to highlight the differences in scale and intensity between China and market democracies is not to demonize China as the Other by returning to Cold War rhetoric. In fact, if the idea of everyday surveillance by the government and tech giants as implicated by China's rising social credit ecology feels dystopian and yet strangely familiar, it is only because we have already seen and experienced fragmentary versions of it. From Brexit to Trumpism, mass surveillance and behaviour modification through digital infrastructures operated by corporations and states has been a vital force in disrupting the old liberal order, unleashing a new wave of popularist and extremist politics that is heavily driven by algorithm-generated disinformation and misinformation. The old sense of the autonomous political subject has thus become increasingly limited if not altogether antiquated. Similarly, instead of creating politically aware citizens, the Chinese one-party security state has now resorted to the production of data subjects susceptible to digital control and manipulation based on pre-inputted parameters and algorithms. Thus, in spite of the many differences between China and market democracies, the two sides converge significantly in how their corresponding surveillance infrastructures have produced a new mode of governing paradigm that, as Zuboff argues, replaces "the engineering of souls with the engineering of behavior" (Zuboff, 2019, p. 376). In short, as human behaviours are increasingly shaped by computer algorithms and feedback loops, we drift toward becoming essentially posthuman (Hayles, 1999; Käll, 2017). If this trend continues, then we may indeed finally (and tragically) reach the end of history.

Acknowledgements I would like to thank participants of the workshops on quantification studies held at the Institut d'Etudes Avancées de Nantes in 2015 and the Ecole Normale Superieure de Cachan in 2016 for their comments on the initial ideas of this paper, especially when there was hardly any attention given to China's emerging social credit system at the time. The paper has also benefited from feedback from the 2018 Historical Studies of East Asia Workshop at the University of Toronto and the 2019 Dialectic of Private and Public Knowledge in Early Modern Europe Workshop at the UCLA Center for Seventeenth- and Eighteenth-Century Studies. I also extend my thanks to the Wissenschaftskolleg zu Berlin, especially members of the Quantification Focus Group, who I worked with during my fellowship residence in 2013–2014.

NOTES

1. In critiquing of the "real" fact as an unqualified concept, I do not mean to promote nihilism or to suggest that reality does not exist but rather to emphasize that facts are always mediated. For a discussion why facts remain important in this context, see Bruno Latour (2004).

2. According to the EU General Data Protection Regulation (GDPR), a data subject is a person whose personal data is subjected to be collected, stored and processed by digital technologies. See Käll's (2017) discussion, especially in the posthuman context.

3. Zuboff also argues that instrumentarianism and totalitarianism are like "two species of power" (2019, chap. 12). While pointing out the cultural, political and institutional differences between China and the West, she nonetheless concludes that the technological trajectories of both are strikingly similar. While I agree that these are two sides of the same coin, my contention is that the Chinese state, which represents a brand of authoritarian neoliberalism with growing global geopolitical ambitions, seems to occupy a space in between these two modes of power. China's Big Other (see Zuboff, 2019, chap. 13), therefore, could be characterized as Big (Br)other.

4. In a sense, it was as if the impulse of liberal governmentality had taken an authoritarian turn. According to Mitchell Dean, "authoritarian governmentality differs from liberalism in that it regards its subjects' capacity for action as subordinate to the expectation of obedience" (Dean, 1999, p. 209).

5. The scope of Stasi's surveillance was nonetheless vast. At the time of the collapse of East Germany in 1989, the agency employed approximately 91,000 full-time staff and 300,000 informants, and it had over six million personal files. See https://wikileaks.org/wiki/Stasi_still_in_cha rge_of_Stasi_files, accessed 1 October 2016.

6. However, it is worth noting that given the way these dossiers were managed in socialist China, the vast rural population who did not work in factories or collectivized farms were generally neglected by the system.

7. Despite the prevalence of these dossiers, very little is known about the operations behind them in the socialist era, and there has not been any in-depth scholarly analysis of them. Nonetheless, some individual dossiers, including high profile ones, have been leaked. Those of high profile individuals provide a glimpse of what was recorded when an individual in question was under intense scrutiny. For example, see Duo (2007).

8. For example, these dossiers were also used by the Public Security Bureau (PSB) for its household registration programme, known as the hukou system, which restricted the mobility of citizens. Household registration determined where individuals were allowed to live or work or attend school, and the dossiers on Chinese citizens contained information that could be used to support or deny any request for transfer and relocation.

9. For example, whereas personal files in the earlier period tried to document the individual's "thought" meticulously, reform-era personal files often contain only simple and generic statements, making differences between individuals indiscernible and hence the files unusable (see Sun, 1994, p. 88).

10. A quick search of the term "social credit" in China Academic Journals, the most prominent and comprehensive database of Chinese publications, is revealing. Throughout the 1990s, there were only about two dozen essays, mostly on the subject of finance, that mentioned the concept of social credit in passing. In 2000 alone, however, there were more than forty articles that did. Moreover, for the first time, social credit appeared in the titles of six articles, suggesting that more in-depth discussions of social credit had begun to emerge. Since then, social credit has become a frequent topic, with several hundred articles either focusing or mentioning the concept each year. Moreover, starting in 2014, there are over a thousand such articles published each year. Many of them were direct responses to the publication of the central government's planning outline.

11. "Nosedive", which is based on a story by Charlie Brooker, was directed by Joe Wright. It was first screened at the Toronto International Film Festival in September 2016 and premiered on Netflix on 21 October 2016, as the first episode of the third season of Black Mirror. See Black Mirror https://www.netflix.com/ca/title/70264888, accessed 5 June 2020.

12. Much has been written about our willingness to feed details of our lives to big data projects. For example, see the discussion of the idea of the quantified self in Swan (2013) and Simanowski (2016).
13. Needless to say, there is no doubt that the system will continue to evolve beyond 2020 based on new requirements and technology. See www.chinalawtranslate.com/en, www.chinalawtranslate.com/social-credit-mou-breakdown-beta, and www.chinalawtranslate.com/en/credit-regula tion. Accessed 10 June 2020.
14. As David Harvey (2005) has observed, the so-called market reform started in 1978 under the late paramount leader Deng Xiaoping has to be understood in the context of the global advance of neoliberalism.

References

Ahmed, S. (2019, May 1). *The messy truth about social credit*. https://logicmag. io/china/the-messy-truth-about-social-credit/. Accessed 25 October 2020.

Asad, T. (1994). Ethnographic representation, statistics and modern power. *Social Research, 61*(1), 55–88.

Bartels, R. (1964). Credit management as a market function. *Journal of Marketing, 28*(3), 59–61.

BBC. (2019, May 2). *China's Xinjiang citizens monitored with police app, says rights group*. https://www.bbc.com/news/world-asia-china-48130048. Accessed 5 May 2019.

Brown, W. (2015). *Undoing the Demos*. Zone Books.

Burke, C. (2020). Digital sousveillance: A network analysis of the US surveillant assemblage. *Surveillance and Society, 18*(1), 74–89.

Cadwalladr, C., & Graham-Harrison, C. (2018, March 17). Revealed: 50 million Facebook profiles harvested for Cambridge Analytica in major breach. *Guardian*. https://www.theguardian.com/news/2018/mar/17/cambridge-analytica-facebook-influence-us-election. Accessed 10 April 2019.

Chai, H. (2015, June 9). Mainland credit-rating network takes shape. *China Daily*. http://www.chinadailyasia.com/business/2015-06/09/content_1527 4221.html. Accessed 1 July 2015.

Chin, J., & Wong, G. (2016, November 28). China's new tool for social control: A credit rating for everything. *Wall Stree Journal*. https://www.wsj.com/articles/chinas-new-tool-for-social-control-a-credit-rating-for-everyt hing-1480351590. Accessed 1 May 2017.

Churchill, O., & Delaney, R. (2019, July 16). *How WeChat users unwittingly aid censorship*. https://www.inkstonenews.com/tech/how-unwitting-users-wechat-aid-chinese-messaging-apps-blacklisting-sensitive-messages/article/ 3018830. Accessed 20 December 2019.

Clarke, R. (1988). Information technology and dataveillance. *Communications of the ACM, 31*, 498–512.

Cqnews.net. (2017, June 17). *A social credit system network is under construction in Chongqing in 2017*. http://cq.cqnews.net/html/2017-06/14/content_4 1933782.htm. Accessed 10 June 2018.

Daum, J. (2017, December 24). *China through a glass, darkly*. https://www.chi nalawtranslate.com/en/china-social-credit-score/. Accessed 23 April 2019.

Dean, M. (1999). *Governmentality: Power and rule in modern society*. Sage.

Derrida, J. (1998). *Archive Fever: A Freudian impression*. University of Chicago Press.

Develle, Y. (2019, May 28). *Time to stop comparing China's social credit to Black Mirror*. https://medium.com/wonk-bridge/time-to-stop-comparing-chinas-social-credit-to-black-mirror-6e54dc98cec8. Accessed 25 October 2020.

Duo, G. (2007). *Youjian yi ce yiluo di dangan* [The surfacing of yet another missing personal file]. http://mjlsh.usc.cuhk.edu.hk/book.aspx?cid=6&tid=157&pid=2989. Accessed 1 October 2016.

Edin, M. (2003). State capacity and local agent control in China: CCP cadre management from a township perspective. *The China Quarterly, 173*, 35–52.

Foucault, M. (1982). The subject and power. *Critical Inquiry, 8*(4), 777–795.

Gordon, C. (1991). Governmental rationality: An introduction. In G. Burchell, C. Gordon, & P. Miller (Eds.), *The Foucault effect: Studies in governmentality* (pp. 1–51). The University of Chicago Press.

Gow, M. (2017). The core socialist values of the Chinese dream: Towards a Chinese integral state. *Critical Asian Studies, 49*(1), 92–116.

Harcourt, B. E. (2015). *Exposed: Desire and disobedience in the digital age*. Harvard University Press.

Harvey, D. (2005). *A brief history of neoliberalism*. Oxford University Press.

Hatton, C. (2015, October 26). *China 'Social Credit': Beijing sets up huge system*. BBC News. http://www.bbc.com/news/world-asia-china-34592186. Accessed 10 May 2017.

Hayles, N. K. (1999). *How we became posthuman: Virtual bodies in cybernetics, literature, and informatics*. University of Chicago Press.

He, H. (2019, February 18). China's social credit system shows its teeth, banning millions from taking flights, trains. *South China Morning Post*. https://www.scmp.com/economy/china-economy/article/2186606/chi nas-social-credit-system-shows-its-teeth-banning-millions. Accessed 10 April 2019.

Horne, E., & Maly, T. (2014). *The inspection house: An impertinent field guide to modern surveillance*. Coach House Books.

Horsley, J. (2018, November 16). China's Orwellian social credit isn't real. *Foreign Policy*. https://foreignpolicy.com/2018/11/16/chinas-orwellian-soc ial-credit-score-isnt-real. Accessed 25 October 2020.

Hu, Y., & Guo, Y. (2020, June 8). *Tencent launches credit scoring system based on WeChat purchases.* https://www.caixinglobal.com/2020-06-08/tencent-launches-credit-scoring-system-based-on-wechat-purchases-101564336.html. Accessed 25 October 2020.

Huang, X. (2002). The birth and development of personal file [Kexue guifan de Zhongguo gongchandang ren de renshi dangan gongzuo]. *Journal of Changsha University, 16*(1), 93–94.

Kai, J. (2014, September 20). The China dream vs. The American dream. *The Diplomat.* http://thediplomat.com/2014/2009/the-china-dream-vs-the-american-dream/. Accessed 2020 May 2015.

Käll, J. (2017). A posthuman data subject? The right to be forgotten and beyond. *German Law Journal, 18*(5), 1145–1162.

Kostka, G. (2019). China's social credit systems and public opinion: Explaining high levels of approval. *New Media and Society, 21*(7), 1565–1593.

Kraus, R. S. (2013). Statistical déjà vu: The National Data Center proposal of 1965 and its descendants. *Journal of Privacy and Confidentiality, 5*(1), 1–37.

Lam, T. (2011). *A passion for facts: Social surveys and the construction of the Chinese nation state, 1900–1949.* University of California Press.

Latour, B. (2004). Why has critique run out of steam? From matters of fact to matters of concern. *Critical Inquiry, 30*(2), 225–248.

Lee, C. S. (2019). Datafication, dataveillance, and the social credit system as China's new normal. *Online Information Review, 43*(6), 952–970.

Leibold, J. (2020). Surveillance in China's Xinjiang region: Ethnic sorting, coercion, and inducement. *Journal of Contemporary China, 29*(121), 46–60.

Lemov, R. (2015). *Database of dreams: The lost quest to catalog humanity.* Yale University Press.

Liu, X. (2016). *Woguo shehui xinyong tixi jianshe wenti* yanjiu [A study of China's social credit system]. Zhishi chanquan chubanshe.

Lu, X., & Perry, E. J. (Eds.). (1997). *The Danwei: Changing Chinese workplace in historical and comparative perspective.* M. E. Sharpe.

Matsakis, L. (2019, July 29). *How the West got China's social credit system wrong.* https://www.wired.com/story/china-social-credit-score-system/. Accessed 25 October 2019.

Naughton, J. (2016). The evolution of the internet: From military experiment to General Purpose Technology. *Journal of Cyber Policy, 1*(1), 5–28.

Rahwan, I. (2018). Society-in-the-loop: Programming the algorithmic social contact. *Ethics and Information Technology, 20*(1), 5–14.

Schwarcz, V. (1986). *The Chinese enlightenment: Intellectuals and the legacy of the May Fourth Movement of 1919.* University of California Press.

Shehui xinyong daima [Social Credit Codes]. (2015, March 13). *China Daily.* http://www.chinadaily.com.cn/opinion/2015-03/13/content_1980 1380.htm. Accessed 4 June 2017.

Shu, C. (2014). *Data from Alibaba's e-commerce sites is now powering a credit-scoring service.* http://techcrunch.com/2015/01/27/data-from-ali babas-e-commerce-sites-is-now-powering-a-credit-scoring-service/. Accessed 10 December 2016.

Simanowski, R. (2016). *Data love: The seduction and betrayal of digital technologies.* Columbia University Press.

Steinberger, M. (2020, October 21). Does Palantir see too much? *New York Times Magazine.* https://www.nytimes.com/interactive/2020/2010/2021/magazine/palantir-alex-karp.html. Accessed 2025 October 2020.

Sun, L. (1994). Qianlun gaigekaifang xingshi xia de renshi dangan guanli gongzuo. *Tianfu xinlun, 5.*

Swan, M. (2013). The quantified self: Fundamental disruption in big data science and biological discovery. *Big Data, 1*(2), 85–99.

The New York Times. (2016, August 6). *China, not Silicon Valley, is cutting edge in mobile tech.* https://www.nytimes.com/2016/08/03/technology/china-mobile-tech-innovation-silicon-valley.html. Accessed 1 May 2017.

Tomba, L. (2009). Of quality, harmony, and community: Civilization and the Middle Class in Urban China. *Positions: Asia Critique, 17*(3), 591–616.

Wang, L. (2011). *Laodong zhengyi caisu biaozhun yu guifan* [Standards for labour dispute arbitration]. Renmin chuban she.

Wood, M. (2018, August 28). *In China, your credit could depend on how often you visit your parents.* https://www.marketplace.org/2018/08/20/china-your-credit-could-depend-how-often-you-visit-your-parents/. Accessed 10 April 2019.

Zak, P. J., & Knack, S. (2001). Trust and growth. *The Economic Journal, 111*(470), 295–321.

Zhang, P. (2019, April 4). Chinese workers could lose social credit for switching jobs too often. *South China Morning Post.* https://www.scmp.com/news/china/society/article/3004704/chinese-workers-could-lose-social-credit-switching-jobs-too. Accessed 10 April 2019.

Zhang, W., & Ke, R. (2003). Trust in China: A cross-regional analysis. *William Davidson Institute Working Paper No. 586* (p. 22).

Zhongguo Guowuyuan. (2014, June 14). *Guowuyuan guanyu yinfa shehui xinyong tixi jianshe guihua gangyao (2014–2020 nian) de tongzhi* [Planning outline for the construction of a social credit system (2014–2020)]. http://www.gov.cn/zhengce/content/2014-06/27/content_8913.htm. Accessed 4 June 2015.

Zhongguo Guowuyuan. (2015, June 11). *Guowuyuan guanyu pizhuan fazhangaigewei deng bumen faren he qita zuzhi tongyi sheihui xinyong daima zhidujianshe zongti fangan de tongzhi* [The state council's approval of the development and reform commission and other departments on the overall plan for the establishment of a unified system of social credit codes on

legal persons and other organizations]. http://www.gov.cn/zhengce/con
tent/2015-06/17/content_9858.htm. Accessed 1 April 2017.
Zuboff, S. (2019). *The age of surveillance capitalism: The fight for a human future
at the new frontier of power*. Public Affairs.

CHAPTER 4

Accounting for Who We Are and Could Be: Inventing Taxonomies of the Self in an Age of Uncertainty

Uwe Vormbusch

Over the last decades, we have witnessed a further advance in quantification. In particular, the rise and spread of digital self-quantification, indicates new taxonomies of the self which (re)frame the human body, everyday practices, emotions and desires. During earlier waves of quantification, particularly from the nineteenth century onwards, accounting and an accompanying "trust in numbers" (Porter, 1995) proliferated at the heart of the economy, the sciences and the state. During the neoliberal era, numbers and calculation have come to fundamentally reframe public services, altering established norms of the common good, "corrupting" the intentions and knowledge of professional actors (Crouch, 2016).

Since the 1980s, calculative tools associated with New Public Management—international educational comparisons (such as PISA), and other

U. Vormbusch (✉)
Institute of Sociology, Faculty of Humanities and Social Sciences, FernUniversität Hagen, Hagen, Germany
e-mail: uwe.vormbusch@fernuni-hagen.de

© The Author(s) 2022
A. Mennicken and R. Salais (eds.), *The New Politics of Numbers*, Executive Politics and Governance, https://doi.org/10.1007/978-3-030-78201-6_4

forms of performance measurement, ranking and rating—have gradually expanded into not yet economized fields of public life, such as education and health, transforming not only the way these work, but also the very objectives they are pursuing. The human body and mind have not been exempted from these developments. Quite to the contrary: these have been a privileged object of quantification from the very beginnings of modern science, most notably in medicine (Foucault, 1973) and statistics.

The early Foucault (1975 [1995]) placed the body centre stage in his studies of power—see here, for instance, Foucault's analysis of Bentham's Panopticon as well as his writings on disciplinary society more generally. Later, Foucault revised the somewhat hierarchical notion of discipline by drawing attention to the interplay of power, knowledge and the self, focusing on "technologies of the self" (Foucault, 1988b). Since newly emerging forms of (digital) self-quantification rely on a quantified *self*-observation far more than earlier practices of self-observation (see for instance diary writing), they seem to be a good case in point for the study of new advances in quantification.

This is not to say that earlier instrumentations from the clinical gaze to statistical classifications and incentive pay systems are not related to subjectification processes and identity politics (see e.g. Espeland & Stevens, 1998). Nevertheless, the new movement in self-quantification indicates a considerable shift in agency. A growing part of the population in western capitalist societies is beginning to engage in new practices of quantified self-observation, thereby moving quantification beyond early aspirations, for instance aspirations aimed at putting a value on humans' competencies through marking (e.g. Hoskin & Macve, 1994).

From the measurement of sleep behaviour, physical and sexual activity, the evaluation of changing moods and labour productivity to the sharing of these data on the Internet, a wide range of calculative self-practices have emerged, validating Miller's (1992) early dictum that accounting as a mode of governing is as much about the calculated, as the actively calculating self. In this context, the Quantified Self movement (in the following: QS) is the most commonly known network of self-trackers and self-quantifiers. The official objective of QS "is to help people get meaning out of their personal data" (http://quantifiedself.com/about/, Accessed 16 July 2019). Patterns and orders of the self are to be discovered, which hitherto have been hidden within the muddy waters of everyday practice. Thereby, the self shall become aware of the hidden undercurrents of everyday practice, precisely those regularities which are governing life without being visible for somebody living in a state of unquestioned familiarity with oneself. The self is called to reconstruct these undercurrents from the aggregated data obtained by systematically observing his or her everyday activities and whereabouts. Lupton (2016, p. 49) rightfully notices that the normative literature about self-quantification and self-tracking is above all pointing to the "ethical responsibility to achieve this authentic self", which "involves delving beneath the surface in order to uncover the hidden desires, drives and motivations that the psyche harbours".

In a first approach, self-quantification can be understood as the attempt to free ineffable corporeal experiences from the sphere of pre-reflexive and pre-predicative knowledge by formally representing and articulating them—in charts, numbers and algorithms, which can be shared, compared, publicly discussed and, eventually, optimized. Therefore, self-quantification presupposes the invention of specific taxonomies targeting body and life: inner sensations bound up with the living body as well as external circumstances and activities that have to be recorded and written down in order to reflect and act upon them. What is more, it is not just individual numbers and calculations that are thereby created. Individual datasets can be, and actually are, linked to other people's datasets, giving birth to entire systems of calculation, or rather "taxonomies of the self".

This chapter explores these taxonomies in the making drawing particular attention to the diversity of representational forms and moral conflicts involved. Digging into exploratory variety, playfulness and ambiguities are important in order not to misunderstand this emerging form of governing the self as a 'juridical' form of power. For what makes power powerful:

> [...] is simply the fact that it doesn't only weigh on us as a force that says no, but that it traverses and produces things, it induces pleasure, forms knowledge, produces discourse. It needs to be considered as a productive network which runs through the whole social body, much more than as a negative instance whose function is repression. (Foucault, 1980, pp. 118–119)

Therefore, the chapter's main focus is on the motives, practices and desires as well as the emerging instrumentation in the field. Showing how something as manifold, ambiguous and unique as the self might have a specific empirical worth requires certain agreements about how to measure and formally represent it, a process commonly dubbed as "commensuration" within the sociology of valuation and evaluation (see Espeland & Stevens, 2008; Fourcade, 2011; Lamont, 2012). From here, some general conclusions about self-quantification and contemporary capitalism are drawn. In doing so, the chapter intends to keep a balance between the economic, cultural and moral dimensions of quantifying the self. This implies a theoretical approach, which is equally sensitive to Foucauldian studies of accounting and governing as well as a more practice-oriented approach related to the "sociology of critique" (Boltanski & Thévenot, 2006 [1991]). In this respect, this chapter might be considered an attempt to simultaneously apply exactly those two research perspectives on quantification that are giving this volume its theoretical appeal. We should not forget that the sociology of quantification always had its roots in both sides of the Channel.

While British critical accounting research, from the 1980s onwards, often followed a Foucauldian trajectory, French conventionalists were simultaneously leaving Bourdieu and Foucault behind by highlighting the practical capabilities of individual actors enmeshed in conventions (Desrosières 2011; Diaz-Bone & Salais, 2011; Diaz-Bone & Didier, 2016; Thévenot in this volume). At the intersection of these two frameworks, self-quantification emerges as a contemporary "institution of the self", not displacing but co-existing with established technologies of the self, such as religious confession, therapeutic and psychoanalytic approaches to identity and authenticity (Noji & Vormbusch, 2018). Consequently, self-quantification is as much a reaction to economic uncertainties and the ambiguities of individual worth as it is a cultural and ethical revolution, offering new foundations for a self which is more or less missing internal principles for action and orientation (see already Riesman, 1950).

While much research quite rightfully stresses the new potential for surveillance that QS-tracking offers (Whitson, 2013) and draws attention to accompanying forms of coercion, alienation and social-psychological pathologies (e.g. King et al., 2018; Lupton, 2015, 2016; Ruckenstein & Pantzar, 2017), it is also worthwhile to consider the ambivalences, ambiguities and contradictions associated with the practice of self-quantification. Sharon, for example, criticizes the polarized nature of the debate about self-tracking for health and asks how following a practice-based approach to self-tracking "can open up new spaces for the enactment of solidarity" (Sharon, 2017, p. 117). Likewise, Nafus and Sherman (2014) stress the systematic tension between autonomy and subordination within the Quantified-Self movement:

> QS also does not escape the constructs of healthiness embodied in the devices that they use, inasmuch as those are the dominant constructs with which participants must wrestle. But wrestle they do. [...] They interact with algorithms not as blind, mindless dupes, but as active participants in a dialogue that moves between data as an externalization of self and internal, subjective, qualitative understandings of what the data means. (Nafus & Sherman, 2014, p. 1793).

In this perspective, self-quantifiers, at least the early adopters within the QS movement, are not uncritically adopting new technologies and data. Instead, they appear to be capable and reflexive actors, deliberately inventing and manipulating technology in order to explore who they are and could become. This sheds light on a more general point highlighted by Diaz-Bone and Didier (2016). Reconstructing the influence of Michel Foucault on the sociology of quantification, they argue that Foucault "did not see that there are actually different statistical techniques and that it makes a difference. He linked statistics, all statistics, mainly to neoliberal Governmentality" (Diaz-Bone & Didier, 2016, p. 15). Alain Desrosières, to the contrary, was very aware "that different modes of quantification are associated with different modes of government" (ibid.), meaning that specific compromises regarding quantification, and thus "investment in forms" (Thévenot, 1984), solidifying the quantitative opportunities as well as related social power relations, would make a difference.

Corporeal Accounting
Within Immaterial Capitalism

The QS movement gained considerable public attention in the U.S. for the first time around 2007. At that time, this movement could be called a kind of "grassroots quantification" movement. Obviously, there must have been more than just new technologies, such as mobile phones and the Internet to let self-quantification as an assemblage of practices unfold. Indeed, the emergence of self-quantification draws heavily on long-established discourses, such as discourses on the "sovereign self" (Miller, 1992) and liberal forms of governing (Foucault, 1981 [1976]), on economic transformations, such as the emergence of the network economy and the rise and spread of self-employment, both closely linked to new orders of justification, such as the "project city" (Boltanski & Chiapello, 2007), and radical political reforms commonly dubbed "neo-liberalism", all of them preceding the QS movement by decades. Therefore, to understand the emergence of self-quantification, we have to take several interlinked processes into account.

Self-quantification is of great interest to the analysis of contemporary capitalism, because it is in this context that the individuals themselves are beginning to transform their body, their idiosyncrasies, their biographical experiences and—particularly important—their imagined futures in terms of quantified and comparable assets. By inventing the very categories and technologies by which an individual's manifoldness is made comparable and measurable, self-quantification constitutes nevertheless an indeterminate and malleable relay between the culture and economy of new forms of capitalism, be it "flexible" (Sennett, 1998), "cognitive" (Boutang, 2012), "emotional" (Hochschild, 1983; Illouz, 2007; Neckel, 2005a, 2005b), "corporeal" (Moore & Robinson, 2016; Smith & Lee, 2015) or "immaterial" (Vormbusch, 2008, 2009, 2012) capitalism. In these new forms of capitalism, immaterial capabilities are the most relevant source for competitiveness and profit, yet, there is still no agreement about how to commensurate subjectivities, let alone reliable methods to empirically measure and evaluate them. Both the economics of conventions as well as actor-network theory (ANT) share the idea that such commensuration requires an active "investment in forms" (Thévenot, 1984) in order to make things common and commensurable. Callon (1998, p. 6) complements this point by asking:

> In order to become calculative, agencies do indeed need to be equipped. But this equipment is neither all in the brains of human beings nor all in their socio-cultural frames or their institutions. What is it then?

For Callon, this equipment can be found in the prostheses rendering actants into calculable and calculating agencies. Some of those prostheses equipping the modern self with calculative powers are outlined later in this chapter. But actor-network theory's assessment might be judged unsatisfactory when it comes to the moral dimension of the "finishing process" by which humans are being made into subjects. If we view contemporary capitalism not only as an economic system but as a life-form, we have to take into account the moral conflicts that arise when human agency is being made up by powerful inscriptions, such as new "taxonomies of the self" provided by practices of self-quantification. Later, we will analyse these conflicts as moral conflicts, rather than merely as conflicts of interest.

Examining the cultural significance of such "corporeal accounting" (Vormbusch, 2015) goes beyond traditional approaches to the study of accounting which have "largely focused on aspects of calculative practices subject to formal organization" (Vollmer et al., 2009, p. 2). It mirrors Didier's interest in "social spheres pretending to remain free from numbers" and in presenting this as a myth no longer suitable within modernity (see Didier's contribution to this volume). In doing so, we have to look for an accompanying shift in agency, since such practices of valuation seem to rely (even) more on the active engagement of the self than others. Whereas accounting in organizations has above all been analysed in its subjectifying capacities (see e.g. Miller & O'Leary, 1994; Mennicken & Miller, 2014), allowing formal organizations to control and to mobilize subjectivity in their favour, self-quantification, at least at its beginnings, has been driven by actors outside the context of formal organization, in their life-world and in the public sphere. One of the constitutive aspects of the QS movement, in particular, is its members' belief in the empowering capacity of self-quantification. As far as I can see, the claim of being recognized as unique as opposed to the way the self is treated within established social institutions (health care is one frequently cited example in this context) is fundamental for the QS movement, leading to the movement's critique of modern institutions as alienating, dispassionate and overall inappropriate for the demands of highly individualized actors within late modernity. Consequently, measuring oneself as being unique ("N=1" is one paramount element of discourse here, indicating that the only relevant reference point for measurement should be the individual) is one crucial promise within the QS movement.

In an unexpected turn in how quantification is regarded by the individuals themselves, it no longer appears to be a threat to how individuality is socially understood, constructed and experienced (such a critique would be in line with classical critical theory). Rather than threatening the integrity and incommensurability of the self, quantification is now warmly embraced as its central source. But it may well turn out that applying metrics to core attributes of one's (and everybody else's) self might as well erode the uniqueness and incommensurability of those who are striving for precisely that. The QS movement may just as well manifest itself as a governor's dream: the dream in which subjects are striving to invent the very categories by which they can be best sorted, managed, activated and moulded in whatever way imaginable. In this sense, the QS may emerge as an exceedingly malleable self; a self always falling short; an unsatisfied self, striving for a better version of him—or herself through calculative means. On the other hand, the subjects engaging in self-quantification are motivated and mobilized by dreams that are just the reverse: namely to evade dispassionate and distorting social institutions which are perceived as being ignorant of and negating these subjects' concrete individuality.

This chapter analyses practices of self-assessment and self-optimization, which have previously been limited to small circles of "self-trackers" and "self-quantifiers" and are currently gaining currency within wider society, last not least, due to the increasing popularity of wearables, the Internet of Things and an ever more digitally connected lifestyle. The initial consideration for our empirical research was that self-quantifiers are, above all, confronted and required to cope with new forms of economic and cultural uncertainty—two fundamental traits of contemporary capitalism.[1] Coping with uncertainty in this context means the calculative quest for discovering the very categories by which the plurality of individual skills and capabilities as well as the plurality of the cultural forms of living can be inscribed into common registers of worth, thereby offering a specific answer to the complexities and ambiguities of life in late modernity (Vormbusch, 2016). The chapter seeks to shed light on some of the contradictions and ambivalences of these new taxonomies of the self: on the one hand, self-inspection through self-quantification might offer new possibilities for self-knowledge, control and emancipation, and could therefore be considered as a form of "enabling accounting". On the other hand, self-quantification threatens to subjugate ever more aspects of individual life by extending instrumental rationality to hitherto incommensurable and incalculable entities: the living body, the self, emotions and desires.

CALCULATION AND THE LIVING BODY

It is not the first time that the body becomes the focus of technologies of the self. Social forces acting upon and through the body are evident at least since the works of Norbert Elias, Michel Foucault and Pierre Bourdieu. In a nutshell, notions of the "civilized body" (Elias), the "disciplined body" (Foucault) or the "body as capital" (Bourdieu) highlight its relevance within historically variable regimes of social domination. In contrast, in early phenomenological thought (Merleau-Ponty, 1962) the "living" or "fleshly" body as belonging exclusively to oneself was perceived to be the only possible approach to the world. Here, the analytical priority is shifted from the body as product and mediator of social practices to the body as the only possible foundation of perception and action. The living body relates *my-self* to everybody and everything else, and simultaneously discerns *my-self* from everybody else, it is "my point of view upon the world" (Merleau-Ponty, 1962, p. 70). It is due to my living body that every possible experience in the world is related to my specific position within this world. The living body is the originator for any possible lived experience and remembrance. It is actively performing, processing and shaping our experiences. In phenomenology the living body is the unavoidable precondition of self-perception as well as the perception of others (Alloa et al., 2012).

Obviously, there is a strong contrast between the concepts of the living body and embodied experience, on the one hand, and the dominant view of calculation as an objectified body of knowledge, on the other. Quantification is intimately related to the instrumental domination of nature and the social world, an observation, which Adorno and Horkheimer (2002 [1944]), drawing on Max Weber, pointedly expressed, and which was later reformulated by poststructuralism. The opposite pole of possible experiential reality represents—at least within the phenomenological school of thought—our living body as "the bearer of the zero point of orientation", as a fundamental way of being in the world. In this perspective, the living body, as the mediator of every possible perception, is impossible to objectify. It cannot be measured and calculated in the same way that other "things" are being measured—not without losing its inherent qualities as an experiencing and experienced living body. The differentiation between "being a living body" and "having a body" (Plessner, 1970) therefore points to the limits of social rationalization. That which cannot be measured, which is always something unique and incommensurable, cannot become the object of formal optimization and instrumental rationality. At least not until now. The current explosion of technically mediated practices of self-quantification points to the historical variability of such a differentiation. It reveals that the distinction between body and living body is nothing ontological as in classical phenomenology, but socially malleable.

Whereas phenomenological thought is built upon the idea that no cognitive representations are possible without the living body actively performing affects, postures and body-environment schemes, the QS movement seems to rely on calculative forms objectively representing the body as a system of determinants. Whereas phenomenological thought regards inner sensations such as emotions, pain and hunger as being without extension, even without any dimension (Schmitz, 2009, p. 71), in the field of QS, measures and measurement procedures are invented for recording, articulating and "writing" them. What has been enclosed within the body shall be formally represented and made operable. But a multitude of transformations must be performed before these can be attributed to the living body. Keeping this in mind and referring back to the seminal works of Elias, Foucault and Bourdieu, the key question that arises is how such a "calculated living body" (a contradiction in itself from a phenomenological point of view) can be brought into existence at all; and how it is related to forms of governing within contemporary society. In what ways is the calculation of the living body making up specific subjects? And, conversely, what does this tell us about our contemporary societies?

The Quantified Self

The QS movement is a global network of self-trackers, self-quantifiers, entrepreneurs, developers and users of mobile and internet-based technologies of self-inspection. It consists of individuals, collective meetings, websites for comparing data and developing metrics, small start-ups and big corporate players from the telecommunications, sports and health industries. It also consists of specific objects that are shaped and introduced into the field by various actors. These objects include material devices, such as mobile phones and wearable sensors and computers, as well as immaterial objects, such as algorithms, apps, and data connections. The self-ascribed motto within the field reads "self-knowledge through numbers" (http://quantifiedself.com/about/, Accessed 16 July 2019).

By systematically quantifying their self-observations, individual users are striving for new insights regarding their bodily, mental, psychological or social status. This includes health data, food records, records of emotional ups and downs, including depressive episodes, sleep behaviour, digestive and sexual habits, the menstrual cycle as well as everyday patterns of movements and whereabouts more generally. Through measurement, the quantified self is exploring his or her possibilities in new ways, opening up new perspectives on who one could be and how to get there: thus, the quantified self is, at least to a large extent, an epistemic self (Noji & Vormbusch, 2018). QS meet-ups are regionally concentrated in western capitalist metropoles (located in the U.S., Western Europe, Australia and New Zealand). Its protagonists—based on our observations, since no reliable data exist—often share a similar educational background and habitus (they are academically educated, technologically apt, prevailingly male, in their twenties and thirties).

Whereas the latest numbers show the active membership of QS (as a social movement and a community of practice) to be somewhere around **40,000** people worldwide, market surveys, such as the study by Grieger (2016), conclude that about 21% of the population in Germany is tracking at least one aspect of their lives on a regular basis. Whereas the latter figure might exaggerate the actual extent of the phenomenon, the first figure is equally misleading, because the social relevance of self-quantification reaches far beyond the inner circle of expert users who actively participate in a global community and who were the primary target group of our research.

Two aspects must be considered here: first, the social relevance of QS is not based on its widespread incidence, but on its character as a global laboratory for inventing new lifestyles and forms of ethics based on technologies and new taxonomies to live them. QS reflects as well as transcends contemporary capitalism by criticizing it. In this sense, today's practices of self-quantification might very well echo the metamorphosis of the Parisian Bohemia at the turn of the century: once despised by bourgeois morality, nowadays a blueprint for the "new spirit of capitalism" (Boltanski & Chiapello, 2007). Second, and directly associated with this, we can already observe a profound transformation of QS from an early "community of practice" (Wenger, 1998) to a mass-market populated by consumers, start-ups and the giant enterprises of the consumer, sports, and telecommunication industries. Self-quantification is on its way to becoming a constitutive part of the digital economy. This latest development is not the focus of this chapter; rather, it is the invention of the taxonomies that preceded it.

For QS-activists, quantification is their method of choice to unveil the undercurrents of corporeal experience and everyday practice. Florian Schumacher, one of the protagonists within the QS movement in Germany, summarizes the main aspects as follows:

We are prevented from monitoring ourselves in a neutral way by protective mechanisms which evolved in the course of our evolution. Therefore, keeping a record of themselves serves for many people the purpose of observing changes or maintaining the motivation to achieve self-defined goals. The externalization of relevant information and its impact on our awareness evolves into a sixth sense allowing us to discover things lying hidden. (Interview with *Die Welt*, 12 October 2013, see http://www.welt.de/gesundheit/article120826726/ Ein-sechster-Sinn-um-Verborgenes-zu-erkennen.html, translated by the author)

This is how one of our interviewees put it:

[…] Having the feedback cycle was really important. Having something to indicate you are stressed at the exact moment when my body was feeling stressed allowed me to see and make connections that I was never able to make before.

Making intangible emotional states visible ("allowed me to see") which are normally hidden to the self implies performative effects, meaning that the represented feeling may to a certain degree be an effect of the representational device or procedures themselves. This is suggested by the following quote from another interviewee, although this chapter will not elaborate on the discussion of performativity any further (but see Callon, 1998):

What I really need is a stress alert system. I need something to tell me when I'm feeling stressed. […] Another thing that was pretty neat about setting up the stress alert system is: I started to learn how my body felt when that light was red.

Self-Quantification relies on technical artefacts, such as activity wristbands, body sensors, smartphones and internet-based diagnosis algorithms. Particularly within sports, the hardware sales of sensors, "smart" (connected) shoes, are on their way to becoming mass-market products and most producers are trying to establish a proprietary world of experience around this form of "connected sport" (see e.g. Nike[plus]). Increasingly, practices of self-quantification are affiliated with gamification applications—partly to address motivational issues, partly in the course of establishing new products and markets. Some observers point to the close relation between gamification applications and surveillance (Whitson, 2013). The integration of self-quantification into larger systems marks a clear break with the original intentions of QS, which surfaced as a form of reflexive monitoring of the self with the objective of healing oneself from chronic diseases and obtaining knowledge about one's own emotions and activities. From the beginning, one of the main topics of the QS movement was the care for the self and the living body.

A large number of the show-and-tell presentations on the global as well as local QS-conferences (https://quantifiedself.com/show-and-tell/) give an account of how people were experiencing long-term suffering without their suffering being institutionally recognized, let alone cured within the established medical system. QS at this stage represented an effort to radically switch from the established procedures of being classified and observed as an object within conventional medicine, where corporeal experiences are residuals or even disturbing variables to technically mediated practices of observation and treatment. The QS presenters, in this context, report healing from diseases commonly considered incurable, such as Crohn's disease. These healings are attributed to an often makeshift kind of self-observation based on numbers and quantification, leading to self-medication and radical redirection of nutrition and other living habits. From a rigorous methodological viewpoint, we are talking not about "big" but rather "dirty data" here: often there is no consistent control of how data are obtained and processed leading to a lack of validity and reliability and a kind of "makeshift-quantification". Nevertheless, these achievements have led to a systematic critique of how people are treated within the established medical systems and to increased calls for including personalized data into the diagnostic process as well as medical treatment (see for example http://quantifiedself.com/2012/04/talking-data-with-your-doc-the-doctors/).

The perceived objectivity and neutrality of calculation (Miller, 1992) as opposed to ineffable corporeal states play an important, even if not uncontested, role in this context:

> And to comprehend myself [...] you can no longer trust yourself; there actually are so many scientific studies such as the Dunning-Kruger-effect from 1999, proving [...] you are having a systematic bias when assessing yourself. That is, one cannot rely on one's feeling any more in different cases. [...] For me, it is beside my subjective sense, I am interested in

an objective perception toward myself, namely facts. There are quantifiable values and I can compare them and I can interpret and judge this completely decoupled from my personal feeling.

Various aspects of what Boltanski and Chiapello (2007) called the "New Spirit of Capitalism"—for instance, autonomy, authenticity, self-realization and networking—are pronounced characteristics of the field. QS in this respect may well be interpreted as being related to a "networked capitalism" built upon flexible networks of auto-entrepreneurs, who are competing and cooperating simultaneously. It is tied up with specific practices of making oneself visible through the web-based sharing of personal, intimate and performance data. It represents a field, which when encompassing the "community of early adopters" had the characteristics of a pioneering network. Meanwhile, there has been an intensified collaboration between users and developers of such self-quantifying technologies. Start-ups, industrial conglomerates and transnationals such as Google, Apple and the likes are investing and building networks in order to create new products and markets, thereby transforming the field.

In the following, we will describe the new taxonomies that are emerging, linking corporeal action and bodily enclosed experiences to accounting procedures. Thereby, the living body as the sensually given, pivotal point of being within the world (Merleau-Ponty, 1962) is being (re)framed and transformed.

Well-Being, Performance and Emotions as Core Issues of *Leibschreiben* (Writing the Body)

How is self-tracking actually performed and what effects does it have on individuals' self-perceptions? In stark contrast to the natural sciences, particularly medical science, the emerging forms of representing the self are to a considerable degree produced by lay actors outside of formal organizations.[2] The emergence of innovative *bodynotations*[3] indicates an entirely new operative scripture for writing the body. We are calling these emerging forms of representing the body *Leibschreiben* (Vormbusch & Kappler, 2018), hereby adapting the basic idea of accounting as a "writing of value" (Hoskin & Macve, 1986) to a certain degree. Alas, within post-structuralist accounting research a resilient concept of the embodied self as well as a concept of human reflexivity is lacking. Unlike poststructuralism, our approach tries to account for both: the sensations of the living body as experienced by concrete individuals, on the one hand, and the emergence of an operative scripture as a form of writing the body related to social discourses, on the other hand. Furthermore, the justification practices constitutive of the actors involved are regarded as a missing link between these two levels of analysis, necessarily preceding the establishment and institutionalization of any operative scripture.

The Foucauldian strand of accounting research (for an overview see Roslender, 1992) investigated how established forms of reading and writing underwent fundamental transformations from the twelfth century onwards. Double-entry bookkeeping in this regard represented one major manifestation of the transformation of writing more generally; more specifically, it represented the "capital form of writing" (Hoskin & Macve, 1986). If we consider accounting as a specific technology within the broader transformation of writing and representing, then self-quantification can be regarded as one form of accounting for the self, as a form of "writing the self", reflecting the above-mentioned changes within contemporary capitalism.

Empirically, there is a wide variety of motives, techniques, programmes, apps, suppliers and objects assembled in the field. We encountered people who are measuring nutrition, physical activity and sleep, depressive periods as well as all kinds of emotional sensations they had throughout the day, some of them tracking their dreams, some of them stressing the importance of sharing their data, some opposing exactly this. As can be expected, there is a fishbowl of narratives, from the *empowerment* discourse (health as a personal "activity" and a "competence") to the *new spirit of capitalism* (sharing data to "connect to people"; sharing as the "new normal" of a new imagined society). In a first step of our analysis at least three distinct discursive and practice-fields within QS emerged: well-being, performance, and emotions (see Kappler & Vormbusch, 2014; Vormbusch & Kappler, 2018).

Well-being refers to the very beginnings of the QS-movement and smoothly connects to contemporary discourses of patient empowerment, public health and, more generally, the "wellness syndrome" (Cederström & Spicer, 2015; Davies, 2015). Many early self-quantifiers were personally affected by chronic diseases, and the public presentation and sharing of their experiences and calculative cure still is a much-appreciated part of every QS gathering. A fundamental critique towards the established medical institutions, types of treatment and forms of knowledge (as expert knowledge distinct from the lived experiences and circumstances of sick people) went along with this. One of the main triggers underlying the movement therefore was a specific approach towards the "care of the self" (Foucault, 1988a) and the search for self-determined ways of healing on the basis of buried linkages between everyday practices and experiences, on the one hand, and the evolution of one's illness, on the other.

The second dimension, *performance*, refers to the ongoing transformation of work, particularly the "delimitation of work" within neoliberal work regimes, its deregulation and subjectification (Bröckling, 2002; Pongratz & Voß, 2003). From this point of view, quantifying the self might be interpreted as a form of subjectifying self-improvement of individual capabilities and human assets with regards to the market and the unrestrained performance requirements that exist within organizations and markets. In this dimension, self-quantifiers are exploring in what specific ways their capabilities might conform to market demands, including moulding themselves with regard to these perceived demands. Critics of these developments have argued that such a delimitation of work is associated with pathological forms of character formation within late modernity, with a tendency of getting "lost in perfection" (King et al., 2018).

The third dimension, *emotions*, refers to several processes within the social world which have been labelled either in terms of a shift of values from material to "postmodern" immaterial values, such as autonomy, self-realization and participation (Inglehart, 1971), or in terms of an "experience society" (Schulze, 1995), or in reference to the "commercialization" of emotions within emotional work (Hochschild, 1983). Neckel (2005b) argues that the modern subject is engaged in a specific form of boundary work caught up between conflicting social requirements: "social discipline", on the one hand, and "social informalization", on the other. Within the field of QS, emotions are not only an important reference point for increased self-awareness, but also a central element in self-presentations ("show and tell!"). Within contemporary capitalism, the awareness and management of emotions has become a major part of these subjects' cultural capital.

THE EMERGING TAXONOMIES OF THE SELF

Inventing Representational Forms

Self-quantifiers are exploring a wide variety of different techniques, representational formats and devices for rendering their selves visible, comparable and manageable. These include narrative formats, such as diaries shared on the web, fully manual or semi-automated forms of measurement and personal feedback, ordinal and metric measures, formal representations and artistically interpreted data (such as graphs[4] or even paintings based on aggregated calculations). In particular, emotions are crucial for self-quantifiers, but only loosely coupled to conventions of how to formally represent them. In contrast to the established fields of writing value (corporate reporting, state statistics, bookkeeping, accounting) the representational forms in the field of *Leibschreiben* are still variable, malleable and non-standardized.

This is why apps such as *Mood Track Diary, T2 Mood Tracker* or *Worry Watch,* all of them easily available on Google Play Store or iTunes, are using quite different ways of "writing" emotions, some of them relying more on graphs, some on colour, some emphasizing the particular context in which specific emotions occur. Currently, the writing of emotions still relies on highly experimental networks of objects, calculations, visualizations and narrations. Following a social-constructivist approach to technology studies (Bijker et al., 1987), we can see that there are quite a lot of social groups participating in the creation of relevant techniques, and there is an equally high interpretative flexibility with regard to these techniques and the objectives of measuring. Similarly, one can also see a wide range of representational practices—starting with simple excel-sheets through the very popular diet apps right up to sophisticated apps demanding agency of their own as to whether and when the human actor is to give data input. In the latter case, the shift in agency from the human subject to internet-based applications is justified by two objectives: first, the elimination of subjective distortions during measuring (particularly a tendency of "measuring only when feeling good"), second, an increase in convenience and a resulting perpetuation of the individuals' motivation for measuring in the course of everyday life.

For example, the application *mood 24/7* (https://www.mood247.com) requires a periodical input of how a person feels by sending him or her an automated message as a call to action, inquiring: "On a scale of

1 to 10 what was your average mood today?" The accumulated longitudinal data are then visualized in a chart which can be shared with other users as well as medical doctors (*mood 24/7* was initially developed in the context of the treatment of depression). Therefore, the application is serving the two-fold goal of objectifying data as well as furthering perpetuation by shifting agency towards the device. Similarly, but more detailed, the application *Track Your Happiness* (https://www.trackyourhappiness. org/) is sending different questions several times a day. Preferably, the individual shall answer to these at once:

> [...] so you get a text and then you go to a little app on the phone and there you have a slider board, with a zero to hundred happiness scale. And then usually they start off with how happy you are, and then it lasts until you answered a series of additional questions. Questions like whether you are inside or not, whether you have to do something, or you want to do it, your actual activity on what you are doing, we have a lot of categories, and then and so on. And so, you do it fifty times now, and you set the parameter to about three or four times a day, minimum. And you are supposed to go through that as responsibly as possible.

Apps such as *Mood 24/7* or *Track Your Happiness* are trying to objectify the measurement of mood and emotions by putting the app in control of the time of measurement and by standardizing stimulus and response. Thereby, the measurement of mood shall be made independent of the mood of the responding person and the context in which this person is located at the time of measurement. But objectification and better comparability have downsides as well:

> I was planning to get rid of all the stuff because I am working and this programme pops up and I think "aaaawww", sometimes I am really annoyed by my own programme, yeah, so sometimes I don't mind and sometimes when you are really into something, but sometimes, if I do not feel like, I don't fill it in.

The obvious problem is that the average answer's quality deteriorates depending on whether the situation seems inappropriate for giving input (such as having lunch with colleagues) or the subject "not feeling like it". As a consequence, devices automatically measuring the emotional state are being developed, such as the *FaceReader* (http://www.noldus. com/human-behavior-research/products/facereader). The *FaceReader* is able to "read" and subsequently write facial expressions using seven basic emotional states. The current combination of these emotional states is entered into a two-dimensional grid, wherein the horizontal axis represents a continuum of emotional valuing, running from pleasant to unpleasant, whereas the vertical axis represents an activity dimension (active to inactive).

Generally, there is substantial disagreement in the field about how to represent the hidden inner state of emotional affairs, hitherto inaccessible to standardized measuring and quantification. The applied representational forms vary to a great degree, combining elements of text with numbers and graphs. The respective advantages and disadvantages are the topic of controversial discussion. Sticking to the topic of emotions, here is a quote from a self-quantifier trying to measure "happiness" and writing a kind of fortune diary which he shares with others on the web:

> [...] I also feel very reductionist if I would do it by numbers, so if I would score it. So I am just curious if other people have experiences with things that are a bit more elaborated than a number, but not as free flow as words or things.
>
> The structure does not help you with emotions, because it is a structure, you do not need a structure but a flow.

In contrast to institutionalized fields of measuring, the absence of a structure is seen here as an advantage for measuring happiness. On the one hand, there obviously is a reluctance to "score" emotions; on the other hand, the interviewee is looking for a kind of middle ground: representational formats not as free as the "free flow of words", but "more elaborated than a number". On the one hand, self-quantifiers are striving for formal knowledge about their emotional experiences and quite often mistrust their own emotional sensations; on the other hand, some of them feel reluctant to formalize it too rigidly. Whatever they are experiencing, it should not be "reduced" or corrupted by the use of numbers. Analytically, emotions within cultural capitalism have to be rationally cultivated. From a participant's perspective, they shall not be simply subsumed to the logic and rigidity of measurement and thereby stripped of their complexity and richness. Such contradictions are well known from other fields of measuring, but they are more pronounced and more difficult to address when it comes to measuring inner state of affairs of the living body which have neither dimension nor extension.

Fortune diaries are another, more text-based approach for representing experiences and emotions. They can be shared via twitter, Facebook or other social media, thereby adding new possibilities of ordering and representing emotions, such as peak moments of happiness or sadness, which are built into the respective platforms:

> [...] then wound up with a lot of private twitter accounts, that has kind of become the closest thing I am doing to journaling now. [...] I had it since spring 2008 and I was doing a twitter study, and I had this whole archive and it was really interesting because the things in the sidebar contains all the years and stuff, it has got little bars of how many tweets there were in each month, and the peaks were [...] when something really sudden was about to happen [...]. And the other peaks were like things that were awful and very sad [...]. And the peaks were when it worsened and when there were changes. And so there is this weird thing it ended up with being a very graphy, mood graph thing, I didn't realize that I was creating it as I did it. It just came out of my user statistics, and it came out of my journaling.

Sometimes, new and innovative forms of representing emotions emerge as an unintended bricolage, composed of different actor-actants (in this case, the user and twitter as a platform providing a graphic representation that was not initially directed towards emotions) and different symbolic systems (narrations as well as graphical representations for the measurement of "peaks" and "changes"). One could argue that QS as a network of post-traditional communities (Hitzler et al., 2008) explores possible ways of measuring and writing health, happiness and performance and thereby forms a global laboratory for doing so. Currently, the most common level of quantification for writing happiness is the use of ordinal scales. Often, for this purpose not only numbers are used, rather these are supplemented by graphical symbols and emoticons such as smileys or visual arguments such as colouration (indicating specific feelings such as red for warmth and tenderness, etc.). The following quote demonstrates that the use of these symbols should not be reduced to a mere assisting function. Quite to the contrary, they are a key means for the inner approval of feelings:

> [...] and then I have this slider, which goes from zero to... I think it is actually divided in the middle, so you get five points to the left and then that is the best mood, for example, and to the right, and it is a good

mood. And I also have this little smiley feedback. So, I put the slider and then I can see the smiley and [it] helps me to adjust, I think "No, not *that* happy, or...", you know, so that gives me kind of feedback to see, if I scored right on the scale. [...] it is just on the continuum happy versus not happy.

Above all, the visualization of an emotion (the smiley) can evoke a sense of coherence between measurement and corporeal experience. In this case, a culturally codified symbol of feeling is serving as a mediator between the inner state of affairs and a metric scale, bridging the missing points of contact between these two. Obviously, this is pointing to questions regarding the epistemological relations between ordinal/metric values and iconic representations ("...No, not *that* happy") as being built into the programme and thereby decontextualized and fixed. While the contribution of formal representations such as graphs and icons to the production of knowledge is an important strand of research within the field of science and technology studies (Jones & Galison, 1998; Latour, 1998; Lynch & Woolgar, 1988), the relationship between formal representations and emotions has not yet been equally explored.

Moral Conflicts in Quantifying the Self

The tentative exploration of the self within QS involves deep moral uncertainties. Drawing on an example of a woman trying to quantify her baby's well-being, the ethical cleavages of self-quantification become apparent. Not entirely convinced by the belief held by some quantifiers that corporeal knowledge compared to quantified metrics should be regarded as inferior knowledge, this person is in an inextricable conflict. She is in deep worry for her baby. She is worried that he might not be sleeping enough ("He must sleep more and that is why I am using this app"). She is worried that she might not be there for him sufficiently ("that he is not getting enough of me"). And she is worried that he may not get enough food ("and when he slept in the meantime, then I know that it CANNOT BE hunger"). Therefore, she began using *Babytracker*, an application that can be downloaded via *itunes*. *Babytracker* is marketed for "busy parents" allowing them to "track everything from your baby's last feeding to that first smile". Parents get various screens showing a summary of events and activities directed toward the baby, in addition to several further screens with personal analytics regarding sleeping and feeding patterns, time-weight graphs, etc., including the possibility to share these data with other parents either via a company-run database or other cloud solutions like Dropbox or iCloud.

Being aware of her concerns, the mother is trying to calm herself down by saying: "Children are self-adjusting somehow". She qualifies her quantifying of the baby quite drastically:

> Such an app is the exact opposite. It is not 'live and let live', trusting that things are just fine and that he will be sleeping and that he is getting enough of me in any case, but it [using the app, added] is above all to control.

Later in the interview she adds:

> In the end everything is getting much more complicated [by measuring it, added]. And much more stressful and it doesn't help you at all. And therefore ... because it gives you the impression you can control it ... but a baby's sleep cannot be controlled.

Despite this latter statement the interviewee continues to give her account about how she is feeling by saying:

> And I hope that when having another baby, I think I will use this [the app, added] definitely again, because there have always been those moments when I was feeling helpless.

On the one hand, the interviewee is acknowledging a baby's general self-sufficient condition by expressing that "a baby's sleep cannot be controlled". With these words, she is referring not only to her child but rather to any baby's sleep or even more to the point: she is referring *first and foremost* to "any baby's sleep" and this should at least in theory include her own. On the other hand, she is drawn to the suggestion of control implied by measuring when saying:

> Data really help next to nothing. It's above all to know, okay, I am in control now and for example, okay, he isn't sleeping throughout the night anymore and he isn't sleeping enough during the day either ... now I am going to take some steps ... yes there is an idea, my plan has just begun ... He must sleep more and that is why I am using this app.

On the one hand, the self-sufficient condition of babies and their practical routines do have a major moral significance for the interviewee. Her statement "a baby's sleep cannot be controlled" does not only tell the obvious; it is not only an observation of a baby's external condition and behaviours. It is also a moral statement about how things *ought to be* in general. That is why she is not addressing her own baby here, but rather every baby in the world. Her firm belief points to a state of affairs that should normally not be touched. On the other hand, her troubles caused by not being able to control what is going on are strong enough to override this feeling and to insert a new kind of device into the situation by measuring, thereby scraping the incommensurability (Espeland & Stevens, 1998) of her baby and her baby's sleep.

Obviously, this is not to say that she does not love her baby as a unique being and hers. But in the course of quantifying new possibilities for evaluating her baby in comparison with other babies (whose parents are also using *Babytracker* or similar devices) emerge, for example assessing his sleep, food intake, and attention. In this, as well as in the case of measuring moods, moral conflicts about if, when and how to measure qualities hitherto unquantified are emerging. The reluctance to score emotions (to "feel very reductionist if I would do it by numbers", see above) and the fear to corrupt one's authentic corporeal sensations as well as the anxiety to interfere with the autonomy of other living beings (as in the last case) are exemplary for what is at stake here. Drawing a line between commensuration and the still incommensurable for self-quantifiers in some crucial areas therefore arises not only as a technical problem, but rather as an everyday moral challenge.

Quantifying Performance: Alternative Measures, Rational Planning and the Deficiency of Corporeal Sensations

Another example draws on the quantification experiments of a passionate triathlete and is situated in the field of performance. Here, we find a variety of measures regarding physical performance. Moreover, this example shows the relationship of these key performance indicators with strategies for not only performing, but rather rationalizing sports performances, in this case triathlon:

> And this is interesting with triathlon. There is ... sounds a little casual, but if you know this threshold value and the distance, you can just as well say,

> I am having a Watt-device here. I am adjusting as if having an autopilot. I
> would like to wind this exact capacity, then you simply wind one, two or
> five hours this capacity and you know that you are not losing too much
> power to reasonably finish the competition.

In a previous section of the interview, the interviewee already characterized the taking of his pulse as being much too imprecise for his
purposes. Unlike taking your pulse, Watt values can simply and directly be
recorded at the bicycle's spindle. In contrast to the generally delayed pulse
values, Watt measurement therefore results in a kind of "instant feedback". In combination with the given distance it is possible to perform
cycling as if being on "autopilot". It is only so that he can "reasonably
finish the competition". Even more than the Watt-value, another performance indicator (VO_{2max}) is allowing him to measure his physical fitness
comprehensively and to make projections, thereby introducing notions of
the time value of performance:

> And what it [a 'smart' running watch from one of the main manufacturers,
> added] also can do, it aggregates everything I do into one measure or key
> performance indicator, one KPI and this is the VO_{2max}. This means okay
> how much oxygen can my body process per minute and per kilogram, and
> this really is the core measure for performance in the field of running.
> And what is really cool, you are provided with projections, straight from
> the watch: okay, how fast can I run this Marathon now and this is quite
> precise. … Thus, how fast I can run is depending on my lung volume.

To summarize, the interviewee is objectifying his bodily experiences
and his sense of effort by framing it, firstly, in terms of the expended Watt-
value during a competition, which, secondly, relies on VO_{2max} as the key
performance indicator aggregating relevant parameters into one master-
measure. This objectified bodily experience is the basis for the reframing
of the body as a rational and improvable machine and for the development
of related rationalization and optimization strategies.

In the following example, another notion commonly held by self-
quantifiers becomes apparent: the notion of the deficiency of embodied
experience.

> What really is absolutely interesting: when I wake up in the morning and
> I feel absolutely whacked and I am about to give up and get me a sick
> leave, I don't feel like working and I don't feel like anything. Then the

device says I shall really take off today and then I am stepping outside and start running and really after some rounds I realize: This is really going to work, the body is really there. But the mind is saying otherwise.

Similar to the above sketched experimental forms of representing emotions, the performative capacity of representing the inner state of bodily affairs is obvious: only by "doing otherwise", that is by ignoring the sensations of his living body, the interviewee arrives at a state of affairs in accordance with the performance projections based on the measurement and evaluation of the collected data. Here, a second line of transformation of inner sensations by calculative means is observable; one that has been discussed above and concerns the translation of inner sensations into numbers and figures in order to formally represent them: the emphasis was on finding adequate, that is at the same time "exact" as well as "rich" and therefore necessarily blurry, indicators for bodily sensations. In the case just discussed the approach is shifting towards an "objectification" by framing the emotional state with the help of calculations which are then taken for granted. At the very least, these calculations are being given more credibility than the interviewee's experienced feelings. The interviewee is following an attitude quite popular with self-quantifiers: that numbers and data are "true" in a deeper way than bodily sensations and feelings. This is also expressed by another interviewee:

> There is a measurable value and I can compare this value and I can interpret and assess this value completely decoupled from my personal feelings.

Such a fundamental "trust in numbers" (Porter, 1995) corresponds with the feeling that "sometimes my body is playing a trick on me". This way, bodily sensations are framed as uncertain and unreliable–in contrast to the capacity of calculations to unequivocally represent and project the true state of affairs. Obviously, there is a great potential for alienation here: the starting point is not to delicately draw out how to translate inner sensations without corrupting them (as in the case of mood tracking), but rather to accredit to numbers and calculations a higher significance when it comes to the most intimate thing humans are made of: their living body. Admittedly, this rather orthodox approach relying on the "mechanical objectivity" of numbers and calculations (Daston & Galison, 2007) is not uncontested within the self-tracking community.

CONCLUSIONS

It is not at all by chance that new forms of calculating and valuing the self are emerging today. Rather, it can be considered a response to the experience of an increasing uncertainty in the culture and economy of advanced capitalist societies. Quantifying the self is as much about the self as a subject competing in markets, as it is about the cultural indeterminacy of today's forms of living. Both aspects are nourishing a comprehensive incertitude. Almost a century ago, Frank Knight (1964 [1921])—assuming that in a dynamic economy there is a great deal of imperfect knowledge of the future–distinguished between "risk" and "uncertainty". The former he reserved for situations where the probabilities for specific outcomes are, at least in principle, calculable. The latter describes "true uncertainty" within settings "not susceptible to measurement" (Knight 1964 [1921], p. 232). Knight, as an economist, believed that only true uncertainty "accounts for the peculiar income of the entrepreneur" (ibid.). Today, in a world where the realm of the calculable and the realm of the incalculable are simultaneously expanding, true uncertainty spreads, not only "ontologically", but empirically. Lifted into public consciousness with the terrorist attacks of 9/11 and the 2008 world financial crisis more recently, it might even be the most fundamental experience for a significant fraction of today's global population, forming their relation to the world, contributing to the rise of anti-modernist movements and political parties, thereby posing existential threats to democratic governing. Against this backdrop, quantifying your self seems to promise one possible answer to the challenges humans are facing today. It is not a random one, but one connecting the social incertitude triggered by Knightean "true uncertainty" with the calculative means provided by classical modernity.

Cultural uncertainty, to be more exact, is related to the principal openness and plurality of forms of living that require ongoing assessments with regard to who I am. Rosa (2016, p. 43) argues that individuals are not able to determine the inner core of their identity, since it has always been elusive. This seems to be even more so under the conditions of an accelerated, permanently shifting modernity. Paradoxically, these ever-shifting conditions solidify into a fairly constant pressure to carve out an authentic and socially recognizable identity. Consequently, we are observing a kind of *identity squeeze*: the more the foundations of a robust identity erode, the more the subjects are occupied with the conditions for establishing it. On "slippery slopes" (Rosa, 2016, p. 691) the self is confronted with the urge not only to be oneself (that is, to be authentic), but also to discover ever more—fundamental and hidden—aspects of oneself in order to carve out what is essential and valuable about oneself.

Thévenot (in this volume) points out that calculation is about the "linkage between counting and counting on". In this sense self-quantification is about the individuals' concerns about what is left to count on when external pillars of the self are deteriorating. Obviously, it is less about what can be *found* as about how the inner pillars of the self can be *negotiated* and stabilized. It is about establishing a calculative truth about oneself which is only true in relation to a world which itself is constituted by numbers (see Salais, 2012, pp. 58–60, on the position of a constructivist realism). Therefore, QS can be seen as a datafied and technically mediated exploration process, whereby individuals try to give meaning to their life under the condition of losing touch with what Berger and Luckmann (1967) called a "natural attitude" towards themselves.

In exactly this sense, self-quantification represents a historically novel "institution of the self" (Hahn, 1982; Noji & Vormbusch, 2018) in the context of an extensive de-naturalization of the familiar world. It supplements established ways of reflecting on and caring for the self, such as the diary, the autobiography, and later various shades of therapeutic intervention. Certainly, its appeal is to be consistent with, if not the logical extension of, the evaluative cultures of contemporary capitalism and modernity itself. Measured and mediated by epistemic objects (see Knorr Cetina, 1999, 2007) such as smartphones, algorithms and apps, ever new angles on the living body and its everyday course of action are created. This ongoing exploration process is not a mere reflex of the actors' social positions and habitus, as could be argued in line with Pierre Bourdieu's sociology. And it would be just as incomplete and misleading to reduce self-quantification to self-optimization, since in many ways there is no fixed relation between ends and means. What self-quantification is for its participants has to be carved out in social practices and is (as of today) open for multiple meanings.

Self-quantification is as much about the actors' position in the social space as it is about defining who they are and who they ought to be. Nevertheless, it is not only about cultural uncertainty within late *modernity*, but as much about economic transformations within modern *capitalism*. It is about the growing importance of self-employment, unfettered and "delimited" work requirements; deregulated and often precarious forms of work, project work and "work on demand". In brief, it is about the deterioration of supporting institutions which had assured long-term security for citizens in Fordist societies. A feeling of economic insecurity has become relevant also for the highly qualified and educated fractions of the workforce—precisely the group investing in new forms of quantifying their selves. "Real" uncertainty in this context manifests itself in particular as uncertainty about the worth of one's immaterial capital, and even more fundamentally about the *notion of worth* applicable to immaterial capabilities.

Institutionalized forms of calculating value in the economy are increasingly undermined by the emergence of so-called immaterial values (Eustace, 2000, 2003), and regular financial crises demonstrate the performative quality of value which is progressively detached from its material basis. This increasing uncertainty concerning the "value of goods" (Beckert & Aspers, 2011) can be regarded as the manifestation of a fundamental shift in the value basis of contemporary capitalism. As knowledge moves to the centre stage of today's economies (as different scholars as Peter F. Drucker and André Gorz argue), and as the "flexible self" (Sennett, 1998), the "enterprising self" (Bröckling, 2002) and the "manpower entrepreneur" (Pongratz & Voß, 2003) are becoming the foundation for competition and profit-making, from a functionalist viewpoint, new taxonomies are needed that are able to frame and calculate living subjectivity.

In earlier works I have argued that the valuation of immaterial capital bound up with the self is performed as a form of quantification that simultaneously relies on objectification as well as subjectification (Vormbusch, 2012). In other words, in order to get a grip on immaterial forms of capital (such as communicative skills, motivation and aspiration) the form of calculation itself has to change. Human Resource Management's latest incarnation, "people analytics" (see Goodell King, 2016; Rasmussen & Ulrich, 2015) and the QS movement have one thing in common: the quest for universally applicable orders of worth for subjectively bound and bodily enclosed forms of capital. It is only by inventing mundane and often conflicting forms of categorization on a micro-level that such new regimes of worth may solidify, and which might then, eventually, traverse the boundaries between the familiar world and the economy.

This is not to say that individuals are consciously striving to make their immaterial capital measurable and correspondingly valuable, or that there is a direct link to "objective" capitalist needs for value realization. This would be functionalist thinking. It is rather argued that a specific social disquiet in advanced capitalism evokes two interlinked exploration problems: explorations regarding the market relevance of the subjects' immaterial capital as well as explorations regarding the hidden undercurrents of their identity. Promising a specific answer to the complexities and contradictions of life in *late modernity* therefore relates closely to the invention of those registers of worth that *capitalism* functionally relies on.

In this sense, self-quantification is an emerging form dealing with the social incertitude constitutive of modern societies. It is about the quest for those qualities of the self, which are regarded as important within the economy and culture of contemporary societies and which cannot be derived from orthodox notions of value. QS therefore is a multifarious social praxis, creating new meaning, which punctuates and shifts the margins of, and boundaries between, economy and culture, and economic and cultural value.

Obviously, this does not simply mean the discovery of subjective qualities already present (and only hidden), but the creation of new forms of representing (and thereby generating) these qualities by creating the context, the observation apparatus (taxonomies) and the normative anchoring which brings them to light as new entities. Making things accountable is bringing them into existence in new ways, and this applies to corporeal accounting, too. In this sense, QS may be seen as a gigantic, globally dispersed laboratory wherein people are investing in new forms, by which the plurality of their individual skills and capabilities, their concrete diversity of living, their uniqueness and incommensurability are being made common and comparable.

Through self-quantification, the human body emerges as a new social entity. Since the turn of the millennium, the living body took centre stage as an object of technological malleability, epistemological deconstruction and social visions to exceed the established boundaries of the human. The living body, far from having ever been something given and uncontested (see the works of Elias, Foucault and Bourdieu), since then became quite a new recipient for questioning, evaluation and improvement. Currently, there is quite a momentum of forging the body into a new object of knowing, as well as the body being one of the core relays for social utopias (see the relevant debates from genetic engineering to transhumanism, see also Lam's contribution to this volume). From a Foucauldian perspective this can be understood as the formation of a new proliferating field of force, suggesting new possibilities for the constitution of a productive subjectivity well suited for the new capitalism—and cutting off others. Here is not the place to discuss in detail the adequacy of a Foucauldian framework when it comes to self-quantification. Obviously, this article is only selectively leaning on such a framework, trying to bypass some of its problems.

Particularly, in order to avoid the equation of discourse and praxis this contribution is drawing more heavily on a participant perspective than Foucault normally did (see also Reckwitz, 2002). In accordance with the sociology of critique (Boltanski & Thévenot, 2006 [1991]), self-quantification can be seen as a deliberative praxis of competent actors. Exploring the cultural and economic qualities of the self by creating an abstract space to compare them, and at the same time extending the margins of accounting in this way, necessarily includes moral conflicts and justifications. Particularly, extending these margins of accounting (Miller, 1998) beyond the hitherto incalculable implies "ethical consequences that are often neglected" (Espeland & Yung, 2019, p. 239). Moreover, judgements about how to do things "right"—or to criticize them as being done the "wrong" way—not only refer to discourses but also to technologies, instrumentations, calculative schemes, formal representations, material (e.g. food) or immaterial (e.g. apps, algorithms, icons) things simultaneously. In this sense, actors are indeed "equipped" (Callon, 1998, p. 6), but this equipment and its practical deployment are in no way normatively neutral.

Both the Foucauldian and the pragmatist approaches have been criticized regarding their stance towards power and domination. Foucault has been accused of ignoring human agency; the sociology of critique has been criticized for ignoring the historically specific restrictions limiting the very possibility for critique (e.g. Celikates, 2006, 2009). We regard QS as an investigative praxis by which new forms of how people relate to each other and new meanings are created *without* neglecting hegemonic discourses (such as empowerment and the hailing of individuality as part of a neoliberal notion of freedom, or activity and connectionism as part of a "network city"). It is only when shifting the analytical angle towards the participants' agency and their capacities of critique that the diversity of their responses to the growing economic and normative uncertainty in today's societies can be acknowledged. By criticizing the shortcomings of how individuals are treated within the established institutions of contemporary societies, and simultaneously embracing some of their central discourses, self-quantifiers are still bringing something new to these societies, hereby confirming the fundamentally dynamic properties of contemporary capitalism (see Boltanski & Chiapello, 2007). Summing up our fieldwork, what kind of critique is then articulated within the QS network of early adopters?

Regarding the *epistemic order*, any form of *subjective* knowledge is rejected, be it bound up with the living body or obscured within the muddy waters of everyday life. Regarding self-trackers' psychological disposition, every form of cognitive abstinence, apathy or naïve familiarity with oneself is rejected. In this regard, self-quantifiers are turning the "project city's" social activity imperative (Boltanski & Chiapello, 2007) inwards, relentlessly exploring what is going on with them. Any idleness and unexamined "business as usual" is dismissed. Above all, a person's *worthiness* is related to the truthfulness and sincerity one has towards him- or herself, towards the meaning of one's personal data, and the consistency with which data are transformed into action, even if this leads to discomfort and considerable strain. The underlying *ontology* is best described in terms of a cybernetic world, within which various entities, be it humans or machines, are connected through feedback loops which are objectified, permanent, preferably immediate, and quantitative.

Self-quantification operates as a relay between the institutional dynamics of capitalist change, on the one hand, and cultural dynamics, on the other. It is varied in its particular empirical shape but consistent in connecting the individuals with newly emerging orders of worth, evaluating their performative, emotional and practical capabilities by establishing new taxonomies of the self. "Accounting for who we could be" surely is no new motive within modern societies' institutional framework. But self-quantification deserves its designation as "accounting" more than the casual "skinny jeans" tracking, or Benjamin Franklin's crude moral bookkeeping. It is deepening the everyday and therefore intimate joints between the economic and the cultural dynamics of modern capitalist societies, highlighting the importance of new forms of creative calculation for capitalist dynamics. As of today, self-quantification is still made up of a diversity of actors, devices, instrumentations and discourses about the self. Considering the growing investments of corporate actors,

start-ups and state agencies, it is not unlikely to turn out as a social innovation "through which something that stands normally outside market
exchange comes to be attributed an economic (monetary) value" (Fourcade, 2011, p. 1723). But quantifying, economizing and marketizing are
quite different technologies (Kurunmäki et al., 2016) with quite different
outcomes regarding participation and democracy. And self-quantification,
as has been shown, is more than just plain economizing. A lot will
depend on if and how "voicing concern and difference" (Thévenot, 2014)
from a plurality of positions will remain relevant when self-quantification
becomes a major component of emerging digital capitalism.

Notes

1. The article draws on the findings of the research project "Taxonomies of
   the Self" (http://www.fernuni-hagen.de/soziologie/lg2/Forschung_Eng
   lish.shtml) funded by the Deutsche Forschungsgemeinschaft (DFG). The
   project follows the methodological principles of Grounded Theory (Strauss,
   1987) and has been conducted by Karolin Kappler, Eryk Noji and Uwe
   Vormbusch. In total, more than 100 different datasets have been collected
   and analysed, from qualitative interviews and participatory observations up
   to group discussions with self-quantifiers, software engineers and start-ups.
2. This holds true at least for the active participants of the QS-movement this
   article is focusing on. However, the balance between professionalized lay
   actors and formal organizations is just about to tip in favour of the latter.
3. The term notation originally refers to varying codifications of how to
   transcribe utterances and gestures of interviewees in the field of qualitative research. In our context, it indicates the various experimental forms
   by which inner sensations as well as physical reactions are "transcribed",
   written down and formally represented by the actors in the field.
4. See, for instance, Alberto Frigo's website: http://2004-2040.com/25_ar.
   htm, Accessed 19 July 2019.

References

Adorno, Th. W., & Horkheimer, M. (2002 [1944]). *Dialectic of enlightenment.*
Stanford University Press.

Alloa, E., Bedorf, Th., Grüny, C., & Klass, N. (Eds.). (2012). *Leiblichkeit:
Geschichte und Aktualität eines Konzepts.* UTB.

Beckert, J., & Aspers, P. (Eds.). (2011). *The worth of goods: Valuation and pricing
in the economy.* Oxford University Press.

Berger, P., & Luckmann, T. (1967). *The social construction of reality.* Doubleday.

Bijker, W. E., Hughes, T. P., & Pinch, T. (Eds.). (1987). *The social construction of technological systems: New directions in the sociology and history of technology.* MIT Press.

Boltanski, L., & Chiapello, E. (2007). *The new spirit of capitalism.* Verso (French edition, 1999).

Boltanski, L., & Thévenot, L. (2006 [1991]). *On justification: Economies of worth.* Princeton University Press (French edition, 1991).

Boutang, Y. M. (2012). *Cognitive capitalism.* Polity Press.

Bröckling, U. (2002). Jeder könnte, aber nicht alle können. *Konturen des unternehmerischen Selbst. Mittelweg, 11*(36), 6-26.

Callon, M. (1998). Introduction: The embeddedness of economic markets in economics. In M. Callon (Ed.), *The laws of the markets* (pp. 1–57). Blackwell.

Cederström, C., & Spicer, A. (2015). *The wellness syndrome.* Polity Press.

Celikates, R. (2006). From critical social theory to a social theory of critique: On the critique of ideology after the pragmatic turn. *Constellations, 13*(1), 21–40.

Celikates, R. (2009). *Kritik als soziale Praxis: Gesellschaftliche Selbstverständigung und kritische Theorie.* Campus.

Crouch, C. (2016). *The knowledge corrupters: Hidden consequences of the financial takeover of public life.* Polity Press.

Daston, L., & Galison, P. (2007). *Objectivity.* MIT Press.

Davies, W. (2015). Spirits of neoliberalism: 'Competitiveness' and 'wellbeing' indicators as rival orders of worth. In R. Rottenburg, S. E. Merry, S.-J. Park, & J. Mugler (Eds.), *The world of indicators: The making of governmental knowledge through quantification* (pp. 83–306). Cambridge University Press.

Desrosières, A. (2011). The economics of convention and statistics: The paradox of origins. *Historical Social Research, 36*(4), 64–81.

Diaz-Bone, R., & Didier, E. (2016). Introduction: The sociology of quantification—Perspectives on an emerging field in the social sciences. *Historical Social Research, 41*(2), 7–26.

Diaz-Bone, R., & Salais, R. (2011). Economics of convention and the history of economies: Towards a transdisciplinary approach in economic history. *Historical Social Research, 36*(4), 7–39.

Espeland, W. N., & Stevens, M. L. (1998). Commensuration as a social process. *Annual Review of Sociology, 24*, 313–343.

Espeland, W. N., & Stevens, M. L. (2008). A sociology of quantification. *European Journal of Sociology, 49*(3), 401–436.

Espeland, W. N., & Yung, V. (2019). Ethical dimensions of quantification. *Social Science Information, 58*(2), 238–260.

Eustace, C. (2000). The intangible economy impact and policy issues. *Report of the European high level expert group on the intangible economy*. European Commission.

Eustace, C. (2003). The PRISM report 2003. *Research findings and policy recommendations*. European Commission Information Society Technologies Programme, Report Series No. 2. European Commission.

Foucault, M. (1973). *The birth of the clinic*. Tavistock.

Foucault, M. (1980). Truth and power. In C. Gordon (Ed.), *Power/knowledge: Selected interviews and other writings* (pp. 1972–1977). Pantheon Books.

Foucault, M. (1981 [1976]). *The history of sexuality, Volume 1: An introduction*. Penguin Books. First published as La Volonté de savoir (Éditions Gallimard, 1976).

Foucault, M. (1988a). *The care of the self*. Vintage.

Foucault, M. (1988b). Truth, power, self. In P. H. Hutton, H. Gutman, & L. H. Martin (Eds.), *Technologies of the self: A Seminar with Michel Foucault* (pp. 9–15). University of Massachusetts Press.

Foucault, M. (1995 [1975]). *Discipline and punish: The birth of the prison*. Vintage Books.

Fourcade, M. (2011). Cents and sensibility: Economic valuation and the nature of "nature". *American Journal of Sociology, 116*(6), 1721–1777.

Goodell King, K. (2016). Data analytics in human resources: A case study and critical review. *Human Resource Development Review, 15*(4), 487-495.

Grieger & Cie. (2016). Quantified Wealth Monitor (2016). *Potenziale für die Monetarisierung von Self Tracking- und Kunden-Daten*. https://www.spl endid-research.com/quantified-wealth.html. Accessed 22 March 2017.

Hahn, A. (1982). Zur Soziologie der Beichte und anderer Formen institution-alisierter Bekenntnisse: Selbstthematisierung und Zivilisationsprozess. *Kölner Zeitschrift Für Soziologie Und Sozialpsychologie, 34*(3), 407–434.

Hitzler, R., Pfadenhauer, M., & Honer, A. (Eds.). (2008). *Posttraditionale Gemeinschaften: Theoretische und ethnografische Erkundungen*. VS Verlag.

Hochschild, A. R. (1983). *The managed heart: Commercialization of human feeling*. University of California Press.

Hoskin, K., & Macve, R. (1986). Accounting and the examination: A genealogy of disciplinary power. *Accounting, Organizations and Society, 11*(2), 105–136.

Hoskin, K., & Macve, R. (1994). Writing, examining, disciplining: The genesis of accounting's modern power. In A. G. Hopwood & P. Miller (Eds.), *Accounting as social and institutional practice* (pp. 67–97). Cambridge University Press.

Illouz, E. (2007). *Cold intimacies: The making of emotional capitalism*. Polity Press.

Inglehart, R. (1971). The silent revolution in Europe: Intergenerational change in post-industrial societies. *American Political Science Review, 65*(4), 991–1017.

Jones, C. A., & Galison, P. (Eds.). (1998). *Picturing science producing art.* Routledge.

Kappler, K., & Vormbusch, U. (2014). Froh zu sein bedarf es wenig ...? Quantifizierung und der Wert des Glücks. *Sozialwissenschaften Und Berufspraxis, 37*(2), 267–281.

King, V., Gerisch, B., & Rosa, H. (Eds.). (2018). *Lost in perfection: Impacts of Optimisation on culture and psyche.* Routledge.

Knight, F. (1964 [1921]). *Risk, uncertainty and profit.* Sentry Press.

Knorr Cetina, K. (1999). *Epistemic cultures: How the sciences make knowledge.* Harvard University Press.

Knorr Cetina, K. (2007). Culture in Global Knowledge Societies: Knowledge Cultures and Epistemic Cultures. *Interdisciplinary Science Review, 32*(4), 361-375. DOI:10.1179/030801807X163571

Kurunmäki, L., Mennicken, A., & Miller, P. (2016). Quantifying, economising, and marketising: Democratising the social sphere? *Sociologie Du Travail, 58,* 390–402.

Lamont, M. (2012). Toward a comparative sociology of valuation and evaluation. *Annual Review of Sociology, 38,* 201–221.

Latour, B. (1998). How to be iconophilic in art, science, and religion? In C. A. Jones & P. Galison (Eds.), *Picturing science producing art* (pp. 418–440). Routledge.

Lupton, D. (2015). Quantified sex: A critical analysis of sexual and reproductive self-tracking apps. *Culture, Health and Sexuality, 17*(4), 440–453.

Lupton, D. (2016). *The quantified self: A sociology of self-tracking.* Polity Press.

Lynch, M., & Woolgar, S. (Eds.). (1988). *Representation in scientific practice.* MIT Press.

Mennicken, A., & Miller, P. (2014). Foucault and the administering of lives. In P. S. Adler, P. du Gay, G. Morgan, & M. I. Reed (Eds.), *The Oxford handbook of sociology, social theory, and organization studies: Contemporary currents* (pp. 11–38). Oxford University Press.

Merleau-Ponty, M. (1962). *Phenomenology of perception.* Routledge and Kegan Paul.

Miller, P. (1992). Accounting and objectivity: The invention of calculating selves and calculable spaces. *Annals of Scholarship, 9*(1–2), 61–86.

Miller, P. (1998). The margins of accounting. In M. Callon (Ed.), *The laws of the markets* (pp. 174–193). Blackwell.

Miller, P., & O'Leary, T. (1994). Governing the calculable person. In A. G. Hopwood & P. Miller (Eds.), *Accounting as social and institutional practice* (pp. 98–115). Cambridge University Press.

Moore, P., & Robinson, A. (2016). The quantified self: What counts in the neoliberal workplace. *New Media and Society, 18*(11), 2774–2792.

Nafus, D., & Sherman, J. (2014). This one does not go up to 11: The quantified self movement as an alternative big data practice. *International Journal of Communication, 8*, 1784–1794.

Neckel, S. (2005a). Die Marktgesellschaft als kultureller Kapitalismus: Zum neuen Synkretismus von Ökonomie und Lebensform. In K. Imhof & T. Eberle (Eds.), *Triumph und Elend des Neoliberalismus* (pp. 198–211). Seismo.

Neckel, S. (2005b). Emotion by design: Das Selbstmanagement der Gefühle als kulturelles Programm. *Berliner Journal für Soziologie, 15*(3), 419–430.

Noji, E., & Vormbusch, U. (2018). Kalkulative Formen der Selbstthematisierung und das epistemische Selbst. *Psychosozial, 41*(2), 16–34.

Plessner, H. (1970). Lachen und Weinen. *Philosophische Anthropologie* (pp. 11–171), S. Fischer Verlag.

Pongratz, H. J., & Voß, G. G. (2003). From employee to 'entreployee': Towards a 'self-entrepreneurial' work force? *Concepts and Transformation, 8*(3), 239–254.

Porter, T. M. (1995). *Trust in numbers: The pursuit of objectivity in science and public life*. Princeton University Press.

Rasmussen, T., & Ulrich, D. (2015). Learning from practice: How HR analytics avoids being a management fad. *Organizational Dynamics, 44*(3), 236–242.

Reckwitz, A. (2002). Toward a theory of social practices: A development in culturalist theorizing. *European Journal of Social Theory, 5*(2), 243–263.

Riesman, D. (1950). *The lonely crowd: A study of the changing American character*. Doubleday (together with Nathan Glazer and Reuel Denney).

Rosa, H. (2016). *Resonanz: Eine Soziologie der Weltbeziehung*. Suhrkamp.

Roslender, R. (1992). *Sociological perpectives on modern accountancy*. Routledge.

Ruckenstein, M., & Pantzar, M. (2017). Beyond the quantified self: Thematic exploration of a dataistic paradigm. *New Media and Society, 19*(3), 401–418.

Salais, R. (2012). Quantification and the economics of convention. *Historical Social Research, 37*(4), 55–63.

Schmitz, H. (2009). *Kurze Einführung in die Neue Phänomenologie*. Verlag Karl Alber.

Schulze, G. (1995). *The experience society*. Sage.

Sennett, R. (1998). *The corrosion of character: The personal consequences of work in the new capitalism*. W. W. Norton and Company.

Sharon, T. (2017). Self-tracking for health and the quantified self: Re-articulating autonomy, solidarity, and authenticity in an age of personalized healthcare. *Philosophy and Technology, 30*(1), 93–121.

Smith, N., & Lee, D. (2015). Corporeal capitalism: The body in international political economy. *Global Society, 29*(1), 64–69.

Strauss, A. L. (1987). *Qualitative analysis for social scientists*. Cambridge University Press.

Thévenot, L. (1984). Rules and implements: Investment in forms. *Social Science Information, 23*(1), 1–45.

Thévenot, L. (2014). Voicing concern and difference: From public spaces to commonplaces. *European Journal of Cultural and Political Sociology, 1*(1), 7–34.

Vollmer, H., Mennicken, A., & Preda, A. (2009). Tracking the numbers: Across accounting and finance, organizations and markets. *Accounting, Organizations and Society, 34*(5), 619–637.

Vormbusch, U. (2008). Talking numbers. *Economic sociology: The European Electronic Newsletter, 10*(1), 8–11.

Vormbusch, U. (2009). *Controlling the future - Investing in people*. Paper presented at the Research Seminar of the Accounting Department at the London School of Economics and Political Science, LSE, 4 February 2009, https://www.academia.edu/9774942/Controlling_the_Future_-_Inv esting_in_People.

Vormbusch, U. (2012). *Die Herrschaft der Zahlen: Zur Kalkulation des Sozialen in der kapitalistischen Moderne*. Campus Verlag.

Vormbusch, U. (2015). *Corporeal accounting and the third advance of quantification*. Paper presented at the IAS-Nantes Workshop on Quantification, IAS-Nantes, April 2015. https://www.google.de/url?sa=t&rct=j&q=&esrc= s&source=web&cd=1&cad=rja&uact=8&ved=0ahUKEwjKsKL3nLjSAhWG nBoKHTBTBHIQFgghMAA&url=https%3A%2F%2Fwww.iea-nantes.fr%2Fr tefiles%2FFile%2FAteliers%2F20150428-Quantification%2Fvormbusch_corpor eal_accounting.pdf&usg=AFQjCNHF4M-3CraSzbRt2MGbTNWoQgEY-A.

Vormbusch, U. (2016). Taxonomien des Selbst: Zur Hervorbringung subjektbezogener Bewertungsordnungen im Kontext ökonomischer und kultureller Unsicherheit. In S. Duttweiler, R. Gugutzer, J.-H. Passoth, & J. Strübing (Eds.), *Leben nach Zahlen* (pp. 45–62). Transcript.

Vormbusch, U., & Kappler, K. (2018). Leibschreiben: Zur medialen Repräsentation des Körperleibes im Feld der Selbstvermessung. In T. Mämecke, J.-H. Passoth, & J. Wehner (Eds.), *Bedeutende Daten* (pp. 207–232). Springer VS.

Wenger, E. (1998). *Communities of practice: Learning, meaning, and identity*. Cambridge University Press.

Whitson, J. R. (2013). Gaming the quantified self. *Surveillance and Society, 11*(1/2), 163–176.

CHAPTER 5

Quantifying Inequality: From Contentious Politics to the Dream of an Indifferent Power

Ota De Leonardis

Inequality is a subject that is being talked about a great deal these days. Large quantities of data and figures provide us with unequivocal evidence of the huge disparities—in income, wealth, and so on—that characterize contemporary global capitalism. This trend towards polarization is all the more evident when we view it from a historical perspective, as Thomas Piketty (2014) has so masterfully done. The figures speak eloquently, but what do they refer to? What do we mean exactly, by "measuring inequality"?

It is a well-known fact that over the course of modern Western history in general, and of capitalism (and anti-capitalism) in particular, the notion of inequality has acquired a relational meaning that refers to power relations between unequal people. It could be argued that the history of the construction of this meaning began with the long

O. De Leonardis (✉)
Department of Sociology and Social Research,
University of Milano-Bicocca, Milan, Italy
e-mail: ota.deleonardis@unimib.it

© The Author(s) 2022
A. Mennicken and R. Salais (eds.), *The New Politics of Numbers*,
Executive Politics and Governance,
https://doi.org/10.1007/978-3-030-78201-6_5

135

struggle against the "high/low" dualism of the mediaeval Christian tradition through the process of secularization.[1] And in this process we might recognize a crucial turn in the "symbolic form" of the Modern Perspective, which provided the cognitive tools for linking the "high" and the "low" together and endowed this link with a political character, through Machiavelli and then Hobbes especially.[2] The French Revolution, which reciprocally gave equality a political status, also represents an important moment. Then, of course, Hegel's *Herr und Knecht Verhältnis* (Master–Slave Dialectic) shaped the framework within which inequality meant a bond of domination constituting both the dominator and the dominated, that conferred an intrinsic dynamism on the social order. There followed the rise of capitalism, from Marx onwards. The capital/labour relationship came to be the main point of reference for inequality, associated as it was with exploitation and the private appropriation of socially produced wealth. For more than a century, inequality became the central critical issue in the labour movement and in the anti-capitalist—and indeed anti-imperialist—conflicts.

My aim in this chapter is to show the shift in the semantic field of inequality that has taken place over the last forty years or so. Owing to this shift, the meaning of inequality as a (historical) bond of dominationand subjection is being obscured. In the current discourse on inequality and its magnitude, the (political) reference to power relations has become weaker and weaker, and inequality as defined by quantification now tends to designate a distributive difference, a gap, a disparity: a distance, and no longer a tie.

To this end, I will, in the following, identify some crucial steps in this semantic shift within the changes that have affected the vocabulary of welfare in Italy (but under pressure from Europe) over the course of some forty years. I will dwell in particular on the *dispositif* of the threshold—first of all the poverty threshold—and on the multiplication of thresholds in the field of welfare. I will look at the influence that the syntax of the threshold has exerted on the reconfiguration of this field, and draw some conclusions regarding the effects on the semantics of inequality—two in particular.

In the first place, "threshold" is a spatial metaphor which particularly shapes social spaces with divisions and separations. This made me focus on space and analyse the spread of space-based technologies of separation in the current governance instrumentation at both local and global levels. Here, I will identify the dynamics of inscriptions of inequality in space, and the emergence of a "spatialized" configuration of the term. "Inequality translates into distance", as Richard Sennett (2006, p. 55) argued. Once inscribed in space, this distance, especially the one between privileged and deprived people, represents the negation of any relationship between them, and all the more so one of domination.

Furthermore, the syntax of thresholds is fundamentally numerical, quantitative, and calculistic. Distances correlate to measurements: they acquire reality to the extent that they are measured. My investigation into the semantic shift from inequality to distance involves numbers. Inequality, being shaped by quantification and measurements, designates the alignment of unequal positions along a linear sequence. Being flattened out, it loses its political significance. Paradoxically, the spotlight on quantitative data illuminates in the tiniest detail how enormous the imbalances are, but the quantitative format reduces inequality to a linear variance, and obscures vertical power relations.

On the way, quantification will emerge as a part, albeit a salient one, of a wider process of symbolic change in which inequality is being reconfigured as distance. In this process, as we will see, words and spatial choreographies come into play together with numbers, as well as interweavings, assonances, and interdependencies amongst these symbolic registers. This perspective on quantification involves two choices of a methodological nature. Here, firstly, I will look at quantification through the influence that it exercises when conferring meanings to the issues it applies to, that is, by analysing numbers as a language—a (situated and historical) *"langage du rapport à la réalité"*, as Desrosières has put it (Desrosières, 2008, ch. 2). Secondly, quantification is framed within a broader context in which various other languages are at work, so that an investigation can be made of its role (its format and uses) in the symbolic institution of society in a given historical-social context: as part of a "thought style", to use Mary Douglas's (and Ludwig Fleck's) expression.[3] When seen from the—indirect, and from the outside, so to speak—perspective adopted here, quantification acquires a significance that is as expressive as it is instrumental: together with tools for knowledge and action, it appears to provide visions as well.

In my conclusions I shall propose some hypothetical remarks about the visions implied by the quantification of inequality, as they seem to express the dream of a domination free from any link with the dominated, morally indifferent and cognitively ignorant towards her/him.

Words: The Semantics of Poverty
and the Syntax of the Threshold

The political history of inequality to which I referred above culminated in the development of the welfare states in Europe in the aftermath of the Second World War. They were the expression of the commitment to reinforcing the social bases of democracy after the devastating experiences of totalitarianism, world wars and mass slaughter, as Alain Supiot has shown in his masterly reconstruction (2010; in a similar vein see also Ken Loach's movie "The Spirit of '45"). Moreover, this development was also stimulated by the challenge posed to European countries by the Soviet Union's collectivist model. Thus labour regulations and systems of social protection were instituted, through different institutional architectures and to different degrees, in the period known as the "*Trente glorieuses*". The welfare state emerged as a political compromise—a deal—between capital and labour (as was still being argued in the literature of the 1970s, for example by Ian Gough [1979]). And the issue of inequality developed within a framework of collective responsibility that called for a redistribution—firstly in relation to labour—not only of goods but also of powers.

The "crisis of the welfare state", officially announced in Italy in the early 1980s, opened a period of welfare restructuring (which is still under way, whether presented as "reforms" or as "modernization"). At that time, my theoretical interests focused on the forms and conditions of institutional change, which I investigated in relation to both the cognitive and normative dimensions, and my main area of research was welfare institutions and policies. It was in this field, therefore, that I began to look more closely at the changes in the vocabulary of welfare used in Italy at both local and national level, from the 1980s onwards, and broadening my focus to include the European Union's social programmes as a crucial source (De Leonardis, 1998, 2000).[4]

Shifting Words

I pointed out that certain words had fallen out of use in the current language of the welfare arenas, whilst new ones were being adopted without encountering any significant resistance. After all, the new words expressed good intentions: "the fight against social exclusion", for instance—how could anyone object to that? As I explored these shifts, I noted (1) the speed of the change and the apparent self-evidence of many of the terms in circulation, with no need for explanation or justification; (2) the sappy rhetoric that dominated discussion; and (3) the reiteration of certain lines of argument that were virtually identical in the various contexts, one highly authoritative source of this being Europe (the European New-speak). This suggested to me that the welfare arena had been subjected to a massive investment in language whilst at the same time "conflicts over the vocabulary" (Fraser, 1989) had attenuated.[5]

We do things with words, as John Austin reminds us: all the more so when they are being used in the official language of a policy or a regulatory system, and therefore have normative force. From this perspective, I focused on the emerging words related to the semantic field of inequality that redefined welfare problems and goals and accompanying discussions on the institutional architecture of welfare (see Fig. 5.1).

Inequality and:

oppression   poverty
exploitation   social exclusion
subalternity

rights   and voices   needs and means

                                              deservingness

                                              responsibility

Those affected:

protesters   victims

An issue for:

contentious politics   humanitarian engagement

antagonism   tolerance

Inequality versus:

Equality   fairness

                                              parity

                                              proximity

Fig. 5.1 Examined semantic shifts

The map (see Fig. 5.1) sums up in an impressionistic fashion the semantic shifts I noted.[6] I paid special attention to the term "poverty". After its eclipse in the golden age of the welfare state, where the term denoted a residual phenomenon, "poverty" reappeared in Italy as a category during the 1980s and acquired a central position in the vocabulary—at least as much as in reality. As I pointed out then, "as the subject of poverty acquires increasing significance in welfare policies, it triggers a change in their vocabulary [...] [P]overty becomes the central framework that shapes policy choices [...] no less than scientific research" (De Leonardis, 2000).

In the semantics of poverty, "needs" replace "rights" as the main referents to identify persons vis-à-vis institutions and policies. Indeed, as the issue of poverty came to the fore, references to rights as essential attributes of (social) citizenship dwindled away, and virtually disappeared from the discourse. The term "needs" took their place, and was aligned with the "means" to satisfy them (and the relative "means tests") provoking an overall reconfiguration of the semantic field of welfare. Moralization and quantification were emerging together as the main drivers of this reconfiguration.

Moralization

As regards moralization, it could be noted that the needs/means pairing was leading to defining and evaluating welfare issues more in moral than in political terms, more in the vocabulary of judgement on personal responsibility than in that of the law regarding the rights one is entitled to. After all, we already know that this is the role that the category of poverty has played in the moral order of capitalism in general and in the history of labour regulations in particular, since the "primitive accumulation" laws against vagrancy, and then recurrently. The moralizing significance of the poverty issue came into play, for instance, in relation to the urban plebs of Haussmann's and Hugo's Paris representing the "*question sociale*". Poverty was seen as being associated with moral degradation, criminality and vice, and approached as the matrix of the "*classes dangereuses*" (Castel, 1995, quoting Chevaliers). The moralizing mark of poverty crops up again and again in the labour movement's battles for social rights, and particularly when the category of the "unemployed worker" was being constructed. Looking at Britain in the early 1930s and 1940s, Noel Whiteside (Whiteside, 2015; but see also Salais et al., 1986) accurately identified the place occupied by "the opprobrium heaped onto the idle poor", whose "demoralization" was assumed to threaten Britain's economic performance (Whiteside, 2015, p. 153). Together with "[...] free enterprise and [...] the efficacy of financial instruments to address risk", this moral stigma constitutes that liberal "collective faith" which is implied in the moral order of capitalism. "All features", Whiteside (2015, p. 153) opportunely adds, "that have proved extremely durable", as shown by the moralizing process of welfare in Italy promoted by the poverty issue from the 1980s onwards. Once again, at its core lies the great divide between the deserving and the undeserving poor, making access to welfare benefits dependent upon assessment of the recipients' deservingness.

It is precisely here, in the assessment operations, that the complementary drive of quantification comes into play in the semantics of poverty reconfiguring welfare matters, values and policies. The deservingness principle operates a division between claimants, which must be justified on scientific grounds promising objectivity. The (moral) judgement on the claimant's deservingness requires scientific evidence grounded on proofs and tests.[7]

The Threshold

The key tool for scientific assessment and measurement of poverty is the threshold, the poverty line dividing the poor and the not-poor. The scientific division it establishes intersects with the moral division between the deserving and undeserving poor. Fixing the threshold is an integral part of the very definition of poverty, as well as of the policy instrumentation in this area. In the widespread debate on the definition of poverty that developed during the 1970s, issues concerning the categorization of poverty were bound up with both measurement and justice issues, concerning criteria and choices: What indicators are pertinent? Indicators related to income, consumer baskets, or "necessities" taking up Rowntree's budgetary approach again? What components should be included, with what scales of equivalence? Whether and how should temporal variations be considered, or the size of the household, or poverty perception, and so on? Discussions hinged on what criteria and what tools should be adopted for determining the poverty threshold, in other words how and where the line should be drawn, how to determine the standard below which life is lived in a condition of poverty.

Of special interest in this respect is the distinction between an "absolute" and a "relative" notion of poverty, which opposed Amartya Sen to Peter Townsend, both key figures in this debate.[8] It was Townsend who, basing the determination and measurement of poverty on the notion of "relative deprivation", dictated the terms of the issue, also in the vocabulary of Europe.[9] Sen has criticized the "relative" definition of poverty arguing that "ultimately poverty must be seen to be primarily an absolute notion" (Sen, 1983, p. 158). There is, says Sen, an "irreducible absolutist core in the idea of poverty" (Sen, 1983, p. 159), which becomes visible from the perspective of capabilities. Capabilities themselves are absolute, if we mean by this term the universal value of the eminently human quality of agency they assess: what constitutes "a derived and variable element" are, instead, the commodities necessary for this quality to flourish. In other words, the determination of poverty based on capabilities raises a question of universal absolutes, whilst taking "a relative form in the space of commodities" (Sen, 1983, p. 161). "Even with exactly the same absolute shortfall [...] a person may be thought to be 'poorer' if the other poor have shortfalls smaller than his. [...] Quantification of poverty would seem to need the marrying of considerations of absolute and relative deprivation even after a set of minimum needs and a poverty line have been fixed" (Sen, 1979, p. 293).

This argument underpins the proposal of compromise Sen advances (Sen, 1983, p. 161): "There is no conflict between the irreducible absolutist element in the notion of poverty (relating to capabilities and the standard of living) and the 'thoroughgoing relativity' to which Peter Townsend refers, if the latter is interpreted as applying to commodities and resources". This proposal, which was rejected by Townsend at the time, was eventually taken up again in an operation that closed the controversy: The Copenhagen Declaration, emerging from the 1995 UN Summit, where Townsend was again a protagonist, makes room for the "absolute poverty" sustained by Sen, and interprets the compromise he proposes by establishing a "two-levels definition". Does this mean everything is settled? Not really. The fact is that in this outcome, the term "absolute" has changed meaning and consequently the compromise in question does not fully correspond to Sen's intentions and reasoning. Here, the term "absolute" defines the manifestation of poverty in extreme forms consisting in "severe deprivation of basic human needs" (UN, 1995, para. 19). "Absolute" has become a synonym of "severe". Assuming the existence of an "absolute poverty" has made it possible, in this official context, to take into account Sen's perspective on poverty in cross-national measurements and comparisons. At the price, however, of a banalization that makes it equivalent to the lowest level on the scale of poverty, that of subsistence.[10] It should, however, be remembered that Sen himself was perfectly clear:

> The characteristic feature of 'absoluteness' is neither constancy over time, nor invariance between different societies, nor concentration merely on food and nutrition. *It is an approach* of judging a person's deprivation in absolute terms... rather than in purely *relative* terms vis-à-vis the levels enjoyed by others in the society. (Sen, 1985, p. 673; first emphasis added; second emphasis in original)

Thus, absoluteness defines not so much a type or degree of poverty but an approach, and specifically an approach that does not consider comparisons to be exhaustive for the purpose of defining poverty. Which is precisely the claim of a relativist approach.[11] The "absolutist" core Sen insists on has nothing to do with its comparative aspect: it is politically determined. The "absolute" opposed to the "relative" by Sen in this definition calls into play a third term to which the relativities of the comparisons are anchored: terms of reference fixed through political compromises on conflictual issues about ends and values. In the comparative perspective on poverty, the terms of reference are instead determined from within, and emerge from the comparisons themselves. The acceptable level of poverty—expressed in the poverty line—is established by means of comparisons between the poorer and the less poor.

Even absolute poverty, banalized as we have seen, has become congruent with this comparative logic. In the end, it is the relative approach that has prevailed, and fixed a comparative frame for the whole set of categorizing, research, measurements, rankings etc. for determining poverty, which meanwhile has continued to develop. This, then, is the format for knowledge that condenses into the figure of the threshold and constitutes it as the central informational basis—Sen's "informational basis of judgement in justice"[12]—on which policy choices in the field of poverty and welfare are based.

Visibility and Obfuscation

Framed by the threshold, attention is focused on what is happening around the borders, and on how to measure marginal differences, variations, transitions, and suchlike, with important consequences for the definition and treatment of poverty. Here, we draw once again on Sen, who discussed this in his latest book co-authored with Drèze on India (Drèze & Sen, 2013, see specifically ch. 7). On the one hand, attention focusing on thresholds entails a bias towards targeted, differential responses to poverty or inequality issues, which tend to fuel "extremely divisive" effects, segmentation, and dynamics of "exclusion and divisiveness" (Drèze & Sen, 2013, p. 191).

On the other hand, in this way attention is diverted from the substance of the issues, from what is happening below, beyond the poverty line. Concerning the poverty line that was officially established in 2011 in India, Sen and Drèze (2013, p. 189) point out that the ensuing public debate has concentrated on the threshold and the dire level that was established, whilst "missing the main point": The fact that, "even with this low benchmark, so many people are below it – a full 30% of the population, or more than 350 million people" (Drèze & Sen, 2013, p. 190). In other words, "the terrifying yet hidden nature of mass poverty – its enormous size – has been quite lost" (ibid.), and, as a result, completely ignored. The threshold is a device that fixes a measurement of poverty. And this measurement, a number in fact, whilst giving poverty great public visibility and attention, equally seems to produce effects of obfuscation. From studies on quantification, we are already aware of this type of effect and, in general, of the selective nature of numbers in giving an account of the phenomena they measure. In the case in question, the measure of poverty shaped by the threshold obscures other crucial information on poverty itself—including, of course, other quantitative data (amongst which the percentage of people who do not have a toilet available: around 50%). From Drèze and Sen's arguments in the following chapters, the impression is given that what is obfuscated and neutralized, is "the grip" of the inequality between "the privileged and the rest" in India today, its enormity, its "outrageousness" (Drèze & Sen, 2013, p. 279). And "the dominance of the privileged" in terms of voice in public reasoning (Drèze & Sen, 2013, ch. 9) is fortified by a systematically diverted attention creating "blind spots" on social failures that are as invisible as they are serious, like the blind spots that are indeed created when attention is focused on the poverty line.

The threshold constitutes a key device for responding to the need, mentioned above, to give scientific grounds to the definition of poverty. By providing measurement criteria and the resulting standards it makes such definition a correlation of its measurement (specifically in comparative terms, as we have seen). As I noted at the time, when reasoning on the growing popularity of this category in the field of welfare, "the bureaucratic passion for categorisation [...] is being replaced by a computational passion that translates all the issues into terms of measurement" (De Leonardis, 2000, p. 95). I shall now add that the poverty line appears to be a central syntactical element for this translation, precisely because it is a line that establishes a binary code—in/out, above/below, yes/no, 1/0—as a basic frame for public knowledge and action. In this way the poverty line aligns the category of poverty—both in its cognitive and normative values, both as a public issue and as the object of policies at all levels of governance—with its quantitative format.

We shall return later to the paths opened up by this investigation, in order to examine the role played by quantification in reconfiguring inequality. To conclude here the argument on the semantics of poverty, we must look back again at the association between the quantitative format established by the threshold and the equally powerful drive of moralization. And we must emphasize the fact that dynamics of division are triggered by both the moral divide between the deserving and the undeserving poor, and the numerical separation enacted by the line's binary code. All in all, the semantics of poverty, when observed at close quarters, implicitly denies the promise that had justified the centrality of the "poverty issue" in the welfare field in Italy. Poverty supposed to be an all-encompassing notion overcoming the—widely criticized—category-based welfare system, was as such surrounded by a universalistic aura, whereas a non-universalistic regime of justice was starting to be established just through the category of poverty. A justice which, according to the quantitative parameter of poverty, subjects the welfare claimant to judgement, weighing up needs and means, selecting and awarding prizes and punishments. In the tangle of moralization and quantification a sort of "bookkeeping justice" is to be glimpsed, weighting the benefits granted against the contributions that people make to society understood as a "shared venture".[13]

Spatial Choreographies:
From Inequality to Distance

The figure of the threshold encourages us to follow another line of investigation into the semantic change in inequality and the role played by quantification: "Threshold" is clearly a metaphor, and a spatial one as it draws a dividing line—in/out, above/below, etc.—according to a binary code. Coming across a spatial metaphor at the heart of the knowledge infrastructure on poverty might not be so important, were it not for its assonance with a more general trend towards forms of "spatialization" in the governance of social issues. All the more so since dividing and separating—as the poverty line does—appear to be, as we shall see, a quite common tendency enacted by these forms.

On the subject of space, space and power, and space as a fundamental lever of governance, a vast body of research is available. Accumulated throughout the history of the disciplines devoted to it,[14] this patrimony has extended its ramifications into the whole corpus of the social sciences. And from Foucault onwards it has been re-investigated and amplified. Because of its symbolic power shaping social organization and conferring an order on it, space emblematically represents a technology of governance that acts indirectly and "at a distance" (Miller & Rose, 1990).

What we learn first and foremost from this background is the basic, preliminary indication that "space" is not "place", as Gieryn (2000, p. 489) has noted. Space is to be "more properly conceived as abstract geometries (distance, direction, size, shape, volume) detached from material form and cultural interpretation [...]. Space is what place becomes when the unique gathering of things, meanings, and values are sucked out" (Gieryn, 2000, p. 465).[15] We should therefore consider firstly that the generative potential of space operates on the territory in the same way, that is by abstraction, as a map does; and secondly that the language of this abstraction is a mathematical, or more precisely geometrical language, once again the language of numbers. As I anticipated, this must be taken into account first and foremost when investigating the current processes of "spatialization", i.e. the diffusion of spatial frames in addressing social matters, in both cognitive and normative terms, and the growing recourse to space-based technologies in governance at both global and local (city) level.[16]

Well in advance Foucault (2001) noted these trends postulating that whilst, "[T]he great obsession of the nineteenth century was, as we know, history [...] The present epoch will perhaps be above all the epoch of space. We are in the epoch of simultaneity: we are in the epoch of juxtaposition, the epoch of the near and far, of the side-by-side, of the dispersed. We are at a moment, I believe, when our experience of the world is less that of a long life developing through time than that of a network that connects points and intersects with its own skein" (Foucault, 2001, p. 1571).

The "network society", investigated and outlined as an emerging social order by Manuel Castells, is the most meaningful expression of spatialization. The network is indeed a geometric figure, the abstract space of a collection of points with its own mathematical laws, which is characterized, Castells maintains, "by the preeminence of social morphology over social action" (Castells, 1996, p. 469). Thus, in my investigation of the semantic changes in "inequality" I will now look at space, spatialization and space-based governance instruments, shifting the focus from words to symbolic artefacts of a spatial nature. My aim is to explore what meanings these artefacts confer on inequality. The analytical background I shall use as a basis comes from research on urban policies, territorial governance, and the transformations that are affecting European cities, mainly.

Cities, which are obviously the preferred environment for space-based policy instruments (first and foremost those based on architecture and urban planning) appear to be affected by two opposing and simultaneous spatial drives. On the one hand, as Françoise Choay argued in her studies of history and anthropology in architecture (Choay, 2006, p. 10), spatialization driven by globalization constitutes an expansive drive that acts upon the extension of the city and produces urban sprawl and thus its disarticulation (and its replacement with the *urbain*). On the other hand, the opposite move towards concentration, which insists on circumscribing the local—the neighbourhoods, typically, in order to control them or increase their value—equally tends to fuel separation, segmentation and disarticulation in a different form.

It is this latter shift that is of interest here. It has its origins in and is fuelled by the issue of "urban (in)security", whose vocabulary of motives was provided by the "fear of the Other". This issue reached Italy (from the Anglo-Saxon world) during the 1990s, establishing itself in public discourse. It marked the beginning of the intense season of policies for urban security, which since then have become a major lever of governance in cities, not only in Italy, and are constantly updated (today also due to the terrorism alert). It has also been noted that in this way problems and solutions are reframed, with attention being shifted from the "social" issue of security, an issue to do with welfare, to its "civil" significance, an issue to do with law and order, and from social protection to police protection.[17] In this frame, whether it is a matter of managing a deprived urban area, of fighting "degradation" by means of "urban décor"[18] interventions, or of preventing social disorder and criminal behaviour by means of surveillance devices, all these policies operate through space-management and space-based instruments. Security is above all territorial, being translated into "securitized territories". And, in its turn, the territory, its borders and identity are increasingly marked by the question of security as the crux of the relationship between citizens and institutions. In the governance of the city (and not only) "territory" becomes a keyword which, whilst giving recognition to the everyday life contexts of people and to local communities, transfers on them the semantic density that the term has accumulated in reference to state sovereignty and its inner security issue.[19] It is in the territory that governance operates by means of area-based instruments, borders, partitions, "*quadrillage*" (grids), and so forth, and thus it is there that clues are to be found to the processes of spatialization, which also affect inequality.

In fact, there is much talk of "territorialization", particularly following in Foucault's footsteps and revealing how powerful it is in the digital and globalized world, just when the virtual is being freed from any spatial bond (see for instance Paul Hirst, 2005, part 3). Anyway, territory and territorialization have become a common yardstick for all policies on social matters—including welfare (Bifulco, 2014), in line with the emphasis placed on the "local" by European policies and programmes. And it is above all in relation to places and spaces, areas, zones, districts, and the like, that policy issues are defined. Even the people's status as citizens—and possibly citizenship itself as a status—is now more directly anchored in their own (local) territory, so that the model of citizen is now provided by the "inhabitant" active in his or her own neighbourhood, as was noted in France by Cathérine Neveu (2011) and Jacques Donzelot (2009).

Territory and territorialization contain the promise of a governance that is closer to the citizens, the promise of a privileged arena for enhancing political participation and democracy. There are experiences and some evidence that give credit to this promise,[20] but these are far from questioning the "territory's" prevalent semantic frame I have just outlined, its intrinsic link to the theme of security and the spatial separation it activates. All in all, on observing forms of territorial governance globally, it can clearly be seen that they tend to produce "exclusion" or "expulsion" (Hirst, 2005; Sassen, 2006, respectively), by making wide use of borders and separations. And we are well aware, with reference to the local scale of governance, that relating people to a given territory may equally be a source of stigma and denied citizenship, as, in the case of France, the young *banlieusards* perfectly know.[21] In our cities[22] the signs by which the territory is marked, consisting of borders and separations, may be slight, yet the well-known phenomena of "relegation", "urban segregation", and the like are important signals and allow us to glimpse drives towards a spatial concentration of homogeneous populations in distinct urban areas. In many of the world's big cities actually, these trends are much stronger, and give rise to spatial concretions of the polarization between privileged and deprived urban populations, such as gated residential communities or luxury areas, on the one hand, and "difficult areas" or slums, on the other. In these forms of spatial inscription of unequal populations significant traces of a "spatialization" of inequality may be detected (Bricocoli & De Leonardis, 2015). Of the space-based technologies employed by territorial governance, many bear the same marks: barriers, enclosures, sensors, checkpoints, (also private) armed police, off-limits areas, fences, walls and, yes, moats. Open spaces, such as public green areas, should also be included in this list when they are designed to create a buffer zone that protects a middle-class residential area from the disorder of the city, and so should the "by-pass roads" that make it possible to skip areas of urban misery, ignore them, and live separate lives. These dynamics of division that slice the city into segments tend to eliminate places, opportunities and reasons for meeting and exchange, and for conflict between unequal populations.

The binary logic of separation finds its most drastic expression in the "walls" that have started to proliferate everywhere, especially as the preferred solution to the threat posed by migratory flows in our globalized and hyper-connected world. The barrier between Mexico and US is the most famous and the longest, and the Israeli "Wall" separating Palestinian people is considered a prototype (Weizman, 2007). In Europe, too, migrants are kept outside its boundaries, as we have had to recognize, and—at its Southern boundary—the armed Mediterranean Sea is now performing a similar function, paradoxically. Starting from Wendy Brown (2009) who first gave an account of the phenomenon, the literature has emphasized several features of this device.[23] Although it is adopted as a quick and easy tool of governance, a ready-made solution, its effectiveness has already proved to be highly dubious. Such barriers perform more of a "theatrical function" (Brown, 2009, p. 122), and what they stage is, in fact, separation. Whilst the walls of total institutions (such as asylums or prisons) which have been familiar in our modern landscape for so long, segregate people for taking charge of, and treating—re-educating, punishing, etc.—them, these new walls produce a separation only, by enacting a spatial division on a territory into two abstract spaces, without any people being taken into charge whatsoever. Through and around these "walls of separation", the powers exercised and the operations performed are directed towards impeding, driving away, rejecting, turning back, establishing a distance, avoiding encounters and denying recognition.[24]

Walls set up a powerful choreography for global inequality. Through spatialization, inequality no longer designates a bond between unequal people, but rather a distance that suspends or denies any relationship between them. "Inequality translates into distance": it is appropriate to recall here what Richard Sennett (2006) argues when discussing how the chain of command changes in the "new capitalism", and observing the enormity of the distances between top managers and workers in the globalized company. "There is nothing like a relationship between a Thai shoe-sticker and a Milanese fashionista; they transact, [...] rather than relate" (Sennett, 2006, p. 55). The translation of inequality into distance, says Sennett, goes hand in hand with the "divorce between command and accountability" (Sennett, 2006, p. 57). To this power, being expressed in a denied bond, and therefore de-responsibilized, corresponds the form of subjection that Sennett himself had described as "the bond of autonomy" (Sennett, 1980). This form of domination consists in the denial of any bond whatsoever with the dominated, resulting in both cognitive ignorance and moral indifference to him/her.

Numbers: Measuring Inequality

What "distance" is—that is the length of a line between two points—may be known and recognized by measuring it. Here we finally focus on the role played by numbers, intended as a language, in the reconfiguration of inequality. As I stressed at the beginning of my discussion, this is the language by which inequality is mostly represented nowadays. In the meantime, however, somewhere along the analytical path followed up to now, the innocence of an objective, neutral description of the phenomenon attached to its measurement has evaporated. We have traced the spread of quantification that goes along with changes in the lexicon of welfare and the grip of the quantitative format on the basic cognitive tools for defining issues and governing them. The measurements of inequality now appear to be involved in a more general reframing, in which inequality is being translated into distance. In which numbers, like words, acquire relevance and demand to be observed for what is made with them, i.e. their performative potential in fabricating a reality. They are "an engine, not a camera" (MacKenzie, 2006), to take up a meaningful image summing up a crucial interpretative key in studies on quantification.[25] And it is in this perspective that we shall proceed to investigate the role of numbers in resignifying inequality, and observe how their virtues are exploited.

Numbers provide the synthesis of a plurality of components, factors and aspects of inequality. Thanks to their parsimony, and the economizing function numbers perform in describing (and assessing), it is simply by means of a few well-constructed figures that the quantitative format provides a precise account of the magnitude of a phenomenon. At the same time, thanks to the related standardization, many qualifications may be translated into figures. And the issue may be split and multiplied into a plurality of inequalities referring to the widest possible variety of assets—inequality in income, wealth, education, health, access to the internet, etc.—and placing one next to another in a linear sequence, from life expectancy at birth to the freedom to choose how one dies.

Nonetheless, we first have to consider that within this multiplicity the same cognitive format is reproduced, made up of quantification, measurements, comparisons, ratings and rankings. As far as inequality is concerned, figures tell us a great deal about variations, distances and unevenness between different positions, but very little about power relations between them. Numbers enact a binary logic, as we have already seen, and are directed to making (horizontal) comparisons rather than talking about (vertical) conflicts. More importantly, it should be considered that the economizing function of numbers is primarily performed on words, qualifications and arguments.[26] And we must ask ourselves what effects this parsimony has on the density of the semantic repertoires that noun "inequality" has accumulated and sedimented, and on its inner contentious meaning. Quantitative data save on qualification and argument, on plural interpretations and representations, on voices and conflicts over vocabulary.

The quantification that establishes what is to be considered relevant knowledge with regard to inequality guides a process of abstraction which—in the same way as space acts on place—"sucks out" plurality, contingency and subjectivity, impoverishing the symbolic repertoire for expressing modes and reasons in talking about inequality. The figures on inequality—so precise and well-founded on the authority of science—efficiently carry out their task of conferring objectivity, the "mechanical objectivity" grounded on calculus and expressed in a "matter-of-fact" format. And of course these are very important results. Nonetheless, when objectivation is exerted on the issue of inequality it comes with high costs. The great variety of other forms of knowledge about inequality is absorbed or replaced by numbers and ends up being neutralized. Costs are high in terms of the naturalization of the inequality issue, with the risk that "an ontological naturalness or essentialism [...] takes up residence in our understandings and explanations" concerning this issue (Brown, 2006, p. 15). As far as it is framed by measurements, inequality also becomes exposed to the effects of their performative potential I have just recalled, when these are taken as a metric for rating operations and assessing performances. This is an aspect that deserves investigation, remembering the research on rankings and the reactivity (or feedback) they produce in the field they measure and order, and more in general the way these types of quantitative instruments function in the "governance by numbers".[27] Indeed, with figures on inequality, their descriptive function may end up being replaced by, or incorporate performance indicators. This is plain to see especially in the area of statistics, comparisons and rankings between countries, as when inequality figures are included in the set of indicators used to evaluate the performances of a country targeted for aid programmes.

In Ousmane Sidibé's contribution on the subject in this volume, the country is Mali, and the indicators to be improved concern inequality in education. The example shows that, under the pressure of ranking (and the incentives correlated to it) an enormous effort is made to raise the rates of education, attaining a numerical objective which, however, obfuscates substantial quality issues of education itself. This is a case in point for the "governance by numbers" where one can see how the inequality issue is being treated within a management by objectives system, formatted by the latter's instrumentation and dynamics, and how widespread practices are for "pushing up the numbers", even by cheating.[28] The outcome in terms of a real reduction of educational inequality is, to say the least, dubious. In any case, the issue of inequality, as it is incorporated into the set of global governance control tools to produce "adjustments" and "alignments", undergoes distortion. Not only is its meaning set as a problem of disparity, a quantitative gap, a distance between positions aligned along a distribution curve, hence losing its relational grounds, its reference to power relations, but it may also happen that inequality itself, as expressed by numbers, becomes a tool for control, discipline and subjection.

The lesson I draw from this case concerns another aspect of the more general reconfiguration of inequality. This paradoxical twist in the meaning of inequality has been made possible by its reshaping and treatment as a matter for management, in accordance with the managerial style of dealing with problems. More precisely, inequality appears to fall within the category of problems that are there to be "managed" instead of being "solved", which Sheldon Wolin identifies as a salient trait of the "managed democracy" (and its "domestication", Wolin, 2008).

However, we also know that numbers may well provide people with strong arguments against power. In the case of inequality figures, too, numbers give phenomena visibility, a visibility grounded on scientific evidence. Inequality, being expressed in the measurement of distances, looks like a "matter-of-fact" issue, and it is in this format that it acquires visibility. It is thanks to this visibility they confer on situations being measured, that quantitative data, their construction and use have often been, and still are, crucial matters at stake in claims for recognition, in political struggles, and in the making of "collectives".[29] Can we say that the same thing is happening today with the figures on inequality?

In a way, yes, it does seem that here, too, figures provide arguments for denouncing disparity and representing a collective, particularly around the polarization they make visible: typically, the "our 99%" against the richest "1%", is a main argument of the "Occupy" movement. The quantitative framework does help denounce an imbalance of power that has grown to the limits of disproportion, giving rise to public protest and collective action. Nonetheless, this power imbalance that numbers highlight appears, in its very disproportion, to be simplified to the extreme, and void of qualification. That 99% remains an aggregate as vast as it is indeterminate, corresponding to, following Robert Castel's (2003) metaphor, a "collection of individuals" sharing a statistical position only, rather than a "collective" of political subjectivities. Whilst attention is focused on the disproportion, the question of what connects the two poles remains obscured. Here again, the issue of inequality is represented more in the form of a gap than of a bond. And as far as domination is concerned, it appears to be a matter of unbalanced quantities. Even when polarization is critically traced back to the new capitalism—as done by Piketty in "Capital in the Twenty-First Century" (Piketty, 2014)—the tendency is to forget Marx's famous warning on how easy it is to pass off capital "as a thing" rather than as the "social relationship" it really is.

CONCLUSIONS: THE DREAM OF AN INDIFFERENT POWER

As scholars of semiotics well know, and as Koselleck's *Begriffsgeschichte* (2007) also teaches us, the meaning of a word has a history that can only be investigated in relation to other words. Inequality is no exception, as it immediately recalls a dense constellation of words and related meanings.[30] Of this constellation, the portion analysed here, however limited in time and place, has provided several leads to explore how the meaning of the word "inequality" has been changing—in a nutshell, it is being translated into "distance"—and what role has been played by quantification, or more precisely by the language of numbers. During this investigation the latter, as a *langage du rapport à la realité*, has come into play together with the language of words and that of spatial artefacts. And I have shown how these languages contribute to reconfiguring inequality as a distance "from", instead of a tie "between".

As far as it is framed as a question of distances and measurements, inequality is captured in a quantitative format. Within this format the vertical configuration, which anchored inequality to burning issues of power, politics, and institutions, is being obfuscated, and inequality appears in the normalized format of a quantitative variance flattening out along a horizontal line. It is established by a comparison between linear positions and intended merely as a matter of plus or minus, more or less, yes or no, according to a binary code. Even though figures can be multiplied, cross-compared and updated in real time, the picture remains flattened out on a one single level. Figures only relate to other figures, whilst the third term that links them tends to disappear. We might say that, because it is only acknowledged as relative, and not as relational, inequality has lost its absolute—that is, its societal and political—dimension (much like Sen's "absolute poverty"). It is worth noting that similar flattening effects of quantification come to light in the making of the "new calculable global world" that Laurent Thévenot discusses in his contribution to this volume, by analysing certification standard setting concerning palm oil. His account, from the perspective of the "smallholders"—farmers and rural communities—involved in the "participative" procedure, vividly shows how in this latter the arguments are both formatted for making things calculable and expressed in a horizontal arena that conceals the (rather obvious) "power imbalance between parties". Any third term between these parties is lacking, aside from the so-called "third parties", whose impartiality appears highly dubious. In that case, too, Thévenot argues, domination bonds get obscured by the "juxtaposition of 'stakes' in a horizontal dialogue around the round table".[31]

In any case, those flattening effects are the mark left by quantification on the semantic field of inequality. However, we should also reverse our view and look at what mark is left on quantification by its involvement in this field and in the more general resignification process affecting inequality. As we have seen, from the vantage point of the inequality issue, quantification is being highlighted in relation to ruling powers, especially those powers that are expressed, since Hobbes, in naming, as well as, obviously enough, in counting. "Depoliticization" may be a first interpretative key when considering that, in quantifying inequality we have seen a power engaged in fabricating a reality without seeming to do so. Of course, as I have already recalled, quantification can also keep political struggle alive, both when it is a matter of constructing data and when these data coagulate political arenas and subjectivities. But the case of inequality bears a quite different mark, as we have seen, which suggests placing quantification amongst the drivers of the depoliticization of political choices characterizing, according to Wendy Brown (2006, 2015) the neoliberal discourse.

Indeed, the issue of inequality, which has for a long time been the crux of political struggles and compromises on the social order, appears to be caught in the grip of the "discourse of depoliticization" as intended (see Brown, 2006, ch. 1) as a discourse that "eschews power and history in the representation of its subject" (Brown, 2006, p. 15),[32] emptying it of political significance. In the end, inequality is transformed, as we have seen, into a problem to be "managed", a terrain on which to apply a managerial logic. "Inequality becomes normal, even normative" (Brown, 2015, p. 38). According to this interpretation, numbers are involved in depoliticization since they provide a language for the economic metric which is at the core of the neoliberal spirit, and for generalizing it within "spheres and activities heretofore governed by other tables of values" (Brown, 2015, p. 21). By providing the "dissimulation of the normative work they do" (Brown, 2015, p. 135), numbers contribute to vanquishing "the already anemic homo politicus", being replaced by "homo oeconomicus" (Brown, 2015, ch. 3).[33]

Our focus on the inequality issue confirms that the quantitative format can be extremely efficient in eschewing power matters in the issues it shapes. But what kind of power operates through the dissimulation provided by numbers in the case in point? As we have already seen the translation of inequality into distance corresponds to a power being exercised through that distance: not only in the sense of "governing at-a-distance"—something we already know numbers may contribute to—but rather in the sense of "keeping at a distance". Numbers provide dissimulation to a power that denies any form of bond with its own object/subject. A domination, I argued or rather conjectured, being enacted through an indifference which takes the form of both ignorance on the cognitive side, and irresponsibility on the moral side.[34] The issue is now how the language of numbers reframing inequality is involved in corroborating this indifference.

We already know how much the neoliberal "bureaucratization" (Hibou, 2012) exploits the performative role of quantitative data in renovating and enhancing the archetypal indifference of the bureaucratic command.[35] As Supiot (2015) shows, the automatisms of the governance by numbers replicate in digital form the long-lasting mechanistic utopia of the *homo automata*, freed from any subjectivity. But the crux of the matter lies in the very virtues of numbers, as we have seen them at work configuring inequality as distance. We have seen how the science-based operations for sparing on words and qualifications, and for "mathematization" (Ogien, 2013) transfigure inequality into an abstract picture. The abstraction that the language of numbers is able to achieve results in a rarefied reality. The latter appears, at the same time, in the spotlight of scientific evidence, and cleared from any other form of knowledge and language, of view and experience, and even of data. It is precisely thanks to the capacity of numbers for abstraction, and thus to the abstract and rarefied picture of inequality they provide, that the inner contentious meaning of inequality, as a power issue, is being sucked out, and may be ignored by the powerful. The "skilled ignorance" grounded on the quantified inequality goes hand in hand with the moral innocence that the reference to destiny or chance (like in a gambling game) in its turn authorizes.

To sum up, the powerful machinery of numerical abstraction transfiguring inequality would seem to support the dream of a power that dominates by pretending to be both cognitively and morally indifferent. But there is a disproportion, here, disturbing the dream. This is the disproportion between the rarefied picture of inequality that numbers achieve—objective, precise, and complete as it claims to be—and the density of the silenced social knowledge on inequality, the related immense sufferings, and the anger growing around it. This disproportion reveals that this dream is just a dream. And when the abstract reality of the quantified inequality grows to the point of irreality, the dream of indifference appears coupled with obtuseness.

Notes

1. See Carlo Ginzburg (1986) on the history of S. Paul's precept "noli altum sapere, sed time".
2. Carlo Ginzburg (1998, p. 180) shows the political implications of Perspective in the Dedication of Machiavelli's Il Principle. As for Hobbes, the focus is on the Leviathan's frontispiece in both Gamboni's and Schaffer's contributions in Latour and Weibel (2005, pp. 162–202).
3. Fleck (1983 [1935]). But I am especially referring to the re-elaboration by Mary Douglas in the framework of her cognitive approach to institutions (Douglas, 1986, 1996).
4. Here, I examined the arguments and justifications put forward in welfare policy arenas, especially those in normative texts, deeds and administrative acts.
5. The word "labour", too, was (and still is) under a process of redefinition subverting its historically sedimented meaning (Salais 2007).
6. It is worth recalling here the "eclipse" of the elite issue from social theory and discourse that Mike Savage and Karel Williams (2008) show to be an effect of, amongst other factors, the quantitative turn of sociological research on stratification and inequality from the mid-1970s onwards. As national sample surveys were unable to highlight the small group at the top, "elites thereby flipped from view" (Savage & Williams, 2008, p. 3). And at the same time, inequality was defined "not as a set of social relations, but as a graduated hierarchy" (Savage & Williams, 2008, p. 5).
7. It is worth pointing out here that a dual semantic matrix of the noun "evidence" converges on poverty, i.e. not only a scientific frame, but also

a legal one, where it is the police who provides the evidence constituting the information on which legal judgement is based. The combination is evoked by the images of the "war on poverty" in the USA, drawn for example from the classic "Regulating the Poor" by Richard Cloward and Francis Fox Piven (1971). I recall the image of the social worker, popularly known as "social police", who visits the home of the single mother applying for benefits in order to check that she is not hiding a husband under the bed. A different perspective on the legal frame was developed by Carlo Ginzburg, recalling Peirce, to illustrate his historiographic approach (Ginzburg, 1986, pp. 159–193).

8. These two qualifications imply two different semantics of poverty, as well as two different grammars of justice. I refer to only a few aspects of this difference here.

9. "Individuals, families and groups in the population can be said to be in poverty when they lack the resources to obtain the types of diet, participate in the activities and have the living conditions and amenities which are customary, or are at least widely encouraged or approved, in the societies to which they belong. Their resources are so seriously below those commanded by the average individual or family that they are, in effect, excluded from ordinary living patterns and activities" (Townsend, 1979, p. 31). The European Commission's definition, adopted in 1984, is similar in tone: "The poor shall be taken to mean persons, families and groups of persons whose resources (material, cultural and social) are so limited as to exclude them from the minimum acceptable way of life in the Member State in which they live" (EEC, 1985).

10. In the meantime, the vocabulary of needs has been reinforced and they have become candidates for the role of "absolutes" in the place of capabilities (Doyal & Gough, 1991).

11. The comparative element is, indeed, part and parcel of the relative concept of poverty, according to which it is only possible to judge whether or not someone is in poverty in relation to other people.

12. As for this crucial question in Sen's capability approach, see Salais, in this volume. See also Salais (2009) and De Leonardis et al. (2012).

13. I am quoting David Schmidtz, a representative of neoliberal discourse on welfare (Schmidz & Goodin, 1998). It is worth noting that something similar to a "bookkeeping justice" seems to underpin the social credit card system instituted in China today, that Tom Lam analyses in this volume. He shows it to be a crucial government's technology to "economize society" and establish a "credit fundamentalism" in pursuing the official dream of a "harmonic society".

14. As for architecture just consider the inner political substance of Vitruvius' and Alberti's "ars edificandi". On geography see the political history of the cartographic reason so wonderfully summarized in Brotton (2012).

15. Gieryn refers, amongst others, to De Certeau: in this case "place" translates the French term "lieu". See also Gregory and Urry (1985).

16. A "spatial- or topological - turn" has also involved social sciences since the 1980s.

17. Following Robert Castel (2003) who points out how this frame–of civil more than social security–benefits from, and in turn feeds, the ghost of "les nouvelles classes dangereuses".

18. In Italian, the corresponding term "decoro" maintains a double meaning, as it refers to both (aesthetical) decoration and (moral) dignity.

19. Whose genealogy Foucault has reconstructed in his lectures at the Collège de France, 1977–1978 (Foucault, 2004). See also Sassen (2006).

20. Also in the light of some case studies I carried out myself (see Bricocoli et al., 2008).

21. Donzelot (2006). On the related ethnicization of social conflict see Castel (2009).

22. The reference is to Europe, as I said. More generally, it should be remembered how different the history of cities in the US is, marked as it was at its very beginnings by "racial" issues and connected dynamics of spatial compartmentalization that are constantly being renewed.

23. See Brown (2009) in her study on today's "porous" states' sovereignty. See also De Leonardis (2013) and the research literature discussed there.

24. According to Saskia Sassen (2014), it is "expulsion" rather than "inequality" that better corresponds to the "predatory" capitalism, which she now sees emerging. "Repulsion" instead is, according to Serge Paugham's recent research with colleagues (Paugham et al., 2017), the common attitude towards the poor amongst the urban élites in Paris, Sao Paulo and Delhi. The aesthetics of these barriers is eloquent testimony to this. Their sheer crudeness transmits brutality, and the hubris of an act of mere force: Consider by contrast how beautiful the Otto Wagner Steinhof Spital in Vienna is. See also Christopher Payne's rich repertoire of the American asylums (Payne, 2009).

25. The image has been taken up by Espeland and Sauder (2016) to study rankings and their social effects, especially in terms of "reverse engineering". On the performative role of quantitative data in general see the concept of "rétroaction" (feedback), regarding statistics (Desrosières, 2011), that of "reactivity", regarding commensuration and ranking (Espeland & Sauder, 2007; but see also Espeland & Stevens, 2008), whilst on performance indicators see the Salais' image of the inverted pyramid, in this volume.

26. On the economizing function, see Guter-Sandu and Mennicken's very rich discussion in this volume, where quantification gets involved in "economizing the social" in three different ways: curtailing, marketizing and financializing. However, I am looking at this a bit differently. As far as

"to economize" may be also intended as the reframing process of "the social" according to (mainstream) economic thinking, also that "parsimony" the latter predicates and the numbers perform so well, is to be taken into account. It's about cognitive economy as well. The two classical essays by Sen (1977) and Hirschman (1985) still represent a relevant background for this question.

27. Here I am quoting the title of Alain Supiot's Lectures at the Collège de France, 2012–2013 (Supiot, 2015), in which he shows how quantification gives rise to a new normativity in which the rule of law is being dismantled and the law itself reduced to an instrument of–he argues–"total market" laws.

28. On similar cases in Africa see also Boris Samuel (2016). On the "government by objectives" see Thévenot (2015).

29. As we know from historical studies especially on labour statistics (notably by Robert Salais) as well as from research on statactivism (Bruno et al., 2014). An example for statactivism, which also is relevant here as it concerns measuring poverty, can be drawn from Appadurai's account on the mobilization of the inhabitants of the slums in Mumbai (now in Appadurai, 2013). When the city government, prompted by this mobilization, wanted to do a statistical survey on their living conditions, they claimed the statistical tools as their own by undertaking the survey themselves and deciding how to measure these conditions. However, see Boris Samuel, in this volume, on some of the limits of statactivism.

30. It would certainly have been appropriate to explore the opposite notion of "equality", as it also was subjected to dynamics of resignification, along with the changes in the words of welfare discussed in the first section of this chapter. One could note that, on the one hand, the noun "equality" was used less and less, or was treated as an equivalent of "homologation" (versus "difference") so that it acquired a negative meaning; and that, on the other hand, the semantic field of equality was breaking up into a plurality of synonyms or substitutes. One of these words is "parity" (parità, parité), which I mention here, because it seems to provide a fertile terrain for studying quantification. "Parity" demands that things be placed on the same level, as peers, it implies a comparative approach, and is correlated to measurements, rankings etc. (as typically the gender parity index). All in all, there is some contiguity to be seen here, with the flattening out we observed in the reconfiguration of inequality as distance. The opposite of parity is "disparity".

31. Quite obviously, this convergence does not imply that flattening effects, occurring in such different contexts, are intrinsic to quantification, which on the contrary can well be a road for voices to travel up vertically (and "en généralité"). Rather, the two opposite moves in which quantification is involved may help to clarify the difference between the two modes

of quantification—"statistics" or "governance-driven quantification"—that Robert Salais has identified in his contribution to this volume (see also Salais, 2010).

32. "Depoliticization involves removing a political phenomenon from comprehension of its historical emergence and from a recognition of the powers that produce and contour it" (Brown, 2006, p. 15).

33. Perhaps, in order to clear the ground of any impression of ideological criticism, it should be remembered that this line of interpretation has a history. In this respect, it is sufficient to recall certain astute observations made by Werner Sombart (2006 [1906]) during his trip in the US. The account of this trip (dating back to 1905) revolves around the question: "Why is there no socialism in the US?" Sombart identifies one of the answers in the American "passion" for figures, or more precisely in the general recourse to quantitative metrics in assessing people and objects. Thus "bigness" has an absolute prevalence in evaluating and appreciating "greatness". Sombart also argues that it is money that, in the end, constitutes the term of reference for these metrics, more precisely money "in the specific capitalist form" (versus Simmel). In the framework of justification theory, see also its comparative-cultural developments in Lamont and Thévenot (2000) on the relative salience of the market as a principle of evaluation in the American (versus French) polity. Similarly, see also Supiot's (2015) research on the historical-cultural matrix of quantification associated to the rise of the "total market".

34. If one follows Richard Sennett 2006, this indifference, seen as a way of exercising domination, and the divorce between command and responsibility it rests upon, may be considered as a salient feature of the culture of the new capitalism. However, this indifference also recalls to me other, disparate, images. First, we might search for the origins of this orientation in "the revolt of the elites" (and "the betrayal of democracy", Lasch, 1995) or in that sort of "class struggle from the top down" that Luciano Gallino (2012) has identified in the dynamics of Italian capitalism since the late 1970s. Second, from the perspective of recognition (starting out from Honneth) indifference towards the "other" may correspond to a lack of recognition, to an identity not imposed–by the "naming" power–but rather denied (De Leonardis, 2013). Third, this indifference of the ruling powers seems complementary to the subjection standing from the threat of, or the condemnation to "uselessness", that is the condition of potential or actual "surnuméraire" in the book-keeping of working, and of living as well, to pertinently quote Robert Castel (2009) once again. Fourth, this indifference of power also looks contiguous with today's cynicism grounded on (both cognitive and moral) relativist positions (Sloterdijk, 2013).

35. Since Webers's "sine ira et studio" disposition corresponding to the bureaucrat's "honour" (see Herzfeld, 1992).

REFERENCES

Appadurai, A. (2013). *The future as a cultural fact*. Verso.

Bifulco, L. (2014). Citizenship and governance at a time of territorialization: The Italian local welfare between innovation and fragmentation. *European Urban and Regional Studies, 23*(4), 628–644.

Bricocoli, M., & De Leonardis, O. (2015). Les protections sociales spatialisées. Rêves et cauchemars. In C. Bianchetti (Ed.), *Territoires Partagés. Une nouvelle ville*. MetisPresses.

Bricocoli, M., De Leonardis, O., & Tosi, A. (2008). L'Italie. Les politiques sociales, du logement et de sécurité: inflexions néo-liberales dans les politiques locales. In J. Donzelot (Ed.), *Villes, violence et dépendance sociale. Les politiques de cohésion en Europe*. La Documentation Française.

Brotton, J. (2012). *A History of the world in twelve maps*. Penguin.

Brown, W. (2006). *Regulating aversion: Tolerance in the age of identity and empire*. Princeton University Press.

Brown, W. (2009). *Murs: Les murs de séparation et le déclin de la souveraineté étatique* (p. 2010). Les Prairies ordinaires (English edition Zone Books).

Brown, W. (2015). *Undoing the Demos*. Zone Books.

Bruno, I., Didier, E., & Prévieux, J. (Eds.). (2014). *Statactivisme: Comment lutter avec les nombres*. Zones.

Castel, R. (1995). *Les métamorphoses de la question sociale : Une chronique du salariat*. Fayard.

Castel, R. (2003). *L'insécurité sociale: Qu'est-ce qu'être protégé?* Éditions du Seuil et La République des Idées.

Castel, R. (2009). *La montée des incertitudes. Travail, protections, statut de l'individu*. Seuil.

Castells, M. (1996). *The rise of the network society*. Blackwell.

Choay, F. (2006). *Pour une anthropologie de l'espace*. Éditions du Seuil.

Cloward, R., & Fox Piven, F. (1971). *Regulating the Poor: The functions of Public Welfare*. Pantheon Books.

De Leonardis, O. (1998). *In un diverso welfare. Sogni e incubi*. Feltrinelli.

De Leonardis, O. (2000). Quel povero abile povero. Il tema della povertà e le culture della giustizia. *Filosofia e Questioni Pubbliche, 6*(2), 89–116.

De Leonardis, O. (2013). Altrove. La configurazione spaziale dell'alterità e della resistenza. *Rassegna Italiana di Sociologica, 14*(3), 351–378.

De Leonardis, O., Negrelli, S., & Salais, R. (Eds.). (2012). *Democracy and capabilities for voice: Welfare, work and public deliberation in Europe*. Peter Lang.

Desrosières, A. (2008). *Gouverner par les nombres II. L'argument statistique*. Presses de l'Ecole des Mines de Paris.

Desrosières, A. (2011). Buono o cattivo? Il ruolo del numero nella città neoliberale. *Rassegna Italiana Di Sociologica, 52*(3), 373–398.

Donzelot, J. (2006). *Quand la ville se défait*. Seuil.

Donzelot, J. (2009). *Vers une citoyenneté urbaine?* Editions Rue d'Ulm.

Douglas, M. (1986). *How institutions think*. Syracuse University Press.

Douglas, M. (1996). *Thought styles: Critical essays on good taste*. Sage.

Doyal, L., & Gough, I. (1991). *A theory of human needs*. Guilford.

Drèze, J., & Sen, A. (2013). *An uncertain glory: India and its contradictions*. Princeton University Press.

EEC. (1985). On specific community action to combat poverty (Council Decision of 19 December 1984), 85/8 EEC. *Official Journal of the EEC, 2*, 24.

Espeland, W. N., & Sauder, M. (2007). Rankings and reactivity: How public measures recreate social worlds. *American Journal of Sociology, 113*(1), 1–40.

Espeland, W. N., & Sauder, M. (2016). *Engines of anxiety: Academic rankings, reputation, and accountability*. Sage.

Espeland, W. N., & Stevens, M. L. (2008). A sociology of quantification. *European Journal of Sociology, 49*(3), 401–436.

Fleck, L. (1983 [1935]). *Genesi e sviluppo di un fatto scientifico*. Il Mulino (English edition: Thaddeus J. Trenn & Robert K. Merton (Eds.), Chicago: Chicago University Press, 1979).

Foucault, M. (2001). Des espaces autres. In M. Foucault (Ed.), *Dits et écrits II, 1976–1988* (pp. 1571–1581, Lecture 1967, Cercles d'études architecturales). Folio Gallimard.

Foucault, M. (2004). *Sécurité, territoire, population*. Seuil et Gallimard.

Fraser, N. (1989). Talking about needs: Interpretive contests as political conflicts in welfare-state societies. *Ethics, 99*(2), 291–313.

Gallino, L. (2012). *La lotta di classe dopo la lotta di classe*. Laterza.

Gieryn, T. F. (2000). A space for place in sociology. *Annual Review of Sociology, 26*, 463–496.

Ginzburg, C. (1986). *"L'alto e il basso" and "Spie. Radici di un paradigma indiziario" in Miti, emblemi, spie. Morfologia e storia*. Einaudi (English edition: Johns Hopkins University Press, 1989).

Ginzburg, C. (1998). *Occhiacci di legno. Nove riflessioni sulla distanza*. Feltrinelli.

Gough, I. (1979). *The political economy of the welfare state*. Macmillan.

Gregory, D., & Urry, J. (Eds.). (1985). *Social relations and spatial structures (Critical human geography)*. Palgrave.

Herzfeld, M. (1992). *The social production of indifference: Exploring the symbolic roots of western bureaucracy*. University of Chicago Press.

Hibou, B. (2012). *La bureaucratisation du monde à l'ère néolibérale*. La Découverte (English edition, Palgrave Macmillan, 2015).

Hirschman, A. O. (1985). Against parsimony: Three easy ways of complicating some categories of economic discourse. *Economics and Philosophy, 1*(1), 7–21.

Hirst, P. (2005). *Space and power: Politics, war, and architecture*. Polity Press.

Koselleck, R. (2007). *Futuro passato. Per una semantica dei tempi storici, Bologna: CLUEB.* CLUEB (Vergangene Zukunft. Zur Semantik geschichtlicher Zeiten, 1979).

Lamont, M., & Thévenot, L. (Eds.). (2000). *Rethinking comparative cultural sociology: Repertoires of evaluation in France and the United States.* Cambridge University Press.

Lasch, C. (1995). *The revolt of the elites and the betrayal of democracy.* W. W. Norton.

Latour, B., & Weibel, P. (Eds.). (2005). *Making things public: Atmospheres of democracy.* ZKM and MIT Press.

MacKenzie, D. (2006). *An engine, not a camera: How financial models shape markets.* MIT Press.

Miller, P., & Rose, N. (1990). Governing economic life. *Economy and Society, 19*(1), 1–31.

Neveu, C. (2011). Just being an 'active citizen'? In J. Newman & E. Tonkens (Eds.), *Participation, responsibility and choice* (pp. 147–159). Amsterdam University Press.

Ogien, A. (2013). *Désacraliser le chiffre dans l'évaluation du secteur public* (Sciences en questions). Quae.

Paugham, S., Cousin, B., Giorgetting, C., & Naudet, J.-D. (2017). *Ce que les riches pensent des pauvres.* Seuil.

Payne, C. (2009). *Asylums.* MIT Press.

Piketty, T. (2014). *Le capital au XXIe siècle.* Editions du Seuil (English edition, Harvard University Press).

Salais, R. (2007). Europe and the deconstruction of the category of unemployment. *Archiv Für Sozialgeschichte, 47,* 371–401.

Salais, R. (July, 2009). *Deliberative democracy and its informational basis: What lessons from the capability approach.* Paper presented at the SASE (Society for the Advancement of Socio-Economics) Conference, Paris.

Salais, R. (2010). La donnée n'est pas un donné. Pour une analyse critique de l'évaluation chiffrée de la performance. *Revue Française D'administration Publique, 135,* 497–515.

Salais, R., Baverez, N., & Reynaud, B. (1986). *L'invention du chômage.* PUF.

Samuel, B. (2016). Macroeconomic calculation and modes of government: The cases of Mauritania and Burkina Faso. In B. Hibou & B. Samuel (Eds.), *Neoliberal reforms and the everyday politics of the state in Africa.* Amalion.

Sassen, S. (2006). *Territory, authority, rights: From medieval to global assemblages.* Princeton University Press.

Sassen, S. (2014). *Expulsion: Brutality and complexity in the global economy.* Harvard University Press.

Savage, M., & Williams, K. (2008). Elites: Remembered in capitalism, forgotten by social sciences. *The Sociological Review, 56*(1), 1–24.

Schmidz, D., & Goodin, R. (1998). *Social welfare and individual responsibility.* Cambridge University Press.

Sen, A. (1977). Rational fools: A critique of the behavioral foundations of economic theory. *Philosophy and Public Affairs, 6*(4), 317–344.

Sen, A. (1979). Issues in the measurement of poverty. *Scandinavian Journal of Economics, 81*(2), 285–307.

Sen, A. (1983). Poor, relatively speaking. *Oxford Economic Papers, 35*(2), 153–169.

Sen, A. (1985). A sociological approach to the measurement of poverty: A reply to Professor Peter Townsend. *Oxford Economic Papers, 37*(4), 669–676.

Sennett, R. (1980). *Authority.* Knopf.

Sennett, R. (2006). *The culture of the new capitalism.* Yale University Press.

Sloterdijk, P. (2013). *Critica della ragion cinica.* Cortina (Kritik der zynischen Vernunft, 1983).

Sombart, W. (2006 [1906]). *Perché negli Stati Uniti non c'è il socialismo?* Bruno Mondadori.

Supiot, A. (2010). *L'Esprit de Philadelphie. La justice sociale face au marché total.* Seuil (English translation: Verso, 2012).

Supiot, A. (2015). *La gouvernance par les nombres.* Fayard (English edition, Verso 2016).

Thévenot, L. (2015). Autorités à l'épreuve de la critique. Jusqu'aux oppressions du 'gouvernement par l'objectif'. In B. Frère (Ed.), *Le tournant de la théorie critique* (pp. 216–235). Desclée de Brouwer.

Townsend, P. (1979). *Poverty in the United Kingdom.* Penguin.

UN. (1995). *World Summit for Social Development 1995.* Final Report. United Nations.

Weizman, E. (2007). *Hollow land: Israel's architecture of occupation.* Verso.

Whiteside, N. (2015). Who were the unemployed? Conventions, classifications and social security law in Britain (1911–1934). *Historical Social Research, 40*(1), 150–169.

Wolin, S. (2008). *Democracy incorporated: Managed democracy and the spectre of inverted totalitarianism.* Princeton University Press.

PART II

The Politics of Evidence

Homo Statisticus: A History of France's General Public Statistical Infrastructure on Population Since 1950

Thomas Amossé

Over the past few decades, views of quantification have changed. Largely seen as accompanying social progress and economic growth during France's post-war boom (1945–1975), quantification is now associated with new forms of domination. There have undoubtedly been differences in the tone, more or less critical, of social–historical work on quantification developed in France and elsewhere, ranging from the more political to the more cognitive (see e.g. Desrosières, 2008, vol. 2, ch. 1). Denunciations of the tools of quantification as reflections of the negative effects of neoliberalism[1] are yet also relatively widespread on an international level, as attested to in practice and by the diverse settings analysed in the contributions of this volume.

How are we to understand this shift? Certainly this is in part a result of differences (semantic, syntactic and pragmatic) in the quantification tools

T. Amossé (✉)
Conservatoire National des Arts et Métiers (CNAM), Paris, France
e-mail: thomas.amosse@lecnam.net

© The Author(s) 2022
A. Mennicken and R. Salais (eds.), *The New Politics of Numbers*,
Executive Politics and Governance,
https://doi.org/10.1007/978-3-030-78201-6_6

emblematic of each period. In France, socio-professional categories—which contributed to making the 1960s and 1970s a golden age in the fight against inequalities (see e.g. Desrosières, 1987 [1977]; Desrosières & Thévenot, 1988; Amossé, 2013)—have little to do with the more recently developed management indicators. Their use likewise varies, as the former was related to knowledge and administration of the national socio-economic situation by the State and by social democratic authorities such as the General Planning Commissioner, while the latter aspire to the self-transformation of micro-economic behaviours (private and public). That said, this evolution is not solely due to a shift in subject-matter, viewing angle, or focal length in perspectives on quantification.

As we shall see, a retrospective examination of the forms taken by the same 'tool' over the course of time—focusing here on changes in the public statistical infrastructure on population—highlights its profound transformation, which attests to a crisis of totalization (Dodier, 1996) as well as a neoliberal inflection (Desrosières, 2014). Moreover, we herein talk of 'statistical infrastructure' rather than 'survey', despite the latter being more common in French and other languages. This choice aims to distinguish the generic subject-matter of our analysis—the statistical infrastructure—from the different forms it takes on over time (the first of which being the 'representative household survey'). Additionally, this term refers to the utilization of administrative registers made up of households or individuals for statistical purposes. We further clarify that company surveys, which focus primarily on accounting and financial data, are excluded.

The statistical survey, a classic tool of quantification, has not disappeared in an era of indicators and benchmarking (Bruno & Didier, 2013). Yet, as much in its themes and technical characteristics as in the questions it poses and categories it retains, it is today much different than in the immediate post-war period. Herein lies the aim of this chapter, to describe in detail its evolution, which we summarize in three models that progressively overlap one another: the 'representative household survey', the 'biographical investigation', and the 'matched panel'. These models articulate, each in a specific way, social science theories, statistical methodologies, and public action conceptions. They summarize a way of seeing and showing the world, of building and acting upon it. From these models emerge three types of *being*[2] that the different statistical infrastructures address—*homo statisticus*[3] they contribute to defining—herein called *subject, person*, and *individual* in reference to the three pillars set out in Alain Supiot's (2007 [2005]) *Homo juridicus*.

These *beings* are, of course, only made of paper or bytes, but they exist: simultaneously as a more or less explicit ideal within the minds of statisticians designing the statistical infrastructure, as a support (abstract but also linguistic) in the interaction between interviewer and interviewee,[4] and as a (more or less visible) unit of analysis in the results. They can be the object of identification or appropriation processes; these *beings* provide a basis for scholarly and ordinary categorizations, reflections of a statistical normativity that, even if weak under the law, is nevertheless very real. Placed under the auspices of the State according to the continental tradition that symbolizes the etymology of the word *Statistik*, this statistical normativity is particularly strong in France, having since its origin been administration-driven (Desrosières, 1998 [1993]).

In what follows, we aim to document the evolution of the relation between the French State and its statistical citizens, in their multiple and successive forms. What can a history of large-scale statistical infrastructure on the general population, of the *beings* that they pre-construct and to whom they are directed, teach us about ways of knowing and governing the economy and society? In addressing this question, we endeavour to go beyond the individualization thesis characteristic of contemporary times. The analysis accordingly shows the diversification of statistical *beings*; the proposed trinity—*subject*, *person*, and *individual*—underlines the non-univocality of the changes at work. It also invites reflection on the tensions that accompany the advent of neoliberalism, from emancipation opportunities to renewal of oppression, without mystification of a past that can be the only bearer of social and economic progress or, on the contrary, synonymous with archaism. Far from any instrumental determinism, this study[5] aspires not only to retain plurality but also the elasticity of quantification forms, of their conceptual base as much as their technical implementation; of their practical uses as much as their political implications. It builds on recent work on the social history of quantification, a subject widely reflected upon within this volume's contributions.

Our three models should not be thought of in purely temporal terms by which the modern is compared to the old. These are 'ideal type' constructions whose characteristics we never find in a 'pure state' in the statistical infrastructure actually carried out. Rather we find various forms of hybridization. If there is a temporal trend, the latter corresponds to a progressive diversification, and not to a replacement of one model for another. Each has passed through, in its own specific way, different methodological changes: the emergence of the individual as observation unit replacing the household; the micro-computerization of collection modes and statistical analyses; the introduction of a longitudinal component in the design. In what follows, we present the three models in detail before turning to a discussion of the resultant forms of *homo statisticus*.

THE REPRESENTATIVE HOUSEHOLD SURVEY

The first classic type of statistical infrastructure examined here has its origins in France in the 1950s. It was imported from the United States by public statisticians working at the INSEE (Institut National de la Statistique et des Études Économiques, the French National Institute for Statistics) in the immediate post-war period, who had spent time across the Atlantic training in the latest methodological innovations.[6] Among other tools, they brought back the random sample survey (then used in the area of employment). This technique gave way to the adoption of the now standard notion of statistical representativeness, providing the necessary conditions for the description of the national population as a whole, as opposed to the hitherto targeting of specific subpopulations (Desrosières, 2008, vol. 2, ch. 8, p. 194). This period was characterized by the strong belief in the scientificity of a 'new statistical language' for economic matters (Desrosières, 2008, vol. 2, ch. 3), which coincided with the desire to depict a post-war society in full reconstruction. Numerous surveys were at the time created, forming a first point of departure for repeated studies still active today. Described as 'structural', these surveys dealt with general themes that both organized administrative action and reported on French daily life: employment (1950), housing (1955), family budgets (1956), health (1960), training and employment skills (1963), time-use (1966). The 'programme of priority surveys on standards of living', adopted in 1965 under the Fifth Plan, reflects the accordance of this type of survey with the knowledge and management of social and economic life objectives pursued at the time by the General Planning Commissioner.

The questionnaires of these surveys are usually short and in paper format.[7] They have only a few filters (technical indications that determine whether to ask one or a set of questions) and the general principle is to use the same questions for the entire population, with identical formulations and response options for all respondents. They follow the model of a social identity card resembling, for example, the census report which de facto defines the principal socio-demographic characteristics of the population. There is thus only space for the 'major variables',[8] or those of an administrative nature, approved by public statistics and assumed to be unanimously and uniformly understood. These variables generally go hand in hand with legal categories or are derived from institutions, such as civil status registers for sex and age, nationality and country of birth for geographic origin, diploma or nationally certified trainings for level of education, administrative subdivisions (departments, regions) for place of residence or work, contractual terms for professional situation (type of contract, working time). As suggested by Michel Gollac (1997), the law saves on construction costs, as there is a shared belief that the categories that refer to the law are solid.

Due to high production costs (up until the introduction of microcomputing)—the survey samples are smaller in size and their analysis constrained by limited automated processing capacities—the results often take the form of tables or charts, with few intersecting variables: on one side there are indicators on employment, housing, health, etc., and on the other side those corresponding to socio-demographic characteristics (sex, age cohort, nationality, region of residence, etc.). Their purpose is to provide 'photographs', thematic snapshots of society, so as to gain an understanding of its organization and functioning. The periodic reissue of the surveys provides insight into macro-social dynamics. The results are produced according to schemas of a structural-functionalist inspiration that govern the elaboration of the surveys. It is in this manner that the demographic, social or economic behaviour of the population and its households are studied: the economy and social matters primarily being the domain of the 'head' (a man)[9]; the domestic and familial reserved for women, their spouses.[10]

Public statistics considers the household to be the central unit of analysis. Indeed, this is the title given, for almost twenty-five years, to the section primarily responsible for designing statistical infrastructure on population at INSEE: the Population and Household Department (from 1966 to 1989). The notion of household separates the interior (private) from the exterior (public) according to a strict gender division of roles. In addition to accommodating a male-dominated vision of the world, this concept also reflects a holistic vision of society and the economy. By law, men and women long had clearly assigned roles within the household. Women could not work without their husband's permission until 1965 in France, and divorce by mutual consent was not introduced until 1975.[11]

In this context, the statistical household is seen as a full-fledged economic actor in terms of income, consumption, savings, or economic expectations. As such, according to the monthly business survey (still in place today), households have opinions, independent of the men and women, parents or children, who compose them.

The statistical nomenclature of socio-professional categories occupies a special place in this survey model. Its success was total during the three decades following their creation in the early 1950s (Desrosières & Thévenot, 1988). Broken down at the level of head of household or father, these categories are systematically used in statistical tables, evincing class inequalities or those of social origin.[12] This 'major variable' is emblematic of the back and forth between public statistics (and its surveys) and the socio-economic administration in France at the time: on the one hand, statistical nomenclature draws on social categories, occupational subdivisions which, backed by the law, are in place within companies and administrations; on the other hand, it is used directly by social actors, whether under the General Planning Commissioner or, to mention just one example, the indexation of the minimum wage which gives rise to national negotiations between labour unions and employers' organizations.

The scope of these surveys is usually households in ordinary accommodations in metropolitan France, which compose the statistical heart of society, an echo of the electoral body. The non-zero probability of selection of households that organize the sampling procedures can, in fact, be thought of as the equivalent of a statistical right to vote. Furthermore, the methods of analysis used are essentially summation techniques, much like the adding up of votes in an election. The notion of representativeness is central here, its statistical meaning lending political acceptance to the term. If, in addition, we consider the particular role played by the law and institutions, this survey model certainly seems emblematic of the representative democracy of intermediary bodies characteristic of France from the 1950s to the 1980s. Indeed, the General Planning Commissioner constituted one of the primary transmission channels in organizing government and social partner participation in the elaboration of medium-term policies based, specifically, on predictions from statistical surveys.

This political-administrative tone is found in the term 'survey' itself. While it has certainly been used in a generic sense since the post-war period within the French statistical community, it also refers, originally, to the search for information within a judicial framework, then by further extension the systematic collection of testimonies or documentation aimed at clarifying an issue or dispute.[13] These different definitions share the implicit meaning of an 'unveiling'; of obtaining private, sometimes secret or hidden, information. The word also recalls the asymmetry of the survey context, which has long been reinforced by the sociological profile of INSEE interviewers (former gendarmes and military personnel) (see Dussert, 1996). This asymmetry is, moreover, particularly significant in that it distinguishes between ordinary households and public agents. The statistical survey, notably that which follows this first model, is marked by the State's seal: official notification letters often accompanied by an obligation to respond, the professional card of the interviewer knocking on the door bearing the colours of the flag—such elements give off an air of an administrative questioning.

This survey model was for many years the only one that existed. Then progressively, starting in the late 1970s and particularly in the 1980s, two other models were developed. That said, this initial form has not disappeared, but rather continues to exist, giving way to hybridization between original and emergent models with the micro-computerization and the integration of longitudinal questionings (see 'matched panels' below). Today they are essentially annual surveys by wave (on the labour force, housing, etc.) which, as a continuation of their antecedents from the 1950s and 1960s, compose the 'back bone' of the INSEE statistical infrastructure.[14] There have, however, been two notable developments. First, they are increasingly governed by European regulations, for the purposes of updating social descriptions at the continental level in the form of reporting indicators or national barometers relying on several 'core variables'. While similar to the 'major variables' mentioned above, the latter differs in not always referring to institutional categories, in the absence of common institutions at the European level.[15] Second, they aim to more fully cover the entire population, surveying segments of the population usually labelled as "outside the scope" such as the homeless, those in institutions (prisons, health or social establishments) or by geographical extension of existent statistical infrastructure (such as for the overseas departments).

THE BIOGRAPHICAL INVESTIGATION

The second type of statistical infrastructure on population examined here has its origins at INSEE during a time of reflection and critique of the social sciences in France starting in the mid-1970s with, for example, the shift towards Pierre Bourdieu's critical sociology, affirmation of the work of Michel Foucault, and the start of Luc Boltanski and Laurent Thévenot's pragmatic sociology. A two-fold movement of diversification thus began to shape large scale public statistical infrastructure. First, in terms of the variables used, with less primacy given to institutional categories in the questionnaires, and more openness towards the social sciences, whereby theoretical advances and methodological observations of ordinary practices provided new ways of questioning the world. Secondly and more broadly, the themes of such infrastructure diversified, going well beyond the economic behaviours of households and the socio-demographic characteristics of their members.[16]

The 'biographical investigation' questionnaires stand out for their length and evident distancing from examinations of a more administrative nature. They follow a linear path much like the biographical interviews of interpretive sociology,[17] and frequently employ retrospective questions and timelines to reconstruct respondents' trajectories. To this regard, two practical protocols were developed. The timelines can, on the one hand, rely on paper chronologies allowing respondents to mark their own points of reference (a birth, a move, a promotion, etc.) and in this way reconstruct different parallel accounts (i.e. familial, residential, professional, etc.). On the other hand, or simultaneously, the timelines can be assessed using resources offered by computerization, allowing to gradually unfold stories in function of past events. Thanks to filters and the configuration of successive questions, the survey fits the life of the respondent like a glove. In this model, the survey situation targets the unit of time and of place in order to ensure the coherence of responses, which rely on memory recall, sensitivity, and the perception of contexts, and thus depend on the interaction between interviewer and interviewee.

The objective of 'biographical investigations' is less about consistency with official categories (specific to the 'representative household surveys') or the pureness of 'matched panels' (see below) than the sincerity and coherence of the responses provided. This statistical infrastructure model has contributed to the development of new questions relative to emotional experiences, whether they be physical or mental, using a subjective (perception or opinion; feeling or emotion) or more objective (ordinary situations, practical experiences) approach. In this way, violence, suffering or hardship, physical ailments or bodily nuisances, satisfaction or happiness, freedom, etc., become 'statisticable' notions. Simultaneously, information on the temporal context or local environment is collected, on different levels or according to different timeframes, often fixed by the respondents themselves. Multiple 'nested' descriptive circles can thus be identified: from the closest members who compose the 'living unit' or 'relations' (terms which invite moving beyond the alleged unicity of the 'household') to the furthest, such as the social class or geographical area to which one feels belonging.

Examples of statistical infrastructure within this model, which share some or all of its features, are as diverse in their themes as in the government departments or administrative bodies that produce them. They were originally carried out mostly by INED (the French Institute for Demographic Studies), where in the 1980s demographers began implementing statistical modelling for the analysis of biographies (Courgeau & Lelièvre, 1989),[18] and then a decade later, multilevel or contextual analyses (Baccaini & Courgeau, 1997).[19] Such statistical infrastructures have, however, also subsequently been used by a number of other public institutions. Two particularly stand out for their attention to biographical nature, to perceptions of past situations, and importance given to the contexts in which personal trajectories unfold. The first, the *Health and Career Path* survey (2006 and 2010; *Santé et itinéraire professionnel*) asks respondents to reconstruct their entire professional careers while also indicating major health events, with the objective of understanding how health and work influence one another over time. The second, the *Life History—Construction of Identities* survey (2003; *Histoire de vie—construction des identités*) combines a complete retrospective timeline (residence, family, employment, economic well-being) with questions aimed at understanding the articulation of the latter with different facets of personal identity (e.g. family, work, friends, hobbies, health, origins, etc.). As explained in the survey's guideline note, the idea is to account for the multiple processes of identifying individuals with places, groups, histories, values: "the individual bears multiple identities" whose "main dimensions must be explored" (see Héran, 1998). These two statistical infrastructures both leave a great deal of freedom to the respondents in how they mark their biographical itineraries and give more room to their subjectivity.

More than one method has been developed to use data collected in this way. Duration models, chronograms and, more generally, life course analysis methods all aim to understand procedural logics, successive choices, bifurcations, potential disruptions or protections of a given trajectory.[20] In a different way, exploratory factorial analyses can both show structural oppositions within the population and the coherence of answers for each respondent. Results can lead to a first, inductive, modelling of areas previously little explored and where a full understanding has not been reached in the absence of structuring 'major variables'.[21] More generally, "biographical investigations" seem to be consistent with the desire to reconcile the holism and individualism we find in the "new sociologies" described by Philippe Corcuff (2007 [1995]).[22] In two different registers and disciplines, the methods derived from Amartya Sen's capability approach in economics and the multilevel analyses used in demography (Baccaini & Courgeau, 1997) have been adopted in efforts to understand the effect of situated interactions and local contexts.[23]

These approaches all have in common the fact that statistical representativeness is not the primary concern. Certainly the statistical infrastructure relies on a random sampling procedure and the subsequent analyses often use weights based on the latter, but this use is secondary in that it is the processes, the consistencies or oppositions, that are of particular interest. Echoing the sociological interview principle, the methods share a reasoning 'by row' at least as much as 'by column'; in the sense that they first follow the logic of the respondents, not that of the variables. This sort of thinking is present from the very conception of the questionnaires—the queries are formulated using verbatim accounts from sociological studies—and of their computerization (filters linked to previous responses, which act like reminders during an interview). The term 'investigation' used here recalls the exploratory dimension of this statistical infrastructure: simultaneously as a study of 'rows' (that is to say, an attempt to reconstruct, for each person, their complex biographical history, their subjectivity, their social inscription) and taken as a whole, in many ways following an exploratory research approach.

THE MATCHED PANELS

The third form of statistical infrastructure responds to a micro-causalist agenda, which differs both from grand narratives which collective entities deploy as historical causes, or biographical narrations where causality is presented in a singular way. The statistical infrastructure that follows this logic was made possible by unprecedented advances in information processing, and maintains a close relationship with micro-econometrics. Long dominated by macro-structural approaches, they were developed along with individual computerization in the 1990s before becoming dominant in the next decade. Like the 'biographical investigation', this statistical infrastructure uses a large number of variables. However, in contrast, their collection of information does not necessarily suppose a specific unit of place and of time. While the information gathered can certainly draw on questions asked by an interviewer during a single interview, the latter is just as likely to derive from subsequent interviews (possibly with different interviewers and respondents for the same panel unit) or from matching with external data of diverse origins. In fact, the origin or the situated consistency of the collected information is not of all that much importance. What counts is their quality, thought of as intrinsic, and the ability of the statistical infrastructure to amass pertinent variables for each respondent so as to be able to saturate the explanatory models with the phenomena under analysis.

When the classic format of the face-to-face interview is used, this form of statistical infrastructure relies on a complex questionnaire including numerous case disjunctions, where computerization plays a crucial role. Multiple filters and parameter settings allow to adjust the questioning and adapt the formulations so as to obtain, for each respondent, the best measure of the targeted variables. Computerized questionnaires are not modelled on either the social identity card of the 'representative household survey', or the carefully tailored 'biographical investigation'. They are foremost routines, computer programmes, difficult to grasp in their entirety. Indeed, the initial structuring requires a paper format translation, necessary in order to give analysts a clear understanding. The current *Labour Force Survey* (*Enquête Emploi*) questionnaire provides a good example: its paper version is more than 80 pages long (compared to the simple DIN-A3 sheet in its original form); its reading, broken up by numerous filters and parameter settings, underlines the complex task of trying to construct a linear, narrative version. It is, more broadly, an intricate set of documents (technical notes, training guide, survey lexicon or glossary, etc.) necessary for using the 'data'. These questionnaires, fragmented and complex, in reality, blend into the object they target, i.e. panel individuals; direct questioning is one way (among others) of obtaining information deemed pertinent. The collected variables rarely, or secondarily, refer to practical experience or to institutional categories, but rather ideally and in principle, to observed facts[24] in accordance with theoretical categories which, besides, derive more from the natural sciences than the social sciences. They are ideally continuous, being categorical only when imposed by constraints (conditioning of reality or resulting from data collection arrangements). It is about 'good variables' (to use a common econometric expression), and not 'ordinary variables' or 'major variables' which are both too dependent on the administrative, spatial or temporal context. Education, for example, is thus ideally measured by number of years of study, and not by diplomas obtained (for which recognition by the State can vary over time) or the level of education reported (susceptible to perception bias). The aim of establishing causalities, scientifically demonstrated and ideally universal in scope, necessitates discarding variables that do not relate to any theory or may be endogenous. The term 'data' is preferred to 'responses'; the constructed and declarative nature of the information must be neutralized in the analysis, following an objectivist plan that aims to get rid of limits associated with the subjectivity of the respondents or with conventions linked to institutions in the development of variables.

This type of statistical infrastructure is also characterized by a specific way of viewing time. Like the 'biographical investigations', longitudinal information is of central importance. In contrast, however, here such information is collected prospectively through data collection or matching of information repeated over time. Retrospective examinations are seen as tainted by bias, echoing the work of Karl Popper in the immediate post-war period, later analysed in detail and criticized by Luc Boltanski (2012), whereby history, the past, memory, are all seen as obstacles to the establishment of scientific truth, which should be timeless. If time plays a role, it is only as a source of variation and turned towards the future; it allows to identify causalities between actual and subsequent events. It is not about understanding historical processes—viewed as impossible to grasp other than as partial constructions (loss of respondents from one examination to another) or biased reconstructions (memory). It is a question of having repeated measures, of independent observations over time of the same panel units, so as to establish statistical links between various changes that have affected them over the period under analysis. At INSEE, a shift in orientation towards prospective panel data occurred in the mid-2000s (Chaleix & Lollivier, 2004), in part explaining our use of the term 'matched panels' to describe this model. The adjective 'matched' emphasizes the matching used to collect various information on the individuals panelled.

Examples of such statistical infrastructure, today numerous, are most often prospective panels. They tend to dominate public statistics and have contributed to transforming 'representative household surveys', leading to a hybridization of the two models. This is the case, for example, with regard to the *Labour Force Survey* (*Enquête Emploi*), whose longitudinal dimension was significantly reinforced in 2003, when it came to rely on a 'continual'[25] statistical infrastructure. Whereas once it was (in its original form) emblematic of our first survey model, two important characteristics now attest to its hybridization. First, its original objective (establishment of the unemployment rate) has shifted from a reporting logic—made possible by the assumed strength of institutions themselves to give meaning to a direct question on labour status, notably unemployment, to a factual logic where only the combination of specific criteria referring to the past week (or reference period) determines unemployed status under the International Labour Office definition (Goux, 2003). Second, the increase in the number of interview waves has allowed for the development of panel analyses.

The *European Community Household Panel* (1994–2001) and its successor, the *Survey on Income and Living Conditions* (2006) provide additional examples of hybridizations. In addition, the Permanent Demographic Sample (*Echantillon démographique permanent*; census subsamples matched with civil status, electoral participation or, since 2011, social-fiscal data) and the *Dads* panel are older examples, which have recently been enhanced and whose exploitation has been strengthened.

Classic themes are addressed in this form of statistical infrastructure (e.g. housing, education, employment, health, etc.). They refer to areas of public action, but in a renewed form compared to the 'representative household survey' model. The emphasis is placed on individual change, its determinants and consequences, rather than on understanding broad macro-structural dynamics in an effort, for example, to organize national planning and accounting. In the shift towards micro-statistics, global management of the economy and society is no longer the central concern, as it was during the time of the planners, but rather scientific expertise, a priori neutral and independent, allowing to evaluate the capacity of public policies to change, within a given domain, individual behaviour. In this regard, incentive theory, sometimes implicitly, plays a determinant role. The political dimension of this form of statistical infrastructure is thus naturalized: their design integrates political objectives in advance, in the definition of eligible populations and target variables. Certain forms of this statistical infrastructure are particularly inventive in their efforts to evaluate public policies, such as the *Panel on State-aided contracts* (*Panel des contrats aidés*) carried out in 2008 and 2014 by the Ministry responsible for labour and employment. In this case, the policies were evaluated using a quasi-experimental model directly imported from the natural sciences (notably medical research).

While specific, this 'matched panel' sub-model (described here as 'quasi-experimental'), is no less important from a symbolic point of view. The INSEE report on longitudinal data (Chaleix & Lollivier, 2004) in fact accords a central place to the latter: panels must be "targeted at specific populations (that is chosen in reference to the evaluation process of a given social policy, which usually means that collection begins before the introduction of the policy in question, so as to be able to observe change over time)", "coupled with information from administrative registers (to limit survey time and assure greater reliability of certain information)" and follow a "sampling plan that includes a control group" (Chaleix & Lollivier, 2004, p. 22).

In 'matched panels', the notion of representativeness is secondary. The descriptive results are usually but a first step, which aims to verify that there is no bias in the structure of the sample excluding an identifiable segment (according to observable characteristics) of the population of interest. Indeed, for panels, maintaining the initial representativeness of the samples is particularly challenging due to attrition. Although dynamic weighting can be developed, helping to guarantee representativeness over time, most studies restrict their analyses to cylindrical samples (i.e. to individuals who responded in all the waves of the panel) and don't use weights. These methodological options highlight the tension between survey specialists and econometricians,[26] and more broadly between the aims of representatively describing populations and estimating factors associated with a specific situation or behaviour. The methods used, most often micro-econometric, neither need nor take into account the representative structure of the samples analysed, as if the 'data' were exhaustive.[27] It is the size of the sample, or the repetition of observations (within a population and/or over time) and the (large) number and ('good') quality of variables that are determinant in demonstrating the 'purified' causalities of composition effects or selection processes.

Econometric modelling generally identifies one or more response variables—a behaviour or a situation that should evolve or improve in the population, such as educational attainment, participation in the labour market, etc.—and a large variety of explanatory variables. Among the latter, an analytical distinction is made between so-called variables 'of interest' and 'control' variables. The former, which correspond to potential public action levers (financial or in-kind support, for example), dominate the reflection while the latter delineate as precisely as possible the socio-economic profile, geographical or professional environmental, etc., of the respondents. Ideally, as mentioned above, control variables should be exogenous (that is, not depend on either the response variables or the variables of interest). They can thus either be an assumed stable property of the respondents (e.g. sex, date of birth, age at end of initial study, etc.) or a characteristic of their environment that they have not chosen.[28] The number and the precision of these control variables condition the degree of purity of the statistical associations between the response variables and the variables of interest. They define what econometricians call observable heterogeneity and are not generally described in published results. They are only occasionally used in analyses of subpopulations (e.g. women or men, those with low levels of education, etc.) aimed at evaluating the differential effectiveness of the policies studied.

Establishing causality relies on even more demanding and sophisticated methods than those leading to controlled correlations. Without going into the technical details of the three main methods employed (instrumental and panel econometrics, experimental matching), we can emphasize that all use the temporal and contextual richness of individual data to eliminate any source of endogeneity (reverse causality, omitted variable, etc.) and to ensure perfect control of individual heterogeneity (observable *and* unobservable[29]). Such types of statistical infrastructure differ from the previous two presented above, as much in their objectives as in their formats and analytical methods. Undoubtedly, these models lean on different social science disciplines and the influence of the latter on French public statistics over the last seventy years. But they are not limited to this influence, in the sense that methods circulate and spread from one discipline to another, and the 'models' that compose the three types of statistical infrastructure identified here take on, in fact, a normative dimension as they are implemented by public statisticians.

HOMO STATISTICUS: THREE TYPES OF BEING CONSTRUCTED BY THE STATISTICAL INFRASTRUCTURE

The description of the three models provides several initial characteristics of the *beings* to which the different statistical infrastructures refer. In what follows, we summarize their features using the same terminology as that employed by Alain Supiot to describe the three pillars of his *homo juridicus* (Supiot, 2007 [2005]), with which the three forms of *homo statisticus* share a surprisingly similar kinship.

We have chosen to call the *being* defined in and by 'representative household surveys', *subject*. This *subject* has a more pronounced political dimension than the *beings* of the other two models. Its tone is administrative, as we have seen, with its official socio-demographic characteristics, the institutional variables defining its identity. From the outset, the characteristics of the *subject* refer to the collective beings with which it is assumed to identify (at a minimum during the survey), according to the process described by Nicolas Dodier (1996) of temporarily accepting assimilation to the pragmatic condition of statistical identity. This identity is borne by the 'major variables' of 'representative household surveys': the *subject* is the representative, among others, of collective beings; it is a member of a class, a category, or a group. Its singularity is not, in a sense, detachable from its properties, which are familial and sexed status within the household ('head' then 'reference person', or spouse, a child, etc.), age group (e.g. youth[30]), social group (e.g. managers or blue-collar workers, farmers, etc.), nationality, or geographical unit of resident (region or department).

Subjects are thus only indirectly represented, by their categories of belonging: institutions such as the family, school, company, or administration, and more broadly the State (as source and guarantor of the law) primarily determine the categories that express, as much to society as to themselves, the surveyed *beings*. It is not, however, solely the range of collective belongings that public statistics uses to characterize its *subject*, but also economic behaviour, and social roles expected of the *subject*. This assignment is particularly clear with regard to the role of women, about whom statisticians ask: how should they assist in the reconstruction of post-war France, by having children (according to a fertility logic whose military inspiration has shifted to a production aspiration) or by providing a source of additional labour-power (Amossé & de Peretti, 2011)? More broadly, analyses of sex and age groups, migratory status or social group, region of residence, are difficult to separate from a planning vision that poses, or rather imposes, behaviours based on a rationale of matching social characteristics to expected socio-economic conditions. According to sex, age, social class, etc. public statistics see, but also and above all, foresee average, or probable, conditions for its *subjects*, in terms of school, work, family, etc.; the training-employment matrices initially developed under the General Planning Commissioner providing a good example.

By their categories of belonging, the *subjects* are integrated into a *holistic* representation of the economy and of society, as cogs in a mechanism. These *beings* are only active *subjects* to the extent that they accept their assigned role of contributing to the collective future. They are actors and acted upon, subjects of and subjected by public statistics, much like the two facets (active and passive) of the notion of 'subject' that Alain Supiot stresses in *homo juridicus* (Supiot, 2007 [2005], p. 17). In descriptions of the tables and graphs that result from these surveys, it is the collective beings that are the subjects of sentences.[31] In this way, they take on a certain realness, becoming common nouns ('women', 'youth', 'managers', etc.), that circulate both in public administration services, in scientific publications and events, in meetings between social partners, and even in ordinary situations. This is a vision of the world made up of *subjects* that are inseparable from their institutional social properties borne by the first survey model.

The parallel between electoral representation and statistical representativeness, central to this type of survey, illustrates in another way the political dimension of these *subjects*. With their non-zero probability of selection, according to the random sampling procedure, they are part of a 'statisticable' population, analogous to the electorate. Linking of the surveyed *being* to the whole is thus achieved in a similar way to the mode of voting, by aggregation-summation or, technically, with the use of survey weights. The *subject* is not only represented by collective entities, but also compared to the whole, to French society. From this point of view, recent change in the 'representative household survey' is not insignificant. Whether in terms of shifts due to processes of European harmonization or extension to margins previously considered 'outside the scope', this statistical infrastructure introduces two new types of *subjects*: French *peri-subjects* and European *proto-subjects*, which join in statistically defining the economy and society without, however, having a political representation that is, as of yet, anything more than marginal or nascent.

Even if sub-models (retrospective, subjective, situated) can be identified within 'biographical investigations', the corresponding statistical infrastructure draw on *homine statistici* with similar characteristics, what we call here *persons*. As highlighted above, such statistical infrastructure shares a certain kinship with sociology's semi-structured interviews in that the *beings* to whom they are addressed are attempts to statistically capture the singularity of personal situations. They are constructed as the questionnaire proceeds: the questions follow a biographical trajectory, and sometimes adapt to the latter. The numerous open-ended questions take into account, in an exploratory fashion, the subjectivity of the respondents, allowing for the expression of opinions or assessments. The co-construction of these *beings* and of the questionnaires underline a critical difference, symbolically, from the *subjects* of the 'representative household surveys'. Indeed, the statistical infrastructure of this second model grants greater freedom to the *beings* it addresses; these beings are not entirely pre-constructed as can be the case for *subjects* (who must accept that their singularity is limited to institutional categories) or *individuals* (whose characteristics must respond to the requisites of scientific theories, as we will see below).

It is neither the 'major' nor the 'good' variables that define the *person*, but the near as possible accounting of their experiences, their practices, their feelings, their wishes. This form of statistical infrastructure seeks, in fact, to reveal the various facets of an identity, possibly plural, constructed over time; the multiple interactions that *persons* have with their loved ones or their environment, their resources and their capacity to take action and plan for the future. The biographical investigation is in itself an experience, during which the *person* surrenders, body and soul, to the interviewer, much like the notion of 'person' described by Alain Supiot.[32] Some, such as the statistical infrastructure underlying the investigation on *Violence Against Women* (2000; *violences envers les femmes*), have moreover raised practical and legal issues for public statistics, in their necessary consideration of the emotional burden and moral commitments faced by interviewees and interviewers.

The practical, sometimes intimate, identity of *persons* considered is thus revealed during the administering of the questionnaire, and the tone of the interviews often remains present in the production and reception of the results: to the personal implication of the respondents corresponds an empathetic recall of statisticians as readers. The process of identification, as described by Dodier (1996), here calls upon the experience, sometimes symbolic, of a common humanity (and not the sentiment of collective belonging as for the *subjects*). From one end of the statistical chain to the other, attention is paid to the body and emotions, to interactions and confrontations with other people, to biographical bifurcations and disruptions. Although there is no one analytical framework used in this type of statistical infrastructure, the capability approach developed by economist-philosopher Amartya Sen (which, for example, has inspired recent forms of statistical infrastructure on educational pathways) corresponds quite closely to the notion of *person*. It describes *beings* who are a priori capable of desiring, of expressing themselves, of making themselves understood, of learning, and of working (when they have good quality jobs), and finally of finding a harmonious balance between work and family (Bonvin & Farvaque, 2007).[33] More broadly, this logic is consistent with political subjectification processes linked to struggles for social rights.

We use the word *individual* for the *beings* corresponding to the 'matched panels'. This *being* can be seen primarily as a way of identifying causalities between two groups of variables (response and of interest, to use econometric terminology). The corresponding statistical infrastructure is thus not really focused on *individuals* per se, either in and of themselves (as *persons*) or as members of the collective entities to which they belong (as *subjects*). From a literal point of view, the term recalls a basic unit that cannot be divided. Certainly, the composite nature of the statistical definition of these *beings*, resulting from an assembling of matched data, brings to mind Arjun Appadurai's (2016) concept of the "dividual" in reference to actors in the finance sector who, like the derivatives they manipulate, are socially divided. Reference might also be made to the algorithmic beings of big data, made up of multiple data marks and imprints (Rouvroy & Berns, 2013). That said, the resemblance is closest to the *individual* described by Alain Supiot in *Homo juridicus* (Supiot, 2007 [2005], p. 13): the individual is at once identical and unique, indivisible and stable (it is the "basic accounting unit *par excellence*"); in its irreducible unicity ("unknowable essence and containing its own end"), a free being ("substantial ego[s] that freely forge social links rather than being fashioned by them").

The *individuals* in this third statistical infrastructure model have these qualities. They are characterized by stable and immutable properties, which neither determine nor are determined externally. These beings are different from *subjects* in that they are not a priori integrated into collective entities. They don't have interactions, connections that they build over the course of time and that contribute in turn to their construction, like *persons*. Analytical methods for 'matched panels' are most often based on so-called 'i.i.d.' hypotheses, meaning that the statistical observations, or the individuals at a given moment in time, are independent and identically distributed. With regard to principles, and beyond specific modelling, dependence is not possible either between *individuals* or between their successive states over time. These *beings* are, in this way, alone and without history. They have neither depth nor belonging. They are a support for statistical identification, this essential heterogeneity on which the models rest, but which they ultimately aim to make disappear. Similar to the creation of data files for the use of heterogeneous information, their characteristics are removable: they are statistical tools (controls), not the objects of analysis, in the absence of a descriptive plan followed by exploitations using 'matched panels'.

An aggregation of situations, individual states or behaviours, results from these statistical models. The statistical *individuals* of the 'matched panels' are not strictly speaking the individuals of standard economics. Although they share numerous traits, their definition is empirical and not theoretical; it is not necessarily assumed that they are driven only by their preferences or utility. There is not, in any event, explicit place for either communities, classes, groups established by belonging, or for local context, familiar environment, close circles. Like atoms without connections to one another, *individuals* cross time and space. Their trajectories are certainly influenced by their environment, which only acts, however, as an external element and is not, a priori, modified in turn. Moreover, this influence does not fundamentally change *individuals*. Time is made up of events, of shocks, that have consequences but don't contribute to building a history, either personal or collective.

As basic *beings* of this type of statistical infrastructure, *individuals* have something almost fictional about them. They are both syntactically central–symbolizing and, especially, allowing the micro-statistical shift that characterizes the analyses—and semantically absent, upstream and downstream of data collection. They are necessary, but must also be overcome: they are seen as obstacles, as reflected by the notion of 'unobservable heterogeneity' and the clear need for the results to be reflected by the notion 'purified'. The polysemy of the term 'identify' characterizes well this present-absent dynamic. It is not about the interviewees, interviewers, researchers, or readers being able to identify with the *individuals* who are produced by these statistical infrastructures (like with *subjects* or with *persons,* but in different ways). Rather it is about being able, thanks to them but also by making them disappear, to statistically identify equations, i.e. to assess statistical associations between variables that should be able to be separated from the observations on which they rely. *Individuals* are supports and not the objects of identification. This sacrifice of diversity is undoubtedly characteristic of statistical methods of analysis more generally, which aim to establish results that are valid beyond single cases. However, a non-trivial observation is that identification resources do exist in the first two statistical infrastructure models, whether in the institutional manner of *subjects* of collective entities to which they belong, or in empathy with *persons* "having declared that...", "suffering from...", "hoping that..." (to take several examples of phrases used in the 'biographical investigations'). These resources disappear for *individuals* who, as we have seen, can be represented as mosaics that, before being assembled, were composed of heterogeneous, fragmented pieces. In analyses produced using this last statistical infrastructure, phrases most often directly link two types of variables: the first, those of interest, reflecting (eventually as consequence) the fact of having benefited from a public policy; the second measuring the evolution of a situation, of a state on a market (housing, education, employment, etc.).

CONCLUSION

Over a period of about seventy years, the French public statistical infrastructure on the general population has diversified. Today, three models co-exist: 'biographical investigations', and the 'matched panels', which were successively added to the original 'representative household surveys'. These models have different specificities, not only in terms of their formats but also their theoretical frameworks, objectives, methods of analysis, and visions of public action. They imply three different types of *beings* who represent three variations of humanity (here in its statistical version) in as many 'ideal types': the *subject*, the *person*, and the *individual*. The last few decades have seen a blossoming of these last two *beings*, who previously were as if restrained by the sole logic of the *subject* in the 'representative household surveys'.

The originality of these two typologies (of statistical infrastructure and of *beings*) is relative. There are echoes of divisions already highlighted and analysed, within the social sciences and its sub-disciplines. Indeed, the *subjects* of the 'representative household surveys' (and the collective entities that represent them) are primarily found in the structural analysis of populations, macroeconomics, structural sociology, and social history of grand narratives. The *persons* of the 'biographical investigations' have instead followed the reorientations of demography (interaction between events, diversity of populations, link to context), sociology (biographical shift, 'new' sociologies, etc.) and history (critical shift and microhistory). The *individuals* of the 'matched panels' are analysed by micro-econometricians, neoclassical economists or rational choice sociologists, with ties to experimental psychology or contract law. Another limit of the proposed typology is that, one might argue, only certain statistical infrastructures possess all the characteristics used to describe the models. Yet, the strength of these proposals is their empirical basis, the way that they precisely aggregate a large number of traits that result in a ternary interplay of oppositions. Following the intuitions of French pioneering work on the social history of statistics, the analysis of quantitative tools seems particularly instructive for organizing the plurality of ways of seeing reality. That said, the transformation of statistical *beings*— the tensions, conflicts, hegemonic drive of notions associated with *subjects, persons* and *individuals*—does not solely correspond to the evolution of social theories in France. There is every reason to think that the elements discussed here have a more general scope, and are relevant to the "politics of statistics" (Desrosières, 1998 [1993]).

What can we conclude, politically, from this diversification of statistical infrastructure and statistical *beings*? First, that the emergence of *individuals*, who occupy both a central and evanescent place in their statistical infrastructure, echoes critiques often formed against neoliberalism, where incentivized *individuals* would be deemed free, but without much power, whereas the *subjects* of a Keynesian planner, admittedly acted upon, were socially protected and retained their capacity for collective action thanks to institutions (Castel & Haroche, 2001). Our third model is no less political, nor any less related to the State than the first. The role of the State is, however, profoundly transformed, in that it aims to establish market instruments entailing feedback loops on individual behaviours, much like the corresponding form of government (neoliberal 'city') described by Alain Desrosières (2014) or the absent State proposed by Robert Salais (2015). The *personal* logic complicates this description, and cannot only be understood as a parenthesis between *subjects* and *individuals*. This logic nuances the 'individualization' move-ment broadly employed (in the social sciences as well as in public debate) to describe contemporary change, but whose ambivalence can be sensed. The real ability to act belongs to *persons*, the only *beings* to which the statistical infrastructure accords a value per se. This evolution leaves some-what open, however, the question of their collective aggregation, or the political forms with which they may be associated.

Acknowledgements This text has been translated from French to English by Maya Judd.

NOTES

1. Given their complexity and range of accepted meanings, both 'neoliber-alism' and 'quantification' could clearly be further discussed, something that is not possible here due to space constraints.
2. The use of italics for the generic term *being* and its various forms (*subject, person, individual*) based on the history of statistical infrastruc-ture, aims to avoid any confusion relative to their general or academic meanings, which vary according to social science disciplines, time-period, and country.
3. Here we take up Jean-Claude Passeron's (1999) expression, in contrast to the *homo singularis* "which can only be recounted on the deathbed of an individual or collective story" (Passeron, 1999, p. 18). We will later see that one of the beings—*person*—in some ways resembles this *homo singularis.*
4. We broadly use the terms 'interviewer' and 'interviewee' here to describe those involved in the collection of information for the various types of inquiries, whether this consists of a 'survey', an 'investigation', or a 'panel'.
5. The analysis uses a variety of material including documentation from more than thirty components of the French statistical infrastructure, as well as different archives of the public statistics system, notably those of INSEE (Institut National de la Statistique et des Études Économiques, the French National Institute for Statistics) collected over the course of the past fifteen years.
6. This experience is recounted, for example, by Raymond Lévy-Bruhl (INSEE, 1977, p. 561).
7. The questionnaire for the first Labour Force Survey, conducted in 1950, provides an example, being composed of just one double-sided DIN-A3 sheet.

8. As observed by Affichard, the first Labour Force Survey allows "experimentation of questions from which a classification of the population into 'major variables' (employed, unemployed, non-active) is produced" (Affichard, 1987 [1983], p. 90).

9. The tables indicate, for example, housing situation, level of income, and consumption patterns based on the characteristics of the 'head of the household' (his age group, employment status, socio-professional category).

10. From 1954 to 1970, the post-censal surveys on family history were only, for example, addressed to married women (Locoh et al., 2003).

11. Note, however, that shortly after the French Revolution of 1789, divorce by mutual consent was legalized for a relatively brief period of time (1792–1803).

12. See, for example, the co-authored book published under the collective name Darras (1966), containing the work of INSEE sociologists and statisticians, including that of Pierre Bourdieu and Claude Gruson.

13. See the entry, 'survey', in Alain Rey (1998), *Dictionnaire historique de la langue française*, Le Robert, vol. 1, p. 1246.

14. According to the medium-term programme objectives of INSEE from 2006 to 2010.

15. The limited number of these variables is less due to technological constraints (as in the past), than to difficulties related to the political harmonization of national categorizations.

16. This is the case, for example, in the area of educational orientation. In a presentation of surveys on these themes, Joëlle Affichard and Michel-Henri Gensbittel (1987) contrast so-called "diagnostic", subjective, or opinion variables, with those of the "State" (Affichard & Gensbittel, 1987, p. 186).

17. The expression "biographical turn" (Chamberlayne et al., 2000) has been used to describe the widespread adoption of comprehensive biographical interviews in sociology in the 1980s and 1990s.

18. The adoption of this approach means no longer using as the unit of analysis an event (marriage, birth, death), examined separately and which is assumed to be the result of homogeneous behaviours within a group (e.g. a social or age group), but instead the "individual biography, considered a complex process" (Courgeau & Lelièvre, 1989, p. 2).

19. Although models of this type were proposed in the 1960s, they were not truly implemented until the 1980s, due to a lack of data (Baccaini & Courgeau, 1997, p. 833).

20. These methods are linked to various analytical frameworks, including the demography of biographies (Courgeau & Lelièvre, 1996), the economy of transitional labour markets, or in a transversal fashion, the analysis of bifurcations or disruptions (Bessin et al., 2010).

21. See, for example, Baudelot and Gollac (2003) on the links between well-being and work and, from a different perspective, that of Amossé and Chardon (2006) on the construction of identities.

22. Removed from quantitative methods, they are not, however, directly implicated in the elaboration of such statistical infrastructure.

23. Multilevel modelling can also be used in a "matched panels" perspective, our final statistical infrastructure model presented here.

24. The difference between described experience and observed fact is important. 'Matched panels' aim, with precise guidelines and formulations (notably in terms of the period of reference concerning an event), to avoid any 'hazy responses', in contrast to the narrative logic employed by 'biographical investigations', where it is expected that the experience will be related as a whole, necessarily subjective to some extent. From the 'biographical investigation' perspective there is no evidence, objectivity, naturalness of facts, but rather the construction of a set of information relevant for describing the experience, feelings and actions of respondents (echoing the 'informational base' theorized by Amartya Sen).

25. The survey comprises, in reality, six waves of quarterly questioning. The longitudinal dimension of the Labour Force survey is old, with a renewal method of the annual sample in thirds which was followed continuously from 1968 to 2002 (Goux, 2003). That said, this design is not in line with longitudinal analysis objectives, and has moreover rarely been used from this perspective.

26. We can highlight two opposing groups of methodological statisticians at INSEE. The battle between them was symbolically lost by the data production specialists, notably of surveys, who in the mid-1990s had to relocate to the laboratory newly associated with ENSAI (training school in Rennes for second rank managers of INSEE, the *'attachés'*), while the specialists of data analysis, mostly econometricians, stayed in Paris in the laboratory associated with ENSAE (one of France's top schools in economics and statistics and in the training of first rank managers of INSEE, the *'administrateurs'*).

27. Often overlooked, the possibility of generalizing 'all things being equal' results on non-representative data remains debatable.

28. The strictest models exclude, at least as exogenous, all variables that could be the result of respondents' past decisions (e.g. region of residence, characteristics of their work organization, etc.).

29. This notion of unobservable heterogeneity aggregates all the non-observed characteristics of the respondents.

30. Laurent Thévenot's (1979) analysis of the lack of clarity concerning this category shows how it had been thought of until then.

31. See Desrosières (2008, vol. 1, ch. 9) for an initial analysis of the evolution of the syntax used in the presentation of statistical results.

32. In their legal version, *persons* are beings endowed with a personality ("generic concept in which body and soul are held together") and historicized, bearers of logic of progress ("the revelation of the human spirit to itself", "history [having] a prophetic dimension") (Supiot, 2007 [2005], p. 21).

33. There is, from this point of view, a tension between two facets of the person; between the passive nature associated with the description of that which they experience, their 'suffering', and a more active dimension linked with their autonomy, with the notion of 'achievement', as in the work, for example, of Amartya Sen.

REFERENCES

Affichard, J. (Ed.). (1987 [1983]). *Pour une histoire de la statistique: Matériaux* (Vol. 2). INSEE & Economica (Première edition: INSEE 1983).

Affichard, J., & Gensbittel, M.-H. (1987). L'entrée des jeunes dans la vie active. In J. Affichard (Ed.), *Pour une histoire de la statistique: Les matériaux, Tome 2* (pp. 177–190). INSEE/Economica.

Amossé, T. (2013). La nomenclature socio-professionnelle: Une histoire revisitée. *Annales, Histoire, Sciences Sociales, 68*(4), 1039–1075.

Amossé, T., & Chardon, O. (2006, November). Les travailleurs non qualifiés: une nouvelle classe sociale? *Economie et statistique* 393–394, 203–229.

Amossé, T., & de Peretti, G. (2011). Hommes et femmes en ménage statistique: Une valse à trois temps. *Travail, Genre Et Sociétés, 26*(2), 23–46.

Appadurai, A. (2016). *Banking on words: The failure of language in the age of derivative finance.* University of Chicago Press.

Baccaini, B., & Courgeau, D. (1997). Analyse multi-niveaux en sciences sociales. *Population, 52*(4), 831–863.

Baudelot, C., & Gollac, M. (2003). *Travailler pour être heureux? Le bonheur et le travail en France.* Fayard.

Bessin, M., Bidart, C., & Grossetti, M. (Eds.). (2010). *Bifurcations: Les sciences sociales face aux ruptures et à l'événement.* La Découverte.

Boltanski, L. (2012). *Énigmes et complots: Une enquête à propos d'enquêtes.* Gallimard.

Bonvin, J.-M., & Farvaque, N. (2007). L'accès à l'emploi au prisme des capabilités: Enjeux théoriques et méthodologiques. *Formation et emploi* (98), 9–24.

Bruno, I., & Didier, E. (2013). *Benchmarking: L'Etat sous pression statistique.* Zones.

Castel, R., & Haroche, C. (2001). *Propriété privée, propriété sociale, propriété de soi. Entretiens sur la construction de l'individu moderne.* Fayard.

Chaleix, M., & Lollivier, S. (2004). Outils de suivi des trajectoires des personnes en matière sociale et d'emploi. *Rapport, 98/B010, Class: 1.5.91.* INSEE.

Chamberlayne, P., Bornat, J., & Wengraf, T. (Eds.). (2000). *The turn to biographical methods in social science: Comparative issues and examples.* Routledge.

Corcuff, P. (2007 [1995]). *Les nouvelles sociologies.* Nathan.

Courgeau, D., & Lelièvre, E. (1989). *Analyse démographique des biographies.* Éditions de l'INED.

Courgeau, D., & Lelièvre, E. (1996). Changement de paradigme en démographie. *Population, 51*(3), 645–654.

Darras (1966). *La partage des bénéfices. Expansion et inégalités en France.* Les éditions de Minuit.

Desrosières, A. (1987 [1977]). Eléments pour l'histoire des nomenclatures socioprofessionnelles. In J. Affichard (Ed.), *Pour une histoire de la statistique* (pp. 155–231). INSEE & Economica.

Desrosières, A. (1998 [1993]). *The politics of large numbers: A history of statistical reasoning.* Harvard University Press.

Desrosières, A. (2008). *Pour une sociologie historique de la quantification.* Presses de l'Ecole des Mines de Paris.

Desrosières, A. (2014). *Prouver et gouverner: Une analyse politique des statistiques publiques.* La Découverte.

Desrosières, A., & Thévenot, L. (1988). *Les catégories socioprofessionnelles.* La Découverte.

Dodier, N. (1996). Les sciences sociales face à la raison statistique (note critique). *Annales HSS, 51*(2), 409–428.

Dussert, F. (1996). Le réseau d'enquêteurs de l'Insee: 50 ans d'histoire. *Courrier des statistiques,* (78), 47–52.

Gollac, M. (1997). Des chiffres insensés ? Comment et pourquoi on donne un sens aux données statistiques. *Revue Française De Sociologie, 38*(1), 1–36.

Goux, D. (2003). Une histoire de l'enquête Emploi. *Economie et statistique* (362), 41–57.

Héran, F. (1998). La Construction des identités: Réflexions et références pour un projet d'enquête. In Département de la démographie (Ed.), *Direction des statistiques démographiques et sociales (Note 44/F101)* (pp. 12). INSEE.

INSEE (1977). *Pour une histoire de la statistique. Les contributions* (Tome 1). INSEE/Economica.

Locoh, T., Hecht, J., & Andro, A. (2003). Démographie et genre, de l'implicite à l'explicite. In J. Laufer, C. Marry, & M. Maruani (Eds.), *Le travail du genre. Les sciences sociales du travail à l'épreuve des différences de sexe* (pp. 299–322). La Découverte.

Passeron, J.-C. (1999). De quel homme les sciences de l'homme parlent-elles? *Revue Européenne Des Sciences Sociales, 37*(113), 5–19.

Rouvroy, A., & Berns, T. (2013). Algorithmic governmentality and prospects of emancipation: Disparateness as a precondition for individuation through relationships? *Réseaux, 177*(1), 163–196.

Salais, R. (2015). Revisiter la question de l'État à propos de la crise de l'Europe: État extérieur, absent, situé. *Revue Française de Socio-Économie,* Special issue (2), 245–262.

Supiot, A. (2007 [2005]). *Homo juridicus: On the anthropological function of the law.* Verso.

Thévenot, L. (1979). Une jeunesse difficile: les fonctions sociales du flou et de la rigueur dans les classements. *Actes de la recherche en sciences sociales* (pp. 26–27), 3–18.

CHAPTER 7

A New Calculable Global World in the Making: Governing Through Transnational Certification Standards

Laurent Thévenot

INTRODUCTION: THE EVOLVING POLITICS OF CALCULABLE WORLDS

The politics of quantification rests on preliminary processes of trans-forming the world to make it quantifiable, by form-giving, formatting, in-forming, codifying and equivalence-making on the basis of a variety of conventions. This chapter concentrates on such transformations that make the world calculable. It first presents the analytical tools of our research agenda on the politics of statistics and quantification. They are used here to characterize the processes of transformation involved in the globaliza-tion of a new mode of governing that operates, away from states, through voluntary certification standards made up of measurable objectives. Initi-ated as a form of communication along the supply chain of major agro-industry products, it enlarged and gained public legitimacy through

L. Thévenot (✉)
École Des Hautes Études en Sciences Sociales, Paris, France
e-mail: laurent.thevenot@ehess.fr

© The Author(s) 2022   197
A. Mennicken and R. Salais (eds.), *The New Politics of Numbers*,
Executive Politics and Governance,
https://doi.org/10.1007/978-3-030-78201-6_7

the implementation of multi-stakeholder governance, while extending the plurality of normative issues it covers, from agricultural good practices to environmental and labour standards, or social accountability. By contrast to other modes of governing, this normative pluralism is entirely encapsulated in the measurable characterization of the product—palm oil in the case we investigated—which is certified to conform with the "sustainability standard" of the Roundtable for Sustainable Palm Oil (RSPO). All regulations of human actors' behaviours are deposited externally in the material product that circulates between them, and its certification. Certification implies a formal statement which is not legal but issued from an accredited third party body. It gives written assurance that the product or service is in conformity with the standard; the certificate being a form of communication between seller and buyer, while the label is a form of communication with the end consumer. While accountants or auditors certify accounting numbers, this third party body certifies product attributes that consumers cannot evaluate even when they use them. It codifies the process of production, its environmental impact and labour conditions—what economists named "credence" attributes of the product. Such governance is not only based on the objectivity of numbers but on a wider expectancy: that the material world of products, with which and through which human beings interact, would turn into a set of objective options, and their certification would guarantee the individuals who choose them that fundamental values or goods are satisfied.

Our analysis deals with the arts of calculating that are at the core of this contemporary mode of governing by certification standards. Calculating has two connected significances in this art: counting to govern with numbers, but also counting on an environment that is designed to be more reliable and offer possibilities to calculate on it. The linkage between counting and counting on is encapsulated in the term "calculable" with its double meaning of quantifiable and dependable. The first points to measurement while the second introduces a broader idea of guarantee and, therefore, an evaluation.[1] In the first sense, the British approach to Foucault's "governmentality" and "the administering of lives" (Mennicken & Miller, 2014) that Nikolas Rose and Peter Miller initiated in the nineties focused early on "governing by numbers" (Rose, 1991) and the invention of "calculating selves", "calculable spaces" and "calculative practices" (Miller, 1992, 2001). It was developed from accounting practices into a major and productive research agenda on modes of governing economic, social and personal life (Miller & Rose, 2008; Rose & Miller, 2010).

Another earlier main research agenda on the politics of numbers originated from the French National Institute of Statistics and Economic Studies (INSEE) at the end of the seventies, and initially dealt with a different domain of practices producing numbers: state statistics. After the pioneering work by Alain Desrosières on the history of socio-occupational categories (Desrosières, 1987 [1977], 1987 [1983]; see also Desrosières & Thévenot, 1988), the historical perspective was reinforced by a symposium in which INSEE brought together historians working on quantitative series and statisticians who were urged to take a reflective and historical look at the surveys they were responsible for (Affichard, 1987 [1977], 1987 [1983]). From the very start, history was a major focus of this research agenda, as shown by the early book co-written by Robert Salais on the social categorization of unemployment (Salais et al., 1986), or the author's research on the genealogy of surveys and social categories that measure social inheritance since Francis Galton's eugenics (Thévenot, 1987 [1983]).

While Desrosières related the history of statistics to the formation of states and their characteristics (Desrosières, 1998 [1993]),[2] I brought together the chain of operations involved in the transformation of personal answers into quantified statistics, and the comparable chain of transformative operations induced by representing and voicing personal concern in the proper format for politics. I early conceptualized the elementary process that makes this possible through the notion of "investment in form". An investment in form "produces equivalence" and "social coding", the term "code" covering "the set of conventions which govern 'regulated' communications between people" (Thévenot, 1983, 1984). On the conceptual basis offered by these "conventional forms of equivalence", the research agenda on the "politics of statistics" (Thévenot, 1990) investigated the various segments of the statistical production chain—survey and data collection, classification, codification and the processing of information—to identify fundamental correspondences between "statistical equivalence forms" and the "political constructions of the bond between members of the same polity" (Thévenot, 1987 [1983]).[3] It extended to policy implementation and evaluation, as well as social, economic and political theories involved in this evaluation (Thévenot, 1983, 2011a, 2016).[4] The initial programme on governing through statistics was expanded to non-state modes of governing through standards and objectives (Thévenot, 1997, 2009, 2015a, 2019b).

The British and French currents lacked opportunities for dialogue.[5] The present book offers such an opportunity, thanks to the interactions facilitated by its co-editors Andrea Mennicken and Robert Salais, and meetings with Peter Miller who provides his views in an afterword, as well as Wendy Espeland.

Far from the politics of state statistics, the transnational—or even a-national—worldwide extension of private voluntary standards has led to a mode of "governing through standards" (Thévenot, 1997, 2009; Ponte et al., 2011; Ponte & Cheyns, 2013) that is intended to make the world not only reliable and countable but even *certifiable* (Thévenot, 2015a). What does it imply in terms of its politics, since this mode of governing has been refurbished in response to criticisms which pointed to the lack of legitimacy that current standard-setting procedures undergo? Several years of a collective research programme I have taken part in focused on investigating the practices of this new calculable world.[6] Observing governing through a certification standard in action has demanded close fieldwork to follow the most vulnerable actors, from their daily life in remote rural areas to "open spaces" of public roundtables, or private confidential negotiations. Proper analytical tools were needed to grasp the wide variety of actors' practices, knowledge, evaluations and voices when they strive to express their concern and criticize.[7] After the next section, which sets out our research programme in more detail, the second part of this chapter examines the normative and regulatory basis of governing through certification standards, which is intended to ensure political legitimacy while taking distance from state legal and political systems. Which alternative to the rule of law does governing through certification standards offer? While the second part deals with the production of regulations, the third one tackles their enforcement. What are the functional equivalents for the judicial system in putting the standard implementation to a critical test? How do actors—particularly those who are most vulnerable—cope with the proper requirement of a certifiable world: Transforming all their concern, from the most personal to the most collectively political, into the format of measurable objectives that the standard enforces?

From State Statistics to Government Through Standards: A Research Programme on the Politics of Conventional Forms and Engagements

Belonging to the "Disobedient Generation" of the "Sixties" (Sica & Turner, 2005), I shared the Marxist critical stance which prompted the reversal of the hierarchical superiority of abstract and formal knowledge in favour of practice. It propped up the significance of practical know-how along the chain of actors which produce statistical data. Following knowledgeable pollsters and coders who were usually downplayed and treated as low-skilled white-collar workers, I turned to the workplace of the statistical chain, and investigated it as an industrial production line. This line creates in-formation not by assembling parts but mostly by trans-forming, changing the format of entities. Manufacturing transforms a personalized matter—currently collected from oral interviews—and shapes it into a standardized public form: the formal format of codified and quantified items. In the case of social statistics, this trans-formation aims at trans-muting "In Person" into "In Common", one of the most intense experiences and learning of the Sixties politicization. The Marxist tension between theoretical and practical knowledge was at stake, but also the exploration of the "Two Bodies" in which every human being invests (Thévenot, 2005) and not only kings and rulers (Kantorowicz, 1997 [1957]). One body is "invested" with a form which ensures communication in the sense of making common and endowing with coordination power. The other formless corporeal, living and mortal body puts into question such conventional "invested forms".

Social Coding *and* Investments in Conventional Forms: *The Prerequisites for the Politics of Quantification*

Each practical step of the transformative chain that creates data was investigated: interviewing respondents at home, filling in questionnaires, coding answers within social classifications (Thévenot, 1981a). The transversal operation of giving form—or formatting—led to conceptualize "social coding" (Thévenot, 1983) which initially focused on the formatting of occupations.[8] "Investment in form" was conceived as the establishment of a conventional form of equivalence such as classification, criterion, code, standard, routine, rule of thumb, house rule, instruction, custom, regulation, right, trademark, model, template, mould (Thévenot,

1984). Certification standards are among those invested forms, and most of the items of the previous list of invested forms were found in our fieldwork on standard-setting and implementation. Three main criteria were initially set out to distinguish various types of invested forms: (1) the lifespan or extent in time—from a short-lived model, up to a perennial custom; (2) the area of validity or extent in space—from a personally and locally attached rule of thumb or house rule, up to international rights; (3) the objectivity or material consolidation—from an ideal mental criterion, up to a solid template.[9] Investing in forms consists of the costly sacrifice of present coordination potentialities to ensure future returns in terms of economies in the cognitive and practical processes of coordinating actions. Formatting into a formal form is a step prior to any quantification, and a basic procedure in making the world calculable. Power relations ensue from invested forms, such as "the power relations between [agents who use very general forms and] agents who make use of more specific forms", this last power being disqualified by Taylorism when formal definitions of tasks phased out rules of thumbs built up by practiced workers (Thévenot, 1984), as did Toyotism later (Charles, 2016).

Our "practice turn" (Schatzki et al., 2001; Thévenot, 2001b) was not initially influenced by American pragmatism. In addition to what I said of the Marxist reassessment of practice, and the legacy of Bourdieu's *Outline of a Theory of Practice* (1976 [1972]), our pragmatic or pragmatist view on invested forms was influenced by research on work and organization, more precisely on the problematic of "coordinating" action.[10] This unusual term in the social and political sciences was taken in the sense of an uncertain process rather than its achievement in order. Such a perspective benefited from the cooperation with François Eymard-Duvernay who elaborated further the economists' notion of "specific investment" to contrast personalized and non-transferable long-term relationships based on codified relationships that can be maintained at a distance (Eymard-Duvernay, 1986). This collaboration was subsequently extended within the founding group of the so-called Economy of Conventions of Convention Theory.[11]

Placing Value on Invested Forms: The Plurality of Orders of Worth *Involved in Justifications and Criticisms Referring to the Common Good*

The definition of invested forms did not explicitly refer to evaluation. Yet, the above-mentioned criteria used to characterize them sustain distinct modes of evaluation (Thévenot, 1984). An early empirical research on the invested forms which respondents and coders used to identify occupations showed three core ways of making one's occupation worthy: the legal qualification or office one fills in (*état*), the art to which one is devoting oneself (*profession*), the traditional trade (*métier*) that one learns by doing (Thévenot, 1981b, 1983, 2016). This first insight into the relation between invested forms and worth—the three kinds of worthiness of occupation roughly correspond to the later identified *civic, industrial* and *domestic* "orders of worth"—was then fully developed through the intense collaboration and co-authorship with Luc Boltanski, which led to the new analytical step of "worth" analysis (Boltanski & Thévenot, 1987, 2006 [1991]).

Boltanski had earlier written on classifications (Boltanski, 1970) and co-authored with Bourdieu a seminal article on classification struggles as a continuation of class struggles (Bourdieu & Boltanski, 1975), giving a classist critical twist to the Durkheim-Mauss legacy on social classifications (Durkheim & Mauss, 2009 [1903]). He had later advanced a more thorough analysis of the representation process in the case of *Cadres* (Boltanski, 1987 [1982]). In our collaboration, we first designed a series of experimental games to investigate the non-expert's modes of classifying occupations and "finding one's way in social space" (Boltanski & Thévenot, 1983).[12] They brought to light the strong connection between bringing occupations together (*"rapprochement"*) in social categories and placing value on them.

Taking part in public debate requires that participants transform their personal concerns—or possibly sacrifice some of them that cannot bear the transformation—to invest in conventional forms and reach a higher level of generality (*"Montée en généralité"*). In the next step of the collaboration with Boltanski, we identified the grammar of *Orders of worth* as the model of the sense of justice that human beings rely on, when they justify and criticize. We initially identified six repertoires of evaluation that correspond to this model, each order of worth seeking legitimacy by claiming to contribute to a distinct conception of common good (Boltanski & Thévenot, 1987, 2006 [1991]). According to this model, considerable aggrandizement of personal concern is needed to demonstrate that one's voice is relevant for the common good. Individual interest does not qualify. Since different orders of worth refer to different constructions of the common good, their confrontation publicizes *difference*—in the sense of *differing* in a dispute—which culminates in severe critical clashes. Each order offers the footing to "denounce" the conception of the common good that another order claims. Unequal commonality entails hierarchical ranking of states of worth. We initially avoided the all-purpose vocabulary of "power", to be more precise about *qualifications* that contribute to both empowerment and domination. Yet, orders of worth relate directly to inequality of power since a higher state of worth provides a higher capacity for coordinating others' actions. Claims of legitimacy strengthen this form of power. Our analysis thus continued Weber's differentiation of orders of legitimate domination.[13]

In addition to distinguishing a plurality of modes of evaluation involved in the justifications and criticisms that aim at public legitimacy, the model of orders of worth differentiates the ways evaluative judgements about worthiness are put to the *test*. The pragmatist realism of the "reality test" (Boltanski & Thévenot, 2006 [1991]) involves items of the material environment, on the condition they *qualify* for the tested order of worth, and may thus be taken into account in the judgement and be granted evidentiary value. Because of the two previous features, this model differs from Actor-Network-Theory (ANT) which does not make it possible to contrast modes of relations between beings—human or not—on the basis of their mode of qualification and thus valuation. With regard to its use by Foucault, the concept of *dispositif* can be made more precise and broken into component parts. The qualification of each of them for an order of worth *disposes* human beings to engage in justification according to this kind of worth. The pragmatist realism that the "reality test" and "qualification" carry contributed to the development of the notion of "quality conventions" that makes more precise the analysis of organizations and markets. Previous research by Eymard-Duvernay on "models of the firm" in diverse industrial sectors (Eymard-Duvernay, 1986, 1989) was refined by differentiating, inside the same organization, a multiplicity of modes of coordination framed by these various "quality conventions" used in the valuation of persons, things and their relations (Eymard-Duvernay, 2002; Storper & Salais, 1997; Thévenot, 2001a).

The Worth of Standards

Standards are valuable for the *industrial* worth of technical efficiency when they primarily contribute to the compatibility of methods and tools of production. This worth remains dominant in the palm oil standard intended to spread a uniform agricultural model that is oriented towards industrial efficiency. Since our pragmatist approach relates valuation to coordination, the temporal orientation of each order of worth impacts on the mode of coordination it governs. *Industrial* worth sustains a forward-looking orientation that is fully reflected in the idea of technical investment, since equipment and methods fabricate the future by providing predictability.[14] In addition, this worth is conducive to quantification, which is currently involved in the reality test for this order of worth.

The standardization of market goods and services also enhances their *market* worth. Companies that engage in the standard-setting process seek a competitive advantage.[15] The two conventions of *industrial* and *market* worth—and their worlds of objects—differ significantly in the spatiality (Cartesian space/free circulation space) and temporality (future/present orientation) they sustain in coordination. These differences stir up critical tensions which are internal to the economy and weigh on standard-setting: fixing for the future the standardized characteristics of products opposes the *market* worth orientation towards an ever-changing present.[16]

The worth of *renown*, or fame, is also significant for standards. It does not rely on prices but on signs of recognition in public opinion. It strongly motivates the commitment of entrepreneurs in standardization procedures when combined with *market* worth in "compromises" such as branding and marketing (Richey & Ponte, 2011). A campaign triggered by Greenpeace in 2010 proposed a devastating parody of a Kit-Kat chocolate bar advertisement, in which a bar was stuffed by the bleeding finger of an orang-utan. Within a few weeks, Nestlé accepted to negotiate with the NGO because the firm was deeply concerned by the drastic consequences of this campaign on the brand image. Like market worth, the worth of renown in opinion orients towards the present and its "trends". This worth is made measurable and quantifiable through opinion polls.

By contrast, *domestic* worth emphasizes traditions and customs, and sustains a temporal orientation anchored in the past. The traditional arts and crafts qualify for this worth and have occupied an important place in the history of quality standards, from the very beginning when they were promoted by corporations. Still very present in today's food certifications (Boisard & Letablier, 1989; Busch, 2011; Cheyns & Ponte, 2018; Diaz-Bone, 2011; Ponte, 2016; Star, 1991), this worth is most often in a critical position in standards of the RSPO type, because of the weight placed on *industrial* worth. This critical position is still weakened by the fact that *domestic* worth is hard to measure and quantify, except through "compromise" variables that combine *domestic* with *industrial* worth.

The reality test of the worth of *inspiration* brings evidence of rupture and revelation, in a temporal orientation on the present and a spatial presence evoking epiphany. Insofar as *industrial* standardization tends to fix things for the future, it generates critical tension with the worth of *inspiration*. Yet, "innovation" processes that create new products and services supposed to reach a compromise that strikes a balance between the conflicting requirements of *inspiration, market* and *industrial* orders of worth. This worth of *inspiration* derives from a genealogy of valuations of religious deeds and spiritual engagements. In our investigation, this *inspiration* worth was involved in forest peoples' denunciation of palm oil standardized plantations practices that ignore sacred places.

Standard-Setting in Search of Legitimacy: The Grammars of Commonality in the Plural

Although orders of worth do play a significant role in some RSPO actors' statements, criticisms or activities, the standard itself, its setting and enforcement processes, thwart actors' attempts to engage in public critiques and justification of large scale. The reason is the following: In response to criticisms pointing to the lack of legitimacy of private, voluntary standards-setting procedures, the RSPO type of transnational standard is built on the "multi-stakeholder governance model", or "multi-stakeholder initiative" (MSI). To understand how MSI governance conflicts with the model of orders of worth, another analytical step is needed: the model of orders of worth has to be situated in a broader analysis of the ways in which actors take part in disputes that claim legitimacy for the whole community.

Grammars of practice which support *pluralist constructions of commonality and difference* can be characterized by basic operations: *communicating*—in its original meaning of making an issue common; *differing*—in the sense of disagreeing; and *composing*—in both the ancient sense of settling a dispute and the wider current meaning of calming (Thévenot, 2014, 2015b). These grammars diverge in the transformation they demand of personal concern into a common format that allows people to agree and disagree. The format of difference channels discrepancies between voices. In addition to evidencing the uses of orders of worth in the United States and France, a comparative research programme (Lamont & Thévenot, 2000; Thévenot & Lamont, 2000) contributed to the identification of the grammar that underlies multi-stakeholder initiatives.

Because of its link to the long and diverse liberal political tradition, I named it *liberal grammar*. As rightly argued by Veikko Eranti (2018), it could be named "grammar of interests". I originally avoided the term "interest" because of its naturalization in the social and political sciences, when it is viewed as an inner force guiding individuals. In the *liberal grammar*, it specifies the format to differ and agree in public. Instead of the large transformation of personal or local concerns into common good issues, and the resulting harsh confrontation when rival conceptions of the common good clash, the transformation of personal concerns into individual choices is lesser, and the confrontation less critical.

Yet, the *liberal* interested individual is also in a state of being formatted for the public. Any personal concern has to be converted into the format of a choice—designated as "preference", or "stake", or "interest"—which an autonomous individual makes between options that should be in a form accessible to all other individuals who constitute the *liberal* public. This format of *opting individuals* cannot express most intimate attachments. Too deeply personal, intimate or emotional concerns are not appropriate for liberal communication (Centemeri, 2015; Stavo-Debauge, 2012). In the grammar of plural orders of worth, differing is strongly critical, resulting from the denunciation of an order of worth in the name of another (Boltanski & Thévenot, 1987, 2006 [1991]), while in this *liberal* grammar differences are less dramatically expressed, because they are presented as individual interests. Criticism is only allowed at a lesser degree and the integration of differences is achieved by "negotiation" and "bargaining" between "stakeholders". Yet, the burden of the transformation weighs on human beings to fit this *liberal* format, and may oppress them as well. This analysis rectifies the current idea that *liberal* politics are "horizontal" as opposed to the vertical hierarchy of *orders of worth*.[17] All *grammars* of commonality are inherently hierarchical because of the gradient they maintain between more and less common formats. The coordinative power of those who engage in the most common format—being able to articulate their concern as choices for common knowledge options, in the *liberal* grammar—entails de facto domination over those who do not and depend on the previous ones.

This distinction of grammars helps to clarify the quantification and evaluation methods used by governing. Fifty years of using the same statistical survey to evaluate social, educational and employment policies showed the dependence on these grammars of: policies, methods of their quantitative evaluations (socio-occupational categories, mobility tables, correspondence analysis, econometrics) and even underlying social and economic theories (reproduction, de-skilling of work, human capital) (Monso & Thévenot, 2010; Thévenot, 2011a). *Civic* and *industrial* orders of worth support the first welfare-social state policies, quantification methods and social theories, whereas *market* order of worth and the *liberal* grammar are backing the more recent policies, quantified evaluations and theories.[18] This dependence is also visible in Emmanuel Didier's contribution to this volume, which examines the relation to quantification that various American sociology currents have. Interactionism and, more broadly, the trends that constitute the "Qualitative sociology" pole influenced by American pragmatism presuppose a *liberal* grammar of opting autonomous individuals, a grammar that also underlies their sociological criticism of the paternalistic welfare state and its categorical statistical treatment of social groups.

Committed to Objects: Valuable Regimes of Engagements with the World Affected by Standardization

Unlike other modes of governing, the one that maintains a calculable world by certification standards intervenes in the surrounding material objects to regulate the relations between human beings. Our analysis has thus to leave discursive public arenas and scrutinize the variety of valued human relations to material objects, whether they are public or not. The concept of *engagement* was crafted to capture such valued relations with the environment, each of them consolidating and empowering the self through a certain mode of coordination with oneself that is secured by this relation (Thévenot, 2006, 2007, 2013, 2019c). This self, or *personality*, has a dynamical identity resulting from the integration of a plurality of modes of engagement.

Each mode of *engaging with* the world rests on a distinct mode of in-formation, if we extend this notion to highly personal knowledge and pay attention to the personal indices or landmarks that give convenient form and disposition to a familiarized background. *Engaging in familiarity* is valuable because of the personal ease it generates, due to familiarization with this environment. This intimate relationship to habituated and inhabited places supports a primordial trust in oneself and is deeply affected by changes in the environment. The industrial agriculture development system that the RSPO standard promotes through "good practices" breaks with the personal relationships of using and inhabiting the land that each member of rural communities has with his or her environment. Highly idiosyncratic, *engaging in familiarity* does not easily lend itself to the commonality and communication.[19] It meets the greatest difficulties to find a place in governing by standards which favours, as we shall see, a completely different regime of *engaging in plan*, or project. More than an instrumental relationship to the world, this engagement aims at the good of being able to project oneself into the future through individual plans, provided that surroundings are seized in a functional format.

Dynamical *regimes of engagement* go through trying moments which provide the opportunity to test landmarks and update them. Engagements are thus polarized by two stances, and the process of certification collapses the distinction between the two in favour of the first. The first stance of static quietude sticks to the fixed form that serves as a marker of the engagement and is tested in the trying moment. It corresponds to the *letter* of the convention or the institutional act when *engaging in public justification*.[20] The second stance of moving inquietude brings the awareness of the sacrifices that this fixed form entails, when the situatedness of the engagement opens up to other possibilities of coordination.

Distinctive Features of a New Calculable World Governed by Certification Standards: Which Substitute for the Rule of Law in the Production of Regulations?

Standards came to constitute a calculable global world through two types of extension of their original technical purpose. Their scope expanded, in terms of the variety of values they take into account. Standard-setting and enforcement procedures enlarged, in search of democratic legitimacy.

Made in Standard: All the Good that Money Can Buy

Standards were originally thought to make technical tools, methods and products more compatible, and to provide economies of scale according to the technical efficiency of *industrial* worth. Quality standards expanded their scope along *market* worth. They developed to bear the burden of the common knowledge identity of market goods that market competition requires against moral hazards, and they backed the market diversification of these goods and services (Busch, 2011; Bowker & Star, 2000; Eymard-Duvernay, 1986, 1989, 2002; Lampland & Star, 2009; Salais & Storper, 1993; Storper & Salais, 1997; Timmermans & Epstein, 2010). As mentioned before, economists use the category of "credence" goods and services when the quality cannot be identified by "search" or "experience"—referring to repair services—and may generate fraud (Darby & Karni, 1973). This category of "credence" good is currently used for the kind of certification that we now consider, such as Fair Trade. Gaëlle Balineau and Ivan Dufeu (2010) rightly contested this categorization, observing that these kinds of goods do not suffer from information asymmetry in the production process, but from another source of uncertainty because the goods' "attributes are seen as means to reach another goal" (Balineau & Dufeu, 2010, p. 335).

This "other goal" actually introduces a most dramatic change in the role of standards. Standards came to carry the responsibility for the satisfaction of basic rights or conceptions of the common good, through certified and measurable properties attributed to goods or services. In democracy, such values are taken into account by legislative bodies of government in the process of making laws. Political public arenas are dedicated to critical debate on this process, with justifications referring to the plurality of these basic rights or conceptions of the common good. The new kind of "standardizing liberalism" (*"libéralisme normalisateur"*) (Thévenot, 1997) and mode of "governing by standards" (Thévenot, 2009, 2015a) arise when such evaluative and normative principles are reduced to measurable characteristics of goods and services, and when individual consumers' choices on the market place replace political debates. This reduction (see Fig. 7.1) can be illustrated on three normative principles: social justice and collective solidarity to struggle against inequalities; environmental concern; tradition and customs. Each of the three corresponds to a separate order of worth (*civic, green, domestic*)

The critical plurality of debated conceptions of the common good turned into the certified qualities of market goods

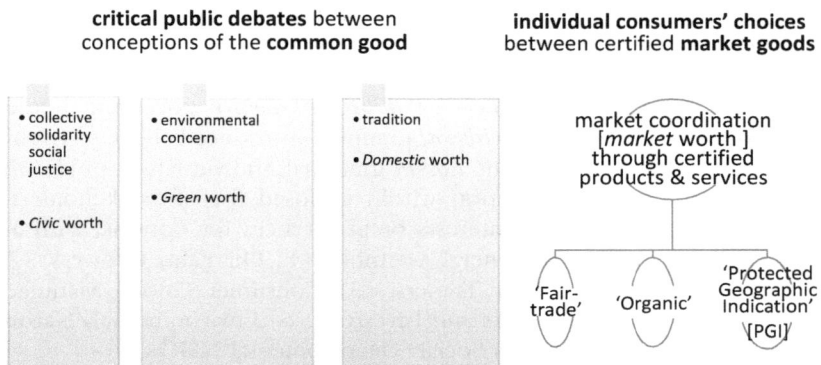

critical public debates between
conceptions of the **common good**

individual consumers' choices
between certified **market goods**

- collective solidarity social justice

- environmental concern

- tradition

market coordination
[*market* worth]
through certified
products & services

- *Green* worth

- *Domestic* worth

- *Civic* worth

'Fair-trade'   'Organic'   'Protected Geographic Indication' [PGI]

Fig. 7.1 From debated conceptions of the common good to certified qualities of market goods

which is involved in public justifications and criticisms that claim legitimacy by referring to various conceptions of the common good (see the left side of Fig. 7.1). Each of these normative principles is formalized and formatted in laws (labour, environmental, customary) and legal rights. In the reduction to certification standards, each is transformed into measurable characteristics of products and services, such as Fair Trade, Organic, Protected Geographic Indication (see the right side of Fig. 7.1).

Multi-Stakeholder Certification: A **Liberal** *Public in Which Opting Individuals Are Formatted as* **Stakeholders** *and Options as Measurable Objectives*

With the expansion of their scope and the range of values they take into account, standards of a new generation had to strengthen their standard-setting and enforcement procedures that were previously criticized for their lack of transparency and legitimacy. This was the case for a series of transnational standards that claim to certify sustainability, independently of the political authority of states, and were built on main value-chains of the agro-environmental business: RSPO (Roundtable on Sustainable Palm Oil) in 2003, today 1000 members; RTRS (Roundtable on Responsible Soy) in 2005; Bonsucro (Better Sugarcane initiative) in 2006; BCI (Better Cotton Initiative) in 2006; RSB (Roundtable on Sustainable Biofuels) in 2008. RSPO began with an agreement that WWF obtained from Unilever in order to be more efficient and faster than states and international legal systems in coping with deforestation. In response to criticisms, these certification standards are ruled by *multi-stakeholder* governing bodies ("roundtables") based on the *liberal* grammar introduced above.[21] In this multi-stakeholder variant of the *liberal* grammar, individuals-in-public are given an additional qualification within a closed list of "stakeholders" corresponding to different interests or preferences for options, with an equal voting right in the general assembly: (1) Oil Palm Growers, (2) Palm Oil Processors and/or Traders, (3) Consumer Goods Manufacturers, (4) Retailers, (5) Banks and Investors, (6) Environmental/Nature Conservation NGOs, and (7) Social/Developmental NGOs.

Standardization alters the *liberal* grammar in that options are to be formatted as elementary *plans* to be engaged in, with the projected output of measurable objectives listed as "indicators" and additional "guidance".[22] Even when they are not quantified, they require codification and formality.[23] In spite of this dominance of the *engaging in a plan*, the RSPO standard progressively included references to conceptions of the good or rights—in terms of "principles of criteria"—that do not fit this format because they overflow the limits of small narrow *plans*. In the 2013 change, four new "criteria" were added, which point to hot issues to be governed: "C1.3 – ethical conduct (Growers and millers commit to ethical conduct in all business operations and transactions)" with the guidance: "A prohibition of all forms of corruption, bribery and fraudulent use of funds and resources"; "C6.12 – forced and trafficked labour"; "C6.13 – respecting human rights"; "C7.8 – minimizing GHG emission from new plantings".

There are six headings of "principles and criteria" under which indicators and guidance are grouped. The first, "Commitment to transparency", concerns information and documentation, including "a written policy committing to a code of ethical conduct and integrity" (indicator 1.3.1) and "a prohibition of all forms of corruption and bribery" (guidance). The second, "Compliance with applicable laws and regulations", states that "2.1. There is compliance with all applicable local, national and ratified international laws and regulations" with extension to other rights:

"2.2. The right to use the land is demonstrated, and is not legitimately contested by local people who can demonstrate that they have legal, customary or user rights". The use of the land is also considered in: "2.3. Use of the land for oil palm does not diminish the legal, customary or user rights of other users without their free, prior and informed consent", with indicator 2.3.1 stipulating the objective of "participatory mapping involving affected parties (including neighbouring communities where applicable, and relevant authorities)", and indicator 2.3.2 stipulating the objective of "copies of negotiated agreements detailing the process of free, prior and informed consent (FPIC)".[24] Each of the last four headings of "principle and criteria" relates to distinct orders of worth that they reduce to objective indicators. The third is *market*-oriented towards economic and financial viability, the fourth is prescribing a certain *industrial* organization of work, the fifth ("Environmental responsibility and conservation of natural resources and biodiversity") raises environmental issues and the sixth ("Responsible consideration of employees and of individuals and communities affected by growers and millers") addresses *civic* labour rights while extending to "community values" and "cultural and religious values" in the guidance.

Participative Technologies and Procedures to Deliberate Over Regulations

The general assembly of members meets annually in a "*convention*" which votes resolutions and changes of the standard. The quest for democratic legitimacy does not rest only on vote but also on direct voices of the constituency through arrangements and procedures that were designed to allow a wide participation in the deliberation over the standard.

Beyond the limitations of voices through exclusion, the concept of a "participation format", developed by Audrey Richard-Ferroudji (Richard-Ferroudji & Barreteau, 2012) in accordance with that of *engagement*, helps to clarify the conditions to take part, and the resulting "burden" that bears on participants as demonstrated by Julien Charles in various domains, from management to politics (Charles, 2012, 2016). In the RSPO general assembly, the requirement to be "pragmatic", "practical", "realistic" and "effective" urges participants to formulate their voice as *engagement in a plan*, and express themselves in public in the format of individual choice between optional plans. This is congruent with *communicating*—making issues common—in accordance with the *liberal* grammar. It explicitly opposes the "absolute or the ideal" of *engaging in public justificatory orders of worth* backed by conceptions of the common good. Prescriptions make explicit the required *liberal* public civility: "understand the stakes" (options) and express your own ("I want that!"), "be not shy", "proactive", "intervene", "make the first move", "take the floor" (Cheyns, 2011).

The large number of participants gathered for the few days of the convention led the organizers of roundtables to provide participatory technologies issued from management. By bringing participants physically closer, the small size of groups is intended to allow more accurate perceptual attention to others. In the 2006 roundtable, the device called "world café" was introduced. It was designed in the end of the 1990s to have participants "spontaneously" formed into "small, intimate table conversations" about shared issues, recording outputs on papers (initially "tablecloths") and periodically switching tables so that ideas might circulate and connect (Brown, 2002; Brown et al., 2005). In the RSPO version, the short-term temporality (twenty minutes) of each session bringing together six unknown people evoked "speed-dating" techniques. This brevity, the circulation from one table to another and the absence of a theme displayed on each table, raised among participants a sense of an arbitrary and poorly significant exchange.

To overcome these shortcomings, another facilitation technique was introduced in the 2013 roundtable, the "open space technology" (OST). Also issued from management and conceived by Harrison Owen in the 1980s, OST was worked out to foster "self-organization" (Owen, 2008). Its device meets the requirements of a *liberal* public of individuals expressing themselves through their choice for options made public. Each individual "convener" takes the responsibility of naming and posting in public an issue for a possible breakout session. Other participants have to choose among the posted themes as options offered to all. A playful staging is intended to turn the implementation into an exciting game. The initiator takes a placard and writes a slogan or objective to gather a discussion group (see Fig. 7.2a, b). Discussions in small groups last for one hour, each taking place in parallel sessions during two days. The output of the conversation is to be written on a flipchart as a list of objectives which will then be collected and transcribed on printed charts. When posted on the walls of the assembly hall, they make possible, in the next step, to produce some proposals to be voted in the plenary meeting.

Fig. 7.2   a, b Photographs of Open Space Technology (OST) in action (*Source* Photographs by Laurent Thévenot)

In addition to the *liberal* matrix, explicit mentions to *market* coordination are given. The "open space" is presented as a "marketplace" where participants "shop" for information and ideas (Owen, 2008). Displaying their respective placards, conveners "sell" their respective slogan-objectives on the competitive "market" of discussion groups. Individuals are urged to "freely" circulate between groups according to the one "law" of OST, the "law of two feet". It urges participants to leave the ongoing conversation of a group of discussion for another, "given both their right and responsibility to maximize their own learning and contribution" when they "lose interest" in a breakout session. Owen affirms that this is the correct civility: moving on is "the polite thing to do". It would be quite rude according to another grammar of commonality more hospitable to attachments and generous hospitality (Thévenot, 2014).

These breakout sessions constitute small-scale and short-lived meetings that do not have the validity of the general assembly and may raise doubt about their legitimacy. Therefore, OST provides "principles" which ensure the validity of the constitution, timing and production of these small groups. The RSPO Open Space kick-off PowerPoint recalls Owen's (2008) four principles that assert the legitimacy of these contingent groups:

> *Whatever happens IS the only thing that could have happened*
> *Whenever it starts IS the right time*
> *Whoever comes IS the right people*
> *When it's over IT IS over*

In the PowerPoint presentation, four other norms introduce additional requisites for communicating in the open space. Most of them specify the conveners' dispositions that are required by the *liberal* public. Norm 4—"[speak your] *voice*, share your opinions and reasons. Do it clearly and briefly"—makes explicit the right mode of *communicating* in the format of one's individual "opinion", stipulating the clarity and brevity already pointed to in the above-mentioned requirement to be "pragmatic". It involves to *engage in a plan*, with a short-term and clear-cut objective. Norm 3—"*respect* [all]"—specifies the disposition of tolerance towards *differing voices*. Mentions to "views" or "styles" ("regardless of whether their views or style are similar to ours") sustain the multicultural extension of the *liberal* grammar.

Norms 1 and 2 complement the *liberal* matrix. Norm 1 (*"listen"*) does not only prescribe the kind of attention to another individual's opinion that the *liberal* grammar requires. It also recommends to "be genuinely curious about their perspective", which points to a distinct regime of *engaging in exploration* (Auray, 2011) aiming at the good of surprising novelty.[25] It echoes another rule of the open space: "Be ready to get surprises". Norm 2 (*"suspend* [judgement]") does not only plead for respect to other individuals' opinions but for avoidance of criticism: "Suspend our agreement or disagreement".

A-liberal Conceptions of Communication and Their Managerial Reductions

If we look into the genealogy of these open space management technologies, we find they were initially designed to go beyond—or below—the *liberal* public space and overcome its limitations, as suggested by the explicit references to "café" style conversations, "backstage", "behind the scene" and "hallway chats". David Bohm, the author of the most influential *On dialogue* (Bohm, 1996) that Open Space Technology draws upon, was concerned by constructions of commonality and difference that significantly depart from the liberal political tradition.

While obtaining his PhD in the theoretical physics group directed by Robert Oppenheimer at the University of California at Berkeley, his engagement in communist organizations prevented him from being integrated in the Los Alamos project, despite Oppenheimer's proposal. When he was an assistant professor at Princeton University, he was called upon by the House Un-American Activities Committee in 1949. He invoked the Fifth amendment right to refuse to testify and give evidence against his colleagues. Princeton suspended him and he left the United States for Brazil and later United Kingdom, as a Professor of Theoretical Physics at the University of London. In addition to the collective spirit, or more precisely the *civic* worth of solidarity that oriented his youth political engagement, his physicist's activity has been a second source of insights into the limitations of the autonomy attributed to entities and even subjects. Pointing to the "fragmentation" that thought processes—and not only theoretical modelling—bring to the perception of the world, he considered that dialogue should shed light on the limitations due to this fragmentation. His contention that "the representation of thought enters the presentation of perception" (Bohm, 1996, p. 57) is not only Kantian but meets Ernst Cassirer's neo-Kantian turn based on Einstein's space–time modelling. A third source of his insights issued from his collaboration on human cognition with Stanford neuroscientist Karl Pribram, the psychological philosopher Jiddu Krishnamurti and the London psychiatrist and practitioner of Group Analysis Patrick de Maré.

All these resources supported Bohm's conception of dialogue which distances itself from the *liberal* composition of differences through "negotiation":

> A great deal of what nowadays is typically considered to be dialogue tends to focus on negotiation [...] People are generally not ready to go into the deeper issues when they first have what they consider to be a dialogue. They negotiate, and that's about as far as they get. Negotiation is trading off, adjusting to each other and saying, 'Okay, I see your point. I see that that is important to you. Let's find a way that would satisfy both of us. I will give in a little on this, and you give in a little on that. And then we will work something out.' (Bohm, 1996, p. 18)

His criticism of negotiation extends to the format of "problem" which occupies a central place in Dewey's pragmatism. He even criticizes the exposition of individual opinion—the mode of taking part in the liberal public—the "pressure [...] to get in there quickly and get your point of view across, particularly if you are one of the 'talkers'. Even if you're not, you have that pressure" (Bohm, 1996, p. 30). Bohm digs into the ground of the *liberal* grammar and illuminates the limitations of *engaging in a plan*: "Now, I'm going to propose that in a dialogue we are not going to have any agenda, we are not going to try to accomplish any useful thing" (Bohm, 1996, p. 17). He refers to what Michael Polanyi has called tacit and personal knowledge (Bohm, 1996, p. 52) pointing to the format we rely on when *engaging in familiarity*.

In spite of these various sources which diverge so strongly from the construction of a *liberal* public of individuals choosing among options, these divergences were blunted in the managerial usages of Bohm's original conception of dialogue and the resulting RSPO Open Space Technology and participatory dialogue mechanisms.

Experiencing Participative Technologies in Practice: "Open Space" and Dialogue Dispositions Put to the Test of "Smallholders" Engaging in Them

All RSPO participants are not equally prepared for the objective-oriented participation format that "Open Space" demands. Managers and international NGO members are well versed in the required techniques and procedures. They demonstrate a skilful utilization of the small group talks, using them strategically as a first move in a sequence of plans expected to extend eventually the objectives listed by the standard in the prescribed format of "criteria, indicators or guidance". We observed a group on labour issues that a member of Oxfam stood ready to offer on the "market" of the open space. He planned that the output of this first strategic step would be the creation of a "working group" that was designed to propose to the vote of the general assembly the revision of the standard and the introduction of new criteria about labour rights. This was an example of a step by step—*plan* by *plan*—process calculated to obtain substantial changes of the standard and introduce links with human rights, once formatted as criteria, indicators and guidance.

From now on, we shall concentrate on RSPO governing devices and procedures as they are practically put to the test by the most vulnerable actors of the palm oil value chain. Since our approach offers a dual analysis based on either *personality* or *community*, we followed both entities: a *personality* of "smallholder" (Arifin) and a "local *community*" the territory of which was severely impacted by industrial plantations (Karang Mendapo). Via a series of surveys (Cheyns, 2011, 2014; Cheyns & Thévenot, 2019a; Silva-Castañeda, 2012; Thévenot, 2018), we were able to observe how such a *personality* and *community* took part in RSPO governing devices and coped with the formats that make this world calculable. They were backed up by a variety of NGOs that we could also observe at different levels of their action.

Reacting to the 2009 plenary "smallholder session" in which presentations were only made by certification companies and agencies while smallholders themselves did not play a part, Arifin took advantage of the "Questions and Answers" session to speak publicly in the general assembly. Introducing the recently created farmers' union "Indonesian Oil Palm Farmers Union" [*Serikat Petani Kelapa Sawit*: SPKS] and promoting the representation of family farmers in RSPO, he emphasized the "inequitable" mechanism used to fix palm bunch prices and the sharing of the value along the chain. As observed by Cheyns, he spoke quietly at first, then his voice became louder and its pitch higher, his movements animated and he often pointed to the dais with the papers in his hand (Cheyns, 2014). As a result, the President asked to "avoid statements" and to provide "shorter questions and comments". A manager of the Indo Oil company commented to a NGO Sawit Watch official: "Why is it that my 'young fruit' makes a noise like that?" The manager referred to a hierarchy within the *Batak* ethnic group to which Arifin and himself belonged. Arifin used the opportunity of a later public meeting to denounce this "young fruit" (i.e. "child") paternalist *domestic* qualification the manager attributed to him. He underlined that he was not taking part in the roundtable as a young *Batak*, but as a representative of the farmers' union, SPKS. He thus claimed for a *civic* qualification (Cheyns, 2014) instead of the *domestic* one that the manager tried to impose.

Commenting on his intervention in an interview by Cheyns, Arifin said that he was angry and felt oppressed because when smallholders "want to fight, they go to jail" while businessmen in RSPO "are the ones who apply this pressure on the ground". Other members of the SPKS family farmers' union actually found his tone "still too soft" because "what is important is to be honest" and Arifin contrasted his full engagement with presentations by a lot of people who "don't speak from the heart" (Cheyns, 2014). Arifin's engagements overflowed in two ways the format of the limited *plan* confused with an objective. He engaged in *public criticism and justification* (in the *civic* denunciation of structural inequalities along the value chain, and promotion of union representation) and also in the *familiarity* of the small farmers' daily life which is severely impacted by industrial "good practices" of farming. Yet, this expansion of the participation formats was harshly criticized and Arifin called to order.

During the 2013 roundtable introducing the Open Space Technology and the marketplace of breakout sessions, Arifin first looked upset by this format (see Fig. 7.3a, left-hand). However, unlike other "smallholders" who were discouraged or eventually disappeared at the moment of welcoming newcomers, he overcame a long moment of embarrassment and hesitation and finally got involved. He lifted up a placard written in Indonesian and English: "Smallholder and replanting. Who to support?" [sic] (see Fig. 7.3b, right-hand). His theme assembled participants who were exclusively smallholders. It produced conclusions presented in Indonesian on a flipchart.

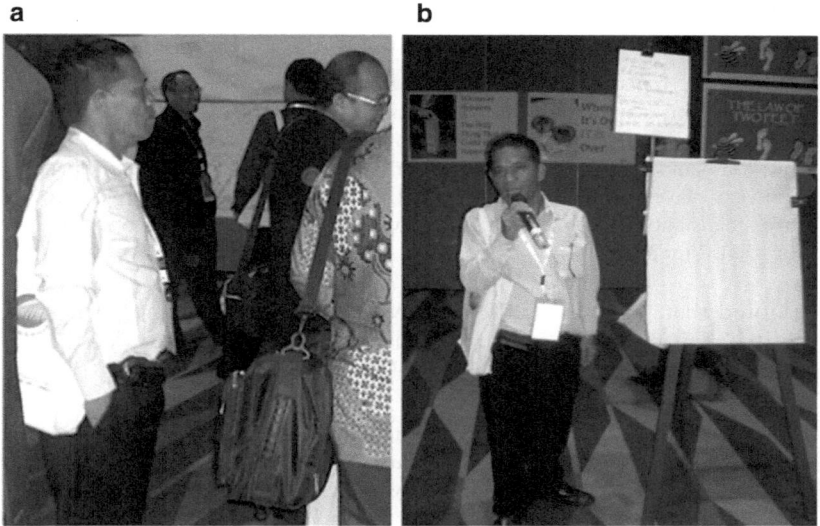

Fig. 7.3 a, b Photographs of Arifin, firstly embarrassed and hesitant in front of the Open Space Technology, then animating his "open space" group (*Source* Photographs by Laurent Thévenot)

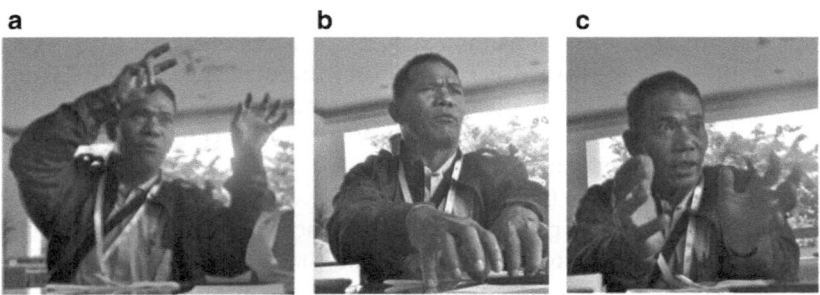

Fig. 7.4 a, b, c Photographs of Arifin, at ease during a conversation at the cafeteria (*Source* Photographs by Laurent Thévenot)

Off the roundtable, the author had a conversation with Arifin in the cafeteria. He was quite at ease, expressing himself in a *familiar* engagement with lively gestures and facial expressions (see Figs. 7.4a, b, and c). He came back to his experience of the Open Space, complaining that:

> [...] on the replanting of palm trees for smallholders who want to remain independent, a major issue for sustainability, because they are numerous, is that there were no representatives of consumers, buyers, major companies or banks. [...] They rather do brand imaging. [...] In Open Space, industry players seem to be passive and wait for conclusions. (Conversation of Cheyns and Thévenot with Arifin, Medan, November 2013, translation by Dani Pradana)

He noted that "NGOs are the ones who open the topic and participate. Industry players don't". Arifin voiced his discontent because of the gap between RSPO discussions and his field experience:

> Initially participants feel satisfied because they expressed their problems, but then they become unsatisfied because they do not see implementation. [...] RSPO is supposed to be a place where various individual interests are combined in a common interest. But once the mutual agreement has been reached, its implementation goes back to individual interest and I am fed up with this situation. It's a waste of time and energy to reach this agreement when it is not implemented. (Conversation of Cheyns and Thévenot with Arifin, Medan, November 2013, translation by Dani Pradana)

To remedy this, Arifin continued, one should "make sure that each topic discussed in these forums reach the executive board. [...] There should be rules to ensure parties enforce mutual agreements". Arifin emphasizes that the costly operation of composing the common interest from individual interests, in a mutual agreement between differing voices, is defeated on the ground.

Another reason for the gap between the discussions framed by the various technologies of participation and the achievements on the ground is that, unlike many actors involved in the formulation of the standard, "parties which have to implement them are common people who don't know much about the procedures of RSPO. [...] They don't know how P&C ['Principles and Criteria' of the standard] came to be; the farmers, workers, businessmen, fields actors who have to implement them" (ibid.). We thus observe that this in-depth criticism of the government by certification standards, informed by Arifin's experience as a small planter, although easily deployed in the *familiar* format of the conversation at the cafeteria, was disallowed by the general assembly because of the mandatory participative format *in a plan.*

Some Lessons Learnt on the "Participative" and "Legislative" Legitimacy of Governing by Certification Standard

In this second part, we considered the principles and effective implementation of the substitute for the legislative function that governing through certification standards offers. Among the various modes of normativity that matter for members of RSPO to express their differing voices, the objective of a calculable world results in drastic selection to fit the standard. Indicators that make it measurable, and the format of *engaging in a plan* which is a prerequisite for the reduction to indicators, heavily constrain the organization of the dialogue and deliberation preparing decisions on the standard. In spite of the explicit domination of the format of the plan, formulated in the imperative of being "pragmatic", some participants go beyond this constraint. They articulate conflicting conceptions of the common good that raise political and economic justifications and criticisms, or air personal suffering because of the violation of their familiar environment. Even if they are regularly called to order, these strongly critical participants, issued from local concerned communities and relayed by NGOs, actively contribute to changes in the standard content and procedures. However, are the standard and procedures enforced?

A-TESTING, PRO-TESTING AND CON-TESTING: SUBSTITUTES FOR THE JUDICIAL SYSTEM IN PUTTING THE STANDARD ENFORCEMENT TO A CRITICAL TEST

In this mode of governing, how are deeds put to a critical test? What are the substitutes for the judicial bodies of a legal system? When criticizing the enforcement of regulations, how do concerned people (1) provide evidence issued from their knowledge and information (A-testing); (2) express objection (Pro-testing); (3) communicate differing views in dispute and conflict (Con-testing)?

Audit Procedures

The participatory setting framed by the "Open Space" operates de facto as an inquiry device since participants use it to *attest* evidence collected in their experience, in support of their argumentations and claims. However, the de jure process of inquiry is the expert auditing integrated in this government. Silva-Castañeda investigated the practical work of four out of the six audit firms that have certified RSPO companies operating in Indonesia (Silva-Castañeda, 2012). In her investigation, she cared to "create a climate of familiarity, taking a walk with villagers in their forest to be sensitive to their familiar engagement with the environment" (Silva-Castañeda, 2012). By contrast, the process of "rendering auditable"—by virtue of designing measurable procedures and performance—is a test of "the quality of internal [...] systems, rather than the quality of the product or service itself as specified in standards" (Power, 1997, p. 84).

Indicators listed in the RSPO standard are codified records of information requests, safety plans, emergency procedures, calendars, monitoring systems, and "companies must develop management systems that will enable them to demonstrate their compliance with the standard" (Silva-Castañeda, 2012). Because auditors regard the document as the ultimate form of evidence, "a lack of 'evidence' – in other words of *documents* – on the side of local communities stands in contrast to the companies' documentary arsenal" (Silva-Castañeda, 2012).

In one of the breakout sessions of the open space that we observed, smallholders and non-smallholders met on the topic of audit. Rather unexpectedly, two auditors joined the group and engaged in criticizing the debated limitations of their work. The self-critical awareness of the auditors allowed an outstanding critical exchange about this core device in the control of the standard enforcement. Because auditors belong to the main inspection body of this mode of governing, their taking part in the critical public debate creates tensions with their official function.

RSPO "Dispute Settlement Facility"

The "RSPO Dispute Settlement Facility" is intended to fulfil some of the jurisdictional functions. Without covering all the aspects of the pre-trial investigation of a case, registering a complaint is part of the critical inquiry we are interested in, in which smallholders are involved. Following our methodological grid, we shall consider the three main operations which compose the chain of transformation between the kind of wrong which the plaintiff suffers and the formal complaint which is duly registered. While *at-testing* begins with attending and points towards providing evidence, *pro-testing* addresses the critical statement to an interlocutor, a step towards the larger public needed in *con-testing*, which is constrained by the strong requirement of togetherness and based on the invested forms needed for a common format. In our fieldwork with Cheyns, we strived to follow the formulation of the complaint from a village in Indonesia close to Jambi, Batu Ampar, up to the RSPO settlement devices.

Contest: Formatting the Complaint in the Right Form for the Public
The needed transformation of the wrong in an official complaint is not a specific feature of this mode of governing. A legal case in court would also require to select "facts" and format them so that they would be taken into account in the procedure. Yet, the plaintiff's frustration, which often results from the formatting, turns out to be particularly severe in RSPO dispute settlement procedures, since complaints have to be strictly formulated in terms of the principles, criteria, indicators and guidelines of the standard. Just as they prepare communities to take the floor at roundtable public sessions—through role-playing exercises among other techniques (Cheyns, 2014)—various NGOs give them a hand in learning and carrying out the transformations of their harm into a standard complaint. It is worth noting that a range of distinct NGOs build up a chain that parallels the needed transformations to turn local knowledge into proper public information formats. Some Indonesian NGOs develop local and close links with smallholders, as Sawit Watch—also a member of the RSPO Executive Board—or Wahli. They help "growing a common cause in proximity" by "accommodating attachments" and emotions within convivial preparatory meetings with farmers and community representatives (Cheyns, 2014). Other international NGOs, such as Oxfam, play their major role within RSPO official bodies, moving forward new rights in the standard. Some of them, such as the Forest Peoples Programme, are able to navigate between the different levels and related *engagements*, from *familiarity* to *public justification* with reference to the common good, because they combine the scientific knowledge and methodology issued from anthropological scholarship and the skilled practice of international organizations (Colchester, 2002; Colchester & Chao, 2013; Cheyns & Thévenot, 2019b).

A first complaint was send to the RSPO DSF (Dispute Settlement Facility) by the village or Batu Ampar and written in Indonesian. Without formal template, it mixed a DSF category ("Effort taken to resolve the issues") with other ones. Gun, a local member of the Indonesian NGO Sawit Watch, helped villagers to write this first version. A second version still written in Indonesian was framed within the DSF Form, after the first complaint had been rejected because of its inappropriate format. The chronological narrative which was so significant for the history of the community was taken out. In addition to "Efforts have been taken to resolve the issues", the other two DSF categories, "Nature of Complaints" and "Supporting Evidence", were filled in. Nauli, who created the local branch of the Indonesian NGO Wahli (Wahli Jambi), gave a helping hand. Already an activist in college, he later organized demonstrations in labour regional unions. As a lawyer, he dealt with cases related to activism, students, labour cases and community cases. We observed this double training in activism and law among other intermediaries who have to compromise between different worlds. Nauli critically comments:

A lot of fuss about "sustainability", catchy phrase. No concern for local communities. No conflict resolution model. Commitment on paper, not in actual practices. Weak involvements of local communities. (Nauli interviewed by Cheyns in Jambi, 6 November 2014)

This second attempt still did not suit the needed format of the DSF complaint. Karlo, who is a less local member of the Sawit Watch NGO, rewrote in English a third version of the complaint. In the "Nature of complaints" category, he transformed the former list which mentioned under the heading "Land dispute" the items: "management system, partnership system, land grabbing". In the new list of grievances, he strictly connected each item to the numbered principle or criteria which were viewed as violated. Under the heading "Land dispute related to violation of the Principles 2.2 and 2.3", he wrote:

The right to use the land is demonstrated, and is not legitimately contested by local people who can demonstrate that they have legal, customary or user rights [2.2]; [...] Use of the land for oil palm does not diminish the legal customary or user rights of other users without their free, prior and informed consent [2.3].

In these successive versions, we see the progressive trans-formation of evidence, from formats that fit smallholders' experienced troubles faced in their engagements, to formats that rigorously stick to the objectives phrased in the headings of the standard.

Attest: Land Appropriation and Appropriate Evidence
The extension of industrial plantations damages a range of farmers' *engagements* with the land and the surrounding nature. Smallholders face great difficulty in bringing into the dispute the acceptable evidence of these damages and rights infringements. The reason is that RSPO Government depends on the formalization of modes of engagement to the land which stay remote from the various relations maintained and valued by villagers. *Liberal* property right is only one among many modes of appropriating land (Silva-Castañeda, 2015) and nature that make them proper to one's use or habitat, and maintain the kinds of dependency that cosmologies or mythologies convey (Breviglieri & Landoulsi, 2016). These modes, which suffer from the radical change provoked by industrial plantations, procedures and normative frameworks, are weakly taken into account by the standard.

The need for exclusive property and associated rights demands mapping and boundaries, between neighbouring communities in particular. A document—in Indonesian—produced by the Batu Ampar village community and distributed at the 2012 roundtable to document the "Case profile" with the questioned company describes the relationship to the land as follows:

> The methods used by the villagers in cultivating their farms and orchards still follow old methods of wise and traditional management, although some have adopted relatively modern tools and framing techniques. Communal work traditions (*gotong royong*) are still practiced on planting and harvest times for agricultural produce, including thanksgiving ceremonies and family parties/celebrations.

In 1916, the community moved across the river to its present location, the document states in this respect:

> [...] the boundaries of Batu Ampar customary territories were delineated in a customary *tembo*, an oral customary *seloko* (poem) which began to be recorded in a written form during the 1940–60s. The written record made in 1985 still forms the guidelines for the current boundaries of the Batu Ampar village.

Table 7.1   The local familiar and customary format of evidence

Tunggul Bungo Besar	Pemayang Tembesu	Tebat Patah	Pematang Belubang
A tree stump with a large flower growing on it/the stump of a large flower	A dry islet in the swamp with a *tembesu* tree on it	A broken/kinked dam or pond	A small island/embankment with a break of mole

Landmarks are based on such customary, local and even personal formats of evidence (see Table 7.1).

We observed in the field the "trans-formatting" chain artfully designed to overcome this handicap via proper devices. Currently engaging in *familiarity* with the natural environment, some villagers—not even the chief—had the familiar knowledge needed to identify local markers of territories, such as plant species or trees—when they were not already destroyed by new industrial plantations. The instrument of global positioning system (GPS) made possible the transformation of these markers into geographic coordinates (see Figs. 7.5a, b). This chain of inquiry translated local and situated formats of information, such as *familiar* landmarks, into formal proofs that would be accepted as publicly *justifiable* "evidence" for the standard. The villager's chief (the woman on the right in the photograph of Fig. 7.5b) was helped by an NGO member with a GPS who calculated "the coordinate points based on the Tembo names obtained from the Batu Ampar village, as well as the agreement between the Batu Ampar and the Karang Mendapo".[26]

Protest: Direct "Private" Interaction

A result of the mapping efforts that Batu Ampar villagers strongly struggled for was a decision [SK] by the local governing authority—the head of district [*Bupati*]—settling the debated boundary between their village (*Batu Ampar*) and the adjacent one (*Karang Mendapu*). It grounded their claim that their rights were infringed because of the faulty map that the company drew in 2001. This new piece of formal evidence was expected to be pivotal in their fight for their rights. In the DSF category "Efforts have been taken to resolve the issues" that was already filled in the first version of the complaint, they recorded three points in the last version: (1) Boundary checking into the field by District and Village Government; (2) Boundary determination by *Bupati* with the issuance

a

b

Fig. 7.5 The transformation of familiar-customary landmarks (**a**) into quantifiable space markers; (**b**) recorded by the NGO technician (left) and the village chief (right) (*Source* Photographs by Laurent Thévenot)

of the decision on the administrate border village; and (3) Asking the company for implementing the decision letter.

The head of the village, Yanti, came from her village to the RSPO annual meeting in Medan. Away from the public arenas, roundtable annual meetings offer the opportunity of "side events" that allow a direct confidential exchange between stakeholders for private negotiation. We managed to attend such a closed meeting between: a director of GAR (Golden Agri-Resources) headquartered in Singapore; the main palm oil company of one of the largest Indonesian conglomerates Sinar Mars[27]; which owns the local subsidiary company KDA in Jambi; the director of this subsidiary company KDA; the village chief Yanti; Karlo from NGO Sawit Watch; Nauli from NGO Wahli local branch in Jambi; Agun, a mediator from the consultant firm TFT (Tropical Forest Trust) who was paid by GAR. Local participants spoke Indonesian while a translator whispered the English translation to the GAR director who did not understand Indonesian. The photographs show the full roundtable (Fig. 7.6a); the GAR director with his assistant listening to the translation in English (Fig. 7.6b); Karlo with the camera recording the meeting (Fig. 7.6c);

Fig. 7.6 Photographs of the direct confrontation with the headquarters company directors. (**a**) (top photograph; right of the table, from left to right): Karlo (NGO Sawit Watch), Bondan (Sawit Watch), Yanti (Batu Ampar village chief), Nauli (NGO Wahli Jambi); (**b**) (bottom left): GAR (Singapore) director of communication; (**c**) (centre): Karlo; (**d**) (bottom right): Yanti (*Source* Photographs by Laurent Thévenot)

Yanti dressed up for the occasion and wearing the hijab (Fig. 7.6d, in contrast with her informal clothing and cap she had on in the field, see Fig. 7.5b).

Karlo (NGO Sawit Watch):
The Batu Ampar villagers have worked hard to resolve the boundary issue during the past year. I present documents describing the effort made to

resolve the boundary issue. [...] we brought this up to RSPO to try and pester the company to implement the government decision so that Batu Ampar can access its lands.

Yanti (chief of the village):

Our last meeting was stuck on boundary issues. In July this was resolved at the government level in favour of Batu Ampar, and I have requested KDA to implement this resolution and allow the Batu Ampar villagers to work the land. The company has not given a satisfactory answer. [...]

Agun (mediator from Tropical Forest Trust, paid by GAR):

The Bupati [head of district] office held the SK [decision] and demanded payment. TFT wouldn't engage with this since it involves bribery and we don't know how the SK [decision] finally got issued. [...]

Nauli (NGO Wahli, local branch in Jambi):

Can't KDA give evidence of progress that can be presented to the Batu Ampar community? Without progress Yanti's efforts may lose community support, so give us something to show. [...]

Karlo:

I don't quite understand the notion of the Bupati asking money for the SK. Does this have anything to do with the Batu Ampar request for funds for boundary mapping? [...]

Yanti:

I'd like to mention that I've borrowed money to perform the mapping and boundary delineation work.

Mediator:

Well, TFT doesn't want to know about bribery.

Yanti:

We're not talking about bribes! It's about the costs of the mapping work.

Yanti (to the director of this subsidiary Company KDA):

Would you dare to walk into the disputed area, since you said you're a Jambinese?

Nasir:

I would as a Jambinese, but that's not the problem.

Some Lessons Learnt on the "Judicial" Legitimacy of Governing by Certification Standards

In this third part dedicated to the ways standard enforcement is put to a reality test, we followed three general procedures for monitoring compliance with the standard and dealing with disputes about it, which have counterparts in the judicial system (complaint, judgement), its periphery (mediation) and other private or public government processes (audit). The three differ due to the formatting of the world they demand to make it calculable, and the disposition of the participants they involve, so that all beings "qualify" for the test and can be taken into account in the judgement. Audit is carried out in the company, most often on the basis of documents, and is therefore the most formal test and the farthest from a direct confrontation between protagonists. The Dispute Resolution process requires of the plaintiffs hard formatting work for the complaint to be admissible. Judgement includes a process of questioning the parties but not the adversarial confrontation organized by the judicial system. Finally, the mediation test is the least formal of the three. The parties in conflict are brought together face-to-face to display their disagreement on their conformity to the standard. It allows them to become involved in *regimes of engagement* that deviate from the format of the plan and the objective set out by the standard. As observed more generally in mediation (Cardoso de Oliviera, 2005), the openness to familiar engagement that this face-to-face relation makes possible is conducive to expressions of suffering or humiliation that complainants consider to be lost in the judicial process that sacrifices them in favour of qualified facts.

In order for the test to have public legitimacy, the bodies that guarantee the procedure should be impartial: auditors, RSPO members issuing a judgement, or mediators. In the three types of tests we examined, this impartiality is highly questionable. Such procedural faults reveal the imbalance of power between parties, which the *liberal* grammar and the *multi-stakeholder* matrix claim to overcome to the benefit of the juxtaposition of "stakes" in a horizontal dialogue around a "roundtable". In the first test, failure comes from the financing of audits by companies with resulting dependency and conflicts of interests. In the second test, a fault results from the choice among the "board" of the "judges" who rule on dispute resolutions, which was recently remedied. In the third test, the flaw ensues from the financing of the mediator by the company that is one of the parties in conflict. The consequences of this flaw are made visible in the meeting above when the mediator himself, and not one of the parties, undertook to disqualify the new major piece of evidence provided by the village chief: the delimitation of the village territory ratified by the local administrative authority, which contradicts the map produced by the company. This disqualification on grounds of "bribery", without any supporting evidence, is an interference by the mediator to which the village chief, Yanti, replied firmly. As a young Muslim nurse, she demonstrated her ability to shift the regime of engagement with great versatility. Although adjusted by her formal dressing and composure to a public space, she was also able, in the presence of managers, to adopt familiar expressiveness and gestures (Fig. 7.6d) of indignation that we also observed on Arifin (Fig. 7.3b).

In the continuation of this move towards familiarity, she challenged the local company's manager to come on site. The request to "come on site", which contrasts most radically with the detachment of the judgement from the contextual situation, and even more with abstraction through numbers, is often expressed by critics to demonstrate that the chain of transformation of their testimony has, in the end, led to the sacrifice of what was most important to them and affected or threatened (Richard-Ferroudji, 2011; Thévenot et al., 2000; Thévenot, 2019a).

Unlike the closure of the judicial system, which is due to its heavy normative equipment, jurisprudence accumulated over time and with support from its professional bodies, the most recent RSPO system is relatively more open to varied modes of normativity. They are either explicitly transferred into the standard (production methods, national and international law, indigenous peoples' rights, customary law, etc.) or advanced via criticism and interventions by legitimate *stakeholders*. The diversity of these stakeholders allows disputes to question the lack—or reduction—of certain modes of normativity even if they eventually undergo a significant transformation to enter the standard.[28] The most vulnerable actors, with the support of an architecture of NGOs that fit the stakeholders' various engagement formats, from local proximity to dealings with international multinationals, seek compromises, through inquiry and critical devices, with such governing through objectives whose limitation they are aware of. With one foot in and one foot out, they also engage in other modes of attesting, protesting and contesting, such as unionization or street demonstration in relation to the 2013 RSPO convention.

Discussion of the Certified Objectivity Sought by a "Standardizing Liberalism": Power-Knowledge and the Enlarged Analysis of Oppression and Criticism

Quantification studies are part of research on objectivity which historians of science nurtured (Daston & Galison, 2007; Porter, 1995). The social sciences have a long tradition of critical deconstruction of claims to objectivity, fostered by social constructivism, ANT (Actor-Network Theory) and STS (Science and Technology Studies). Just as STS authors concerned with the coordination power of "regulatory objectivity" (Cambrosio et al., 2009), our research agenda adopts a pragmatist approach to the study of the politics of quantification. More precisely, it investigates the relations between quantified objectivity and modes of governing that make the world calculable. In contrast to a range of politics that govern by quantifying the individuals themselves (Thévenot, 2011a, 2019b), be they state policies based on survey statistics, or organizational ones that digitally track individuals, or the quantified "social credit" which Tong Lam introduces in this volume and which combines the central control of the Chinese state with the data mining of digital individual tracks, or even the "quantified self" movement that Uwe Vormbusch examines in this volume, the politics considered in this chapter govern human beings indirectly through objects, via the transnational voluntary certification of market goods. In spite of its material basis and "private" character, this form of government takes on values that are generally under the responsibility of the "public" government of people based on the rule of law. As pointed out by Andrei Guter-Sandu and Andrea Mennicken in their contribution to this volume, "the boundaries between the public and private are blurred and/or reworked". The *liberal* grammar that stays in the background of these new modes of governing indeed facilitates the link with *market* coordination and contributes to this blurring. Yet, research attention should be given to the consolidation of this link by the standardization of goods—a process which is, in itself, neither *liberal*, nor implicated in *market* competition. Rather it evolved into a new kind of "standardizing liberalism" (Thévenot, 1997) which has expanded on a world scale.

Marc Breviglieri analysed at large this expansion in the domain of "the guaranteed city" (Breviglieri, 2018), showing the formatting it brings about in the environment that stands close to a person's singular intimacy—the habitat—and supports the construction of commonality built on cohabitation. The array of labels, accreditations and certifications that guarantee a "smart city", "inclusive city", "global active city", "what works cities", etc., transforms what Breviglieri calls the "architectures of usages" which, by contrast, rely on familiarized dependencies between inhabitants and the spaces they dwell in. This transformation produces certified options that allow the choice of opting liberal individuals, and market coordination as well. In her comprehensive research on "halal" certification (Bergeaud-Blackler & Kokoszka, 2017; Bergeaud-Blackler et al., 2016), Florence Bergeaud-Blackler demonstrated that the investments in forms and measurement conventions required by this standardization led to a *dispositif* that combines *inspiration*, *market* and *industrial* orders of worth to reach a stabilized compromise and create a certifiable halal quality of goods and services. She showed the effect of this reduction of faith to a measurable and certifiable quality: it reinforces, on a large scale and insidiously via the *market*, a literalist and fundamentalist conception of religion (see also Stavo-Debauge, 2018).

The European government of education (see also Corine Eyraud's contribution in this volume), health, social work and employment—i.e. core policies of the former welfare state—also display this process of "standardizing liberalism". What are the similarities and differences with the certification we studied? Although standards are central (Landri, 2016; Normand, 2016a; Timmermans & Berg, 2003; Normand, 2010), as well as good practices (Normand & Derouet, 2016) in these post-welfare policies, both are grounded in "politics of evidence" (Normand, 2016b, 2020) which do not have the same role in the RSPO certification standard. Decision-making processes also diverge, since the European policies combine elements of state legal systems with experts from influential transnational think-tanks and private firms, all of them being connected in networks through which normative and evaluative tools "travel" (Lawn, 2013). Yet, in spite of the links with representative democracy, this combination does not appear to be much more favourable to critical questioning (Bruno, 2016) than the RSPO standard.

Research on these different policies, and "standardizing liberalism" more generally, provides an important meeting place for the two research approaches on standardization and quantification brought together in this volume (see also Hansen, 2016, 2017; Normand, 2016a). This volume enables a dialogue between two long-standing research agendas on quantification, which until hitherto, with some notable exceptions, particularly more recently, have not interacted that much: Foucault inspired studies of quantification, on the one hand, and French works on the "Economie des conventions", modes of justification and orders of worth, on the other hand. The dialogue engendered here helps elucidate convergences and divergences between these approaches.[29]

The first point of convergence results from the significance of words-things connections already unfolded in Foucault's *Les mots et les choses* (Foucault, 1966), before STS and ANT paid attention to materiality. Being concerned with the notion of "qualification" in our analysis, these became also important in Foucault oriented studies of quantification which extended from accounting to material arrangements, such as those studied in the "spatial reordering of the manufacture" by Peter Miller and Ted O'Leary (1994); Foucault's dispositif being a shared inspiration for both.

In contiguity with the previous connection through Foucault's micro-power of dispositifs, the power-knowledge connection (Foucault, 1985 [1984], 1995 [1975]) is a second central issue in both research streams, although grasped differently. In Foucault inspired governmentality studies, the "administering of lives" (Mennicken & Miller, 2014) and managing at a distance through the "conduct of conduct" are key issues. In the other approach, in-forming, trans-forming and formatting through invested conventional forms are central operations, because they sustain coordination power under uncertainty.

A third shared concern—clearly visible in this chapter—is the move beyond the state, but also beyond the "neoliberal" as an all-encompassing notion. But also here the main categories used to achieve this move differ. Compared to the Foucauldian historical-genealogical approach aimed at studying and unpacking regimes of political rationality, the "Economie des conventions" approach is interested in the critical pluralism of modes of evaluation constituted by conventions, orders of worth and valued regimes of engagements. While "programmes" (of governing) are the main empirical objects in the Foucauldian tradition to scrutinize management reforms, breaking down policies and politics into a variety of valued modes of coordination is at stake in the other tradition, with special attention given to the tensions between most public conventions and most personally convenient modes of engaging, and the chain of transformation involved when shifting from familiar attachments to public qualifications (Nielsen, 2015).

Shared concern and dialogue might intensify on research objects that bring to light changes in governmentality and modes of governing in response to former waves of criticism. Mennicken and Miller[30] noted that research on modes of "exercising voice" is all the more needed today, as for instance the language of NHS health policy guidelines they study shifted from an earlier focus on "satisfaction", to "choice" and now towards "voice", with even attention paid to "formal and informal knowledge ('soft intelligence')" and "relationship knowledge". In their contribution on quantification in the English Prison Service in this volume, Guter-Sandu and Mennicken also come to take into account a diversity of values that quantification gives room to, including the rise of measures of decency in response to former criticisms.

Accounting for differing voices, their evaluative orientations, the formats of the evidence that qualifies for claims according to grammars of commonality, are a main strength of the extended Convention Theory research programme. Following the processes of attesting, protesting and contesting brought new insights into politics, the ways people take part in the polity and express a differing voice (Luhtakallio & Thévenot, 2018). Many contributors to this volume share a concern for critical voices and what quantification does to them. They follow the process of investigation, construction of the categories and the information used to quantify and govern, considering both structural domination and the participation of the persons in the organization and criticism of this domination, with a possible enhancement of citizens' capabilities (De Leonardis et al., 2012). Fostering on this point the dialogue between the two research agendas nurtured by this volume, Vormbusch recalls in his chapter that, while the sociology of critique has been criticized for overlooking the historically specific restrictions limiting the very possibility for critique, Foucault has been accused of neglecting human agency. Regarding self-quantifying practices as "an investigative praxis [...] *without* neglecting the discourse of power", Vormbusch shifts the analytical angle towards "the participants' agency and their capacities of critique".

In her contribution to this volume, Ota de Leonardis brings to light a "semantic shift" towards "spatialization" of inequality that obscures the "political issues of power" and the "vertical political architecture of modernity", because of a new ideological "dream of a domination free from any bond with the dominated". This statement converges with the changes that result from the new calculable global world that this chapter is dedicated to. Sharing the concern of the governmentality agenda for a more precise analysis of what is usually covered by the extensive term "neoliberal", we identified three distinct components whose combination obscures dependency between individuals and domination bonds.

The first component is political liberalism, currently presented as "horizontal" politics that free autonomous individuals from hierarchies and the kind of hierarchical bonds that we modelled in orders of worth.

Yet, the analysis of the *liberal* grammar shows that, as any grammar of commonality, it creates a dependency between human beings because of the coordination power provided by the format of commonality, and the dominated situation of those who are not in a position to engage in this format. The undeniable emancipatory empowerment of the liberal grammar has a price: all engagements are to be transformed into individual choices—or preferences—for common knowledge options. Even intimate bodily engagements involved in sexual and love relationships, for which this liberal emancipation is widely recognized, have to be transformed into common knowledge options designated as "sexual preferences".

The second component is the *market* competition convention of coordination. It is distinct from the *liberal* grammar because of its two dominating common forms: money and price. Yet the connection between liberal politics and market competition is easily made possible by confusing options with market goods, and "choosing" with "buying".

The third component strengthens this connection by encompassing in market goods a wide variety of conceptions of the common good, or fundamental rights, as certifiable and assessable properties of these goods. In an unprecedented extension of the commodity fetishism that Marx exposed, the combination of these three components reinforces the illusion that human beings are freed from dependency and hierarchy bonds and only connected through a world of things, becoming independent opting individuals facing an array of secured options. Yet, unless critical capacity recovers from the illusions of an assessable world, as we have seen among some "smallholders" and NGOs, under the pressure of the new calculable world people's temper might become incalculable.

Acknowledgements In addition to discussions with other authors of this volume, I am grateful to Ota de Leonardis and Robert Salais for their constructive comments on an earlier version of this chapter, and Andrea Mennicken for her careful reading and relevant suggestions.

NOTES

1. Because of our attention to conflicting valuations in calculability, we depart from other uses of the term in literature, such as Callon and Muniesa (2006).

2. This approach is particularly suited to relate statistics with the history of the new state of the Soviet Union, as demonstrated by Martine Mespoulet in her chapter in this volume (see Chapter 2, "Creating a Socialist Society and Quantification in the USSR").

3. Wendy Espeland developed an important and influential research agenda on "commensuration" (Espeland & Stevens, 1998), without linking this programme explicitly to the study of the modes and politics of producing equivalence.

4. For recent comparative research, conducted in the same spirit, on changes in lifelong learning systems among five European countries (Germany, Denmark, France, Sweden, UK), see Verdier (2017). Magnus Paulsen Hansen (2017) pragmatically followed the processes of justification and critique, emancipation and coercion, as well as quantified modes of "trying the unemployed", in comparative research on measurement, measures and evaluation that compose contemporary unemployment reforms in Denmark and France aimed at the "Active Society".

5. For a recent overview on the sociology of quantification, see Diaz-Bone and Didier (2016).

6. Directed by Emmanuelle Cheyns (CIRAD), this research project was funded by the Agence Nationale de la Recherche (ANR) (grant No. ANR-11-CEPL-0009). See Cheyns (2014, 2011), Cheyns and Thévenot (2019a), Thévenot (2018). See also the research carried on by Laura Silva-Castañeda (2012).

7. I do not refer here to the precisely delimited concept of "voice" that Albert Hirschman distinguished from "loyalty" and "exit" (Hirschman, 1970), but to the sequence of operations of attesting, protesting and contesting (see section "A-testing, Pro-testing and Con-testing: Substitutes for the Judicial System in Putting the Standard Enforcement to a Critical Test" of this chapter).

8. A large part of the original article on "social coding" (Thévenot, 1983) has been made available in English (Thévenot, 2016) with additional comments benefiting from thirty years' experience with the analysis of investments in qualification and quantification, unfolded in Convention Theory.

9. On the relations with Bourdieu's approach to [symbolic] forms, see Thévenot (2011b, 2016); to Simmel's forms, see Thévenot (2017).

10. Foucault's *Les mots et les choses* ("Words and things", translated as *The order of things* (see Foucault, 1971 [1966])) also exerted a significant influence over our initial research, as evidenced in the title of the article presenting our research on classifications, "Words and numbers [*les mots et les chiffres*]: socio-professional nomenclatures" (Desrosières &

Thévenot, 1979). Because of the close attention Foucault paid to power-knowledge and the variety of forms of bringing together [*rapprochement*], his influence departed from that of Bourdieu.

11. For more detailed presentations of this avenue of research in English and its continuation in Convention Theory and French Pragmatic Sociology, see Amossé (2013), Desrosières (2011), Diaz-Bone (2016), Thévenot (2016).

12. A new generation of sociologists well versed in quantitative methods replicated and elaborated further the experimental games: Amossé (2013), Deauviau et al. (2014), and Penissat et al. (2015).

13. Any unequal ordering or ranking among human beings threatens an ideal of a *common humanity* which has been endorsed by various moral, religious and political principles. We found that two distinctive features—shared by all orders of worth but not all repertoires of evaluation—are intended to appease this threat: (1) the "superiority" of more *worthy* should benefit all, as far as their *worth* is linked to a construction of the *common good*; and (2) this "superiority" or unequal qualification for worth should not be permanently ascribed to their persons but regularly *put to the reality test* of effective coordinated actions.

14. For a wide view on the future and expectations, see Beckert (2011, 2016).

15. Robert Salais and Michael Storper devised a pluralist approach to the organization of economic activity which has some family resemblance with orders of worth but aims at integrating production and market. It differentiates "worlds of production" through institutions, social practices and conventions, which coordinate economic actors by defining specific "frameworks of economic action" (Storper & Salais, 1997). Quite independently, Roger Friedland and Robert Alford have developed another pluralist approach of organizations in terms of "institutional logics" (Friedland & Alford, 1991) which became influential in the English literature on organizational studies, and only recently came into dialogue with Convention Theory and Pragmatic Sociology (Brandl et al., 2014; Cloutier & Langley, 2013; Patriotta et al., 2011).

16. François Hartog underlined the contemporary "presentism" by situating it in a succession of "regimes of historicity" (Hartog, 2003).

17. On the criticism of "flatland", see De Leonardis (2008). See also our concluding critical discussion in this chapter on this point.

18. In his contribution to this volume, Thomas Amossé examines the continuation of this history and the recent rise of "matched panels" techniques involved in a quasi-experimental evaluation of incentive policies targeted at specific populations, and intended to establish "purified causalities" of "good variables" on individuals' behaviours.

19. *Engaging in familiarity* hardly achieves a mutuality which remains partial. It is obtained by communicating through personally and emotionally

invested "common-places" of extremely unequal scales, from the closest level of intimacy in love or friendship to wide communities. The good of ease in familiarity can turn into evil when oppressing other engagements, or made instrumental in populist politics. On these issues, see Thévenot (2014, 2020).

20. The sociological use of the "taken-for-granted" of "lifeworld" mainly sticks to this first side of *engaging in familiarity* usually termed "routine".

21. A meta-norm principle borrowed from the "International Social and Environmental Accreditation and Labeling" (ISEAL), which enacts standards of standards and defines "what credibility looks like for standard systems", is called "engagement" and introduces this governance principle with the following definition: "Engagement. Standards-setters engage a balanced and representative group of stakeholders in standards development. Standards systems provide meaningful and accessible opportunities to participate in governance, assurance and monitoring and evaluation. They empower stakeholders with fair mechanisms to resolve complaints" (see https://www.isealalliance.org/credible-sustainability-standards/iseal-credibility-principles, accessed 16 June 2020).

22. On the reduction of European Community government and evaluation to "guidelines" which are themselves reduced to quantitative "indicators", with the detailed case of the guideline "Ensure inclusive labour markets", see Salais (2006, 2017). More on the distance between the political rhetorical justification one can see in the wording of guidelines and the effective policies that are driving their monitoring through performance indicators can be found in Salais' contribution to this volume. On benchmarking for state policies, see also Bruno and Didier (2013).

23. In the 2013 changes, forty new indicators were added, such as the quantified "4.4.4. mill water use per ton of Fresh Fruit Bunches (FFB)", "4.6.2. records of pesticide use", "5.4.1. renewable energy use and fossil fuel use per ton or Crude Palm Oil (CPO)".

24. On the normative principle of "Free, Prior and Informed Consent", its anchoring in the *liberal* grammar of individual choice, and possible or contingent extension to peoples' collective consent, see Cheyns and Thévenot (2019a, 2019b).

25. This sense of curiosity to others is nowadays frequently observed in norms and practices of US liberal communities (Berger, 2012; Eliasoph, 2011). Curiously *engaging in exploration* converges with the key insistence on "experience" that was at the heart of Dewey's pragmatism.

26. Our young Javanese translator who accompanied us in the field, although moved by a curiosity as strong as ours, was not able to understand and thus translate without ambiguity the wording of the landmarks. He was obviously lacking the *familiar engagement* with the places thus named and pointed to by the villagers.

27. After a Greenpeace campaign in 2010, Burger King, Unilever and Nestlé cancelled their supplier contracts with GAR subsidiaries due to unsustainable farming practices. GAR adopted afterwards a *zero-deforestation policy* which required Greenpeace, GAR and a consultancy firm to develop a tool to codify and quantify forest called the High Carbon Stock (HCS) approach. Because of the pre-eminence of the *liberal* grammar, it still rests on the format of relations to the land and bargaining negotiations that raise criticisms similar to those reported here (Cheyns et al., 2020).

28. On the opening of legal studies to a wider variety of modes of normativity, see the special issue in English of *La Revue des Droits de l'Homme* dedicated to "Modes on normativity and normative transformations", edited and introduced by Véronique Champeil-Desplats, Jérome Porta and Laurent Thévenot (Champeil-Desplats et al., 2019). On the "'transformation' of social rights [transferred] under modes of normativity other than those of human rights", such as objectives, programmes, indicators, standards, see in this special issue Porta (2019).

29. In addition to the numerous meetings that were held in connection with the production of this book, Andrea Mennicken and Peter Miller hospitably organized an additional side-meeting at the LSE in April 2017, with Uwe Vormbusch, initiator and go-between, and myself. This meeting nurtured my comments that still do not pretend to synthetize the generous and considerate conversations we then had.

30. Exchange during the London meeting.

REFERENCES

Affichard, J. (Ed.). (1987 [1977]). *Pour une histoire de la statistique: Contributions (Vol. 1)*. INSEE & Economica (première édition: INSEE 1977).

Affichard, J. (Ed.). (1987 [1983]). *Pour une histoire de la statistique: Matériaux (Vol. 2)*. INSEE & Economica (première edition: INSEE 1983).

Amossé, T. (2013). La nomenclature socio-professionnelle: une histoire revisitée. *Annales, histoire, sciences sociales, 68*(4), 1039–1075.

Auray, N. (2011). Les technologies de l'information et le régime exploratoire. In P. Van Andel & D. Boursier (Eds.), *La serendipité dans les arts, les sciences et la décision* (pp. 329–343). Editions Hermann.

Balineau, G., & Dufeu, I. (2010). Are fair trade goods credence goods? A new proposal, with French illustrations. *Journal of Business Ethics, 92*, 331–345.

Beckert, J. (2011). The transcending power of goods: Imaginative value in the economy. In J. Beckert & P. Aspers (Eds.), *The worth of goods: Valuation and pricing in the economy* (pp. 106–128). Oxford University Press.

Beckert, J. (2016). *Imagined futures: Fictional expectations and capitalist dynamics*. Harvard University Press.

Bergeaud-Blackler, F., Fischer, J., & Lever, J. (2016). *Halal matters: Islam, politics and markets in global perspective*. Routledge.

Bergeaud-Blackler, F., & Kokoszka, V. (2017). La standardisation du religieux. La norme halal et l'extension du champ de la normalisation. *Revue du MAUSS, 49*(1), 62–91.

Berger, M. (2012). La démocratie urbaine au prisme de la communauté. Effervescence, emphase et répétition dans la vie civique à Los Angeles. *Participations, 3*(4), 49–77.

Bohm, D. (1996). *On dialogue*. Routledge.

Boisard, P., & Letablier, M.-T. (1989). Un compromis d'innovation entre tradition et standardisation dans l'industrie laitière. In L. Boltanski & L. Thévenot (Eds.), *Justesse et justice dans le travail* (pp. 209–218). Cahiers du Centre d'Etudes de l'Emploi n°33, Presses Universitaires de France.

Boltanski, L. (1970). Taxinomies populaires, taxinomies savantes; les objets de la consommation et leur classement. *Revue française de sociologie, 11*, 34–44.

Boltanski, L. (1987 [1982]). *The making of a class: Cadres in French society*. Cambridge University Press.

Boltanski, L., & Thévenot, L. (1983). Finding one's way in social space; a study based on games. *Social Science Information, 22*(4/5), 631–679.

Boltanski, L., & Thévenot, L. (1987). *Les économies de la grandeur*. Presses Universitaires de France et Centre d'Etude de l'Emploi.

Boltanski, L., & Thévenot, L. (2006 [1991]). *On justification: Economies of worth*. Princeton University Press (French edition, 1991).

Bourdieu, P. (1976 [1972]). *Outline of a theory of practice*. Cambridge University Press.

Bourdieu, P., & Boltanski, L. (1975). Le titre et le poste: rapports entre le système de production et le système de reproduction. *Actes de la recherche en sciences sociales, 1*(2), 95–107.

Bowker, G. C., & Star, S. L. (2000). *Sorting things out: Classification and its consequences*. MIT Press.

Brandl, J., Daudigeos, T., Edwards, T., & Pernkopf-Konhäusner, K. (2014). Why French pragmatism matters to organizational institutionalism. *Journal of Management Inquiry, 23*(3), 314–318.

Breviglieri, M. (2018). The guaranteed city: The ruin of urban criticism. In J. M. Resende, A. C. Martins, M. Breviglieri, & C. Delaunay (Eds.), *The challenges of communication in a context of crisis* (pp. 220–227). Cambridge Scholars Publishing.

Breviglieri, M., & Landoulsi, I. (2016). Gestes publics d'appropriation et phénoménalisation progressive d'une ambiance. Autour d'un source sacrée intégrée à la réhabilitation d'une médina (Tiznit, Maroc). In N. Rémy & N. Tixier (Eds.), *Ambiances, tomorrow: Proceedings of the 3rd International*

Congress on Ambiances (pp. 105–110). International Network Ambiances, University of Thessaly.

Brown, J. (2002). *The world café: A resource guide for hosting conversations that matter.* Whole Systems Associates.

Brown, J., Isaacs, D., & World Café Community. (2005). *The world café: Shaping our futures through conversations that matter.* Berrett-Koehler.

Bruno, I. (2016). Silencing the disbelievers: Games of truth and power struggles around fact-based management. In R. Normand, & J.-L. Derouet (Eds.), *A European politics of education: Perspectives from sociology, policy studies and politics* (pp. 140–154). Routledge.

Bruno, I., & Didier, E. (2013). *Benchmarking: L'Etat sous pression statistique.* Zones.

Busch, L. (2011). *Standards: Recipes for reality.* MIT Press.

Callon, M., & Muniesa, F. (2006). Les marchés économiques comme dispositifs collectifs de calcul. *Réseaux, 122,* 189–223.

Cambrosio, A., Keating, P., Schlich, T., & Weisz, G. (2009). Biomedical conventions and regulatory objectivity. *Social Studies of Science, 39*(5), 651–664.

Cardoso de Oliviera, L. R. (2005). *Droit légal et insulte morale. Dilemmes de la citoyenneté au Brésil, au Québec et aux États-Unis.* Les Presses de l'Université Laval.

Centemeri, L. (2015). Reframing problems of incommensurability in environmental conflicts through pragmatic sociology: From value pluralism to the plurality of modes of engagement with the environment. *Environmental Values, 24*(3), 299–320.

Champeil-Desplats, V., Porta, J., & Thévenot, L. (2019). Introduction: A cooperative and transversal research experience between law and the social sciences. *La Revue des droits de l'homme, 16,* 1–14. https://journals.openedition.org/revdh/7132.

Charles, J. (2012). Les charges de la participation. *SociologieS.* https://sociologies.revues.org/4151.

Charles, J. (2016). *La participation en actes. Entreprise, ville, association.* Desclée de Brouwer.

Cheyns, E. (2011). Multi-stakeholder initiatives for sustainable agriculture: Limits of the 'inclusiveness' paradigm. In S. Ponte, P. Gibbon, & J. Vestergaard (Eds.), *Governing through standards: Origins, drivers and limitations* (pp. 210–235). Palgrave Macmillan.

Cheyns, E. (2014). Making 'minority voices' heard in transnational roundtables: The role of local NGOs in reintroducing justice and attachments. *Agriculture and Human Values, 31*(3), 439–453.

Cheyns, E., & Ponte, S. (2018). Convention theory in Anglophone agro-food studies: French legacies, diffusion and new perspectives. In G. Allaire & B.

Daviron (Eds.), *After the great transformation: Between ecology and capitalism* (pp. 71–94). Routledge.

Cheyns, E., Silva-Castañeda, L., & Aubert, P.-M. (2020). Missing the forest for the data? Conflicting valuations of the forest and cultivable lands. *Land Use Policy, 96*, 1–10.

Cheyns, E., & Thévenot, L. (2019a). Government by certification standards: The consent and complaints of affected communities. *La Revue des droits de l'homme, 16*(online since 5 July 2019). https://doi.org/10.4000/revdh.7156.

Cheyns, E., & Thévenot, L. (2019b). Interview with Marcus Colchester, founder of the NGO Forest Peoples Programme, on the 'Free, Prior and Informed Consent' of communities. *La Revue des droits de l'homme, 16*. https://doi.org/10.4000/revdh.6894.

Cloutier, C., & Langley, A. (2013). The logic of institutional logics: Insights from French pragmatist sociology. *Journal of Management Inquiry, 22*(4), 1–21.

Colchester, M. (2002). Indigenous rights and the collective conscious. *Anthropology Today, 18*(1), 1–3.

Colchester, M., & Chao, S. (Eds.). (2013). *Conflict or consent? The oil palm sector at a crossroads.* FPP, Sawit Watch and TUK INDONESIA.

Darby, M. R., & Karni, E. (1973). Free competition and the optimal amount of fraud. *Journal of Law and Economics, 16*(1), 67–88.

Daston, L., & Galison, P. (2007). *Objectivity.* MIT Press.

De Leonardis, O. (2008). Nuovi conflitti a Flatlandia. In G. Grossi (Ed.), *I conflitti contemporanei. Contrasti, scontri e confronti nelle società del III millennio* (pp. 5–21). UTET.

De Leonardis, O., Negrelli, S., & Salais, R. (Eds.). (2012). *Democracy and capabilities for voice: Welfare, work and public deliberation in Europe.* Peter Lang.

Deauvieau, J., Penissat, E., Brousse, C., & Jayet, C. (2014). Les catégorisations ordinaires de l'espace social français. Une analyse à partir d'un jeu de cartes. *Revue française de sociologie, 55*(3), 411–457.

Desrosières, A. (1987 [1977]). Eléments pour l'histoire des nomenclatures socio-professionnelles. In J. Affichard (Ed.), *Pour une histoire de la statistique* (pp. 155–231). INSEE & Economica.

Desrosières, A. (1987 [1983]). Des métiers aux classifications conventionnelles: l'évolution des nomenclatures professionelles depuis un siècle. In J. Affichard (Ed.), *Pour une histoire statistique* (pp. 35–56). INSEE & Economica.

Desrosières, A. (1998 [1993]). *The politics of large numbers: A history of statistical reasoning.* Harvard University Press.

Desrosières, A. (2011). The economics of convention and statistics: The paradox of origins. *Historical Social Research, 36*(4), 64–81.

Desrosières, A., & Thévenot, L. (1979, April). Les mots et les chiffres: les nomenclatures socio-professionnelles. *Economie et statistique, 110*, 49–65.

Desrosières, A., & Thévenot, L. (1988). *Les catégories socioprofessionnelles*. La Découverte.

Diaz-Bone, R. (2011). Discourse conventions in the construction of wine qualities in the wine market. *Economic Sociology: The European Electronic Newsletter, 14*(2), 46–53.

Diaz-Bone, R. (2016). Convention theory, classification and quantification. *Historical Social Research, 41*(2), 48–71.

Diaz-Bone, R., & Didier, E. (2016). Introduction: The sociology of quantification—Perspectives on an emerging field in the social sciences. *Historical Social Research, 41*(2), 7–26.

Durkheim, E., & Mauss, M. (2009 [1903]). *Primitive classification*. Routledge (translated with an introduction by Rodney Needjam).

Eliasoph, N. (2011). *Making volunteers: Civic life after welfare's end*. Princeton University Press.

Eranti, V. (2018). Engagements, grammars, and the public: From the liberal grammar to individual interests. *European Journal of Cultural and Political Sociology, 5*(1–2), 42–65.

Espeland, W. N., & Stevens, M. L. (1998). Commensuration as a social process. *Annual Review of Sociology, 24*, 313–343.

Eymard-Duvernay, F. (1986). La qualification des produits. In R. Salais, & L. Thévenot (Eds.), *Le travail. Marché, règles, conventions* (pp. 239–247). INSEE-Economica.

Eymard-Duvernay, F. (1989, March). Conventions de qualité et pluralité des formes de coordination. *Revue Economique, 2*, 329–359.

Eymard-Duvernay, F. (2002). Conventionalist approaches to enterprise. In O. Favereau & E. Lazega (Eds.), *Conventions and structures in economic organization: Markets, networks and hierarchies* (pp. 60–78). Edward Elgar.

Foucault, M. (1966). *Les mots et les choses*. Editions Gallimard.

Foucault, M. (1971 [1966]). *The order of things: An archaeology of the human sciences*. Pantheon Books.

Foucault, M. (1985 [1984]). *The history of sexuality. Volume 2*. Penguin.

Foucault, M. (1995 [1975]). *Discipline and punish: The birth of the prison*. Vintage Books.

Friedland, R., & Alford, R. R. (1991). Bringing society back in: Symbols, practices, and institutional contradictions. In W. W. Powell & P. J. DiMaggio (Eds.), *The new institutionalism in organizational analysis* (pp. 232–263). University of Chicago Press.

Hansen, M. P. (2016). Non-normative critique: Foucault and pragmatic sociology as tactical re-politicization. *European Journal of Social Theory, 19*(1), 127–145.

Hansen, M. P. (2017). *Trying the unemployed: Justification and critique, emancipation and coercion towards the 'active society'. A study of contemporary reforms in France and Denmark*. Copenhagen Business School.

Hartog, F. (2003). *Régimes d'historicité. Présentisme et expériences du temps*. Seuil.

Hirschman, A. O. (1970). *Exit, voice and loyalty: Responses to decline in firms, organizations and states*. Harvard University Press.

Kantorowicz, E. H. (1997 [1957]). *The king's two bodies: A study in mediaeval political theology (with a new preface by William Chester Jordan)*. Princeton University Press.

Lamont, M., & Thévenot, L. (Eds.). (2000). *Rethinking comparative cultural sociology: Repertoires of evaluation in France and the United States*. Cambridge University Press.

Lampland, M., & Star, S. L. (Eds.). (2009). *Standards and their stories: How quantifying, classifying, and formalizing practices shape everyday life*. Cornell University Press.

Landri, P. (2016). Standards and standardisation in European politics of education. In R. Normand & J.-L. Derouet (Eds.), *A European politics of education: Perspectives from sociology, policy studies and politics* (pp. 13–30). Routledge.

Lawn, M. (2013). Voyages of measurement in education in the twentieth century: Experts, tools and centres. *European Educational Research Journal, 12*(1), 108–119.

Luhtakallio, E., & Thévenot, L. (2018). Politics of engagement in an age of differing voices. *European Journal of Cultural and Political Sociology, 5*(1–2), 1–11.

Mennicken, A., & Miller, P. (2014). Foucault and the administering of lives. In P. S. Adler, P. du Gay, G. Morgan, & M. I. Reed (Eds.), *The Oxford handbook of sociology, social theory, and organization studies: Contemporary currents* (pp. 11–38). Oxford University Press.

Miller, P. (1992). Accounting and objectivity: The invention of calculating selves and calculable spaces. *Annals of Scholarship, 9*(1–2), 61–86.

Miller, P. (2001). Governing by numbers: Why calculative practices matter. *Social Research, 68*(2), 379–396.

Miller, P., & O'Leary, T. (1994). Accounting, 'economic citizenship' and the spatial reordering of manufacture. *Accounting, Organizations and Society, 19*(1), 15–43.

Miller, P., & Rose, N. (2008). *Governing the present: Administering economic, social and personal life*. Polity.

Monso, O., & Thévenot, L. (2010). Les questionnements sur la société française pendant quarante ans d'enquêtes *Formation et Qualification Professionnelle*. *Economie et statistique, 431–432*, 13–36.

Nielsen, M. H. (2015). Acting on welfare state retrenchment: In-between the private and the public. *International Journal of Sociology and Social Policy, 35*(11/12), 756–771.

Normand, R. (2010). Expert measurement in the government of lifelong learning. In E. Mangenot & J. Rowell (Eds.), *What Europe constructs: New sociological perspectives in European studies* (pp. 225–242). Manchester University Press.

Normand, R. (2016a). *The changing epistemic governance of European education: The fabrication of the Homo Academicus Europeanus?* Springer International Publishing Switzerland.

Normand, R. (2016). What works?': From health to education, the shaping of the European policy of evidence. In K. Trimmer (Ed.), *Political pressures on educational and social research* (pp. 25–40). Routledge.

Normand, R. (2020). The politics of metrics in education: A contribution to the history of the present. In G. Fan & T. Popkewitz (Eds.), *International handbook on educational policy studies* (pp. 345–361). Springer and Shanghai Education Press.

Normand, R., & Derouet, J.-L. (Eds.). (2016). *A European politics of education: Perspectives from sociology, policy studies and politics.* Routledge.

Owen, H. (2008). *Open space technology: A user's guide* (3rd ed.). Berrett-Koehler.

Patriotta, G., Gond, J.-P., & Schultz, F. (2011). Maintaining legitimacy: Controversies, orders of worth and public justifications. *Journal of Management Studies, 48*(8), 1804–1836.

Penissat, E., Brousse, C., & Deauvieau, J. (2015). Finding one's way in social space: genèse, postérité et actualité d'une enquête originale. *Sociologie, 6*(1), 31–42.

Ponte, S. (2016). Convention theory in the Anglophone agro-food literature: Past, present and future. *Journal of Rural Studies, 44*(4), 12–23.

Ponte, S., & Cheyns, E. (2013). Voluntary standards and the governance of sustainability networks. *Global Networks, 13*(4), 459–477.

Ponte, S., Gibbon, P., & Vestergaard, J. (Eds.). (2011). *Governing through standards: Origins, drivers and limitations.* Palgrave Macmillan.

Porta, J. (2019). Social rights' transformations in global context. *La Revue des droits de l'homme, 16.* https://journals.openedition.org/revdh/7161.

Porter, T. M. (1995). *Trust in numbers: The pursuit of objectivity in science and public life.* Princeton University Press.

Power, M. (1997). *The audit society: Rituals of verification.* Oxford University Press.

Richard-Ferroudji, A. (2011). Limites du modèle délibératif: composer avec différents formats de participation. *Politix, 24*(96), 161–181.

Richard-Ferroudji, A., & Barreteau, O. (2012). Assembling different forms of knowledge for participative water management: Insights from the Concert'eau game. In C. Claeys & M. Jacqué (Eds.), *Environmental democracy facing uncertainty* (pp. 97–120). Peter Lang.

Richey, L. A., & Ponte, S. (2011). *Brand aid. Shopping well to save the world.* University of Minnesota Press.

Rose, N. (1991). Governing by numbers: Figuring out democracy. *Accounting, Organizations and Society, 16*(7), 673–692.

Rose, N., & Miller, P. (2010). Political power beyond the state: Problematics of government. *British Journal of Sociology, 61*(Special Issue: The BJS: Shaping sociology over 60 years), 271–303.

Salais, R. (2006). Reforming the European social model and the politics of indicators: From the unemployment rate to the employment rate in the European employment strategy. In M. Jepsen & A. Serrano (Eds.), *Unwrapping the European social model* (pp. 189–212). Policy Press.

Salais, R. (2017). *Social investment: From European conception towards a human rights-capabilities approach* (Working Paper). Europe H2020 Project Re-Invest, grant number 649447.

Salais, R., Baverez, N., & Reynaud, B. (1986). *L'invention du chômage.* Presses Universitaires de France.

Salais, R., & Storper, M. (1993). *Les mondes de production.* Ed. de l'EHESS.

Schatzki, T. R., Knorr Cetina, K., & von Savigny, E. (Eds.). (2001). *The practice turn in contemporary theory.* Routledge.

Sica, A., & Turner, S. (Eds.). (2005). *The disobedient generation: Social theorists in the sixties.* University of Chicago Press.

Silva-Castañeda, L. (2012). A forest of evidence: Third-party certification and multiple forms of proof—A case study on oil palm plantations in Indonesia. *Agriculture and Human Values, 29*, 361–370.

Silva-Castañeda, L. (2015). In the shadow of benchmarks: Normative and ontological issues in the governance of land. *Environment and Planning A, 48*(4), 681–698.

Star, S. L. (1991). Power, technology and the phenomenology of conventions: On being allergic to onions. In J. Law (Ed.), *A sociology of monsters? Power, technology and the modern world* (pp. 27–57). Basil Blackwell.

Stavo-Debauge, J. (2012). *Le loup dans la bergerie. Le fondamentalisme chrétien à l'assaut de l'espace public.* Labor et Fides.

Stavo-Debauge, J. (2018). Le Divin marché (de dupes). Un fondamentalisme qui ne paie pas de mine mais rapporte gros. *SociologieS.* https://journals.ope nedition.org/sociologies/8230.

Storper, M., & Salais, R. (1997). *Worlds of production: The action frameworks of the economy.* Harvard University Press.

Thévenot, L. (1981a). Les catégories socioprofessionnelles et leur repérage dans les enquêtes. *Archives et documents 38*. Institut National de la Statistique et des Etudes Economiques.

Thévenot, L. (1981b). Un emploi à quel titre: l'identité professionnelle dans les questionnaires statistiques. In L. Thévenot (Ed.), *Les catégories socio-professionnelles et leur repérage dans les enquêtes* (Vol. Archives et documents 38, pp. 5–39). Institut National de la Statistique et des Etudes Economiques.

Thévenot, L. (1983). L'économie du codage social. *Critiques de l'Economie Politique, 23–24*, 188–222.

Thévenot, L. (1984). Rules and implements: Investment in forms. *Social Science Information, 23*(1), 1–45.

Thévenot, L. (1987 [1983]). Les enquêtes Formation Qualification Profession-nelle et leurs ancêtres français. In J. Affichard (Ed.), *Pour une histoire de la statistique* (Vol. 2, pp. 117–165). INSEE & Economica.

Thévenot, L. (1990). La politique des statistiques: les origines sociales des enquêtes de mobilité sociale. *Annales E.S.C., 45*(6), 1275–1300.

Thévenot, L. (1997). Un gouvernement par les normes; pratiques et politiques des formats d'information. In B. Conein & L. Thévenot (Eds.), *Cognition et information en société* (pp. 205–241). Ed. de l'EHESS (Raisons Pratiques 8).

Thévenot, L. (2001a). Organized complexity: Conventions of coordination and the composition of economic arrangements. *European Journal of Social Theory, 4*(4), 405–425.

Thévenot, L. (2001b). Pragmatic regimes governing the engagement with the world. In T. R. Schatzki, K. Knorr Cetina, & E. von Savigny (Eds.), *The practice turn in contemporary theory* (pp. 56–73). Routledge.

Thévenot, L. (2005). The two bodies of May '68: In common, in person. In A. Sica & S. Turner (Eds.), *The disobedient generation: Social theorists in the sixties* (pp. 252–271). University of Chicago Press.

Thévenot, L. (2006). *L'action au pluriel. Sociologie des régimes d'engagement*. La Découverte.

Thévenot, L. (2007). The plurality of cognitive formats and engagements: Moving between the familiar and the public. *European Journal of Social Theory, 10*(3), 413–427.

Thévenot, L. (2009). Governing life by standards: A view from engagements. *Social Studies of Science, 39*(5), 793–813.

Thévenot, L. (2011a). Conventions for measuring and questioning policies: The case of 50 years of policies evaluations through a statistical survey. *Historical Social Research, 36*(4), 192–217.

Thévenot, L. (2011b). Powers and oppressions viewed from the perspective of the sociology of engagements: A comparison with Bourdieu's and Dewey's critical approaches to practical activities. *Irish Journal of Sociology, 19*(1), 35–67.

Thévenot, L. (2013). The human being invested in social forms: Four extensions of the notion of engagement. In M. Archer & A. Maccarini (Eds.), *Engaging with the world: Agency, institutions, historical formations* (pp. 162–180). Routledge.

Thévenot, L. (2014). Voicing concern and difference: From public spaces to commonplaces. *European Journal of Cultural and Political Sociology, 1*(1), 7–34.

Thévenot, L. (2015a). Certifying the world: Power infrastructures and practices in economies of conventional forms. In P. Aspers & N. Dodd (Eds.), *Reimagining economic sociology* (pp. 195–223). Oxford University Press.

Thévenot, L. (2015b). Making commonality in the plural, on the basis of binding engagements. In P. Dumouchel & R. Gotoh (Eds.), *Social bonds as freedom: Revising the dichotomy of the universal and the particular* (pp. 82–108). Berghahn.

Thévenot, L. (2016). From codage social to economie des conventions: A thirty years perspective on the analysis of qualification and quantification investments. *Historical Social Research, 41*(2), 96–117.

Thévenot, L. (2017). Simmel et la mise en forme de l'humain. In D. Thouard & B. Zimmermann (Eds.), *Simmel, le parti-pris du tiers* (pp. 135–153). Ed. du CNRS.

Thévenot, L. (2018). Droits et biens pris en compte par les engagements volontaires d'entreprises dans des standards internationaux. La 'sustainable palm oil' certification au regard des plus défavorisés. In A. Jeammaud, M. Le Friant, P. Lokiec, & W. Cyril (Eds.), *Le droit sans frontières* (pp. 852–868). Dalloz.

Thévenot, L. (2019a). Endiguer les transes d'un péril prochain. La transition d'un Plan de Prévention de Submersion et la vague de réactions locales contre l'État. In F. Padovani & B. Lysaniuk (Eds.), *Les gestions des transitions* (pp. 31–66). L'Harmattan.

Thévenot, L. (2019b). Measure for measure: Politics of quantifying individuals to govern them. *Historical Social Research, 44*(2), 44–76.

Thévenot, L. (2019c). What engages: The sociology of justifications, conventions, and engagements, meeting norms. *La Revue des droits de l'homme, 16*(online since 3 July 2019). https://doi.org/10.4000/revdh.7114.

Thévenot, L. (2020). How does politics take closeness into account? *International Journal of Politics, Culture, and Society, 33*(2), 221–250.

Thévenot, L., & Lamont, M. (2000). Exploring the French and American polity. In M. Lamont & L. Thévenot (Eds.), *Rethinking comparative cultural sociology: Repertoires of evaluation in France and the United States* (pp. 307–332). Cambridge University Press.

Thévenot, L., Moody, M., & Lafaye, C. (2000). Forms of valuing nature: Arguments and modes of justification in French and American environmental disputes. In M. Lamont & L. Thévenot (Eds.), *Rethinking comparative cultural sociology: Repertoires of evaluation in France and the United States* (pp. 229–272). Cambridge University Press.

Timmermans, S., & Berg, M. (2003). *The gold standard: The challenge of evidence-based medicine and standardization in health care.* Temple University Press.

Timmermans, S., & Epstein, S. (2010). A world of standards but not a standard world: Toward a sociology of standards and standardization. *Annual Review of Sociology, 36,* 69–89.

Verdier, E. (2017). How are European lifelong learning systems changing? An approach in terms of public policy regimes. In R. Normand & J.-L. Derouet (Eds.), *An European politics of education: Perspectives from sociology, policy studies and politics* (pp. 194–215). Routledge.

CHAPTER 8

Do Performance Indicators Improve the Effectiveness of Development Aid?

Ousmane Oumarou Sidibé

Performance indicators, inspired by benchmarking techniques invented in the private sector, are increasingly used in public management, according to the notion that in order to survive in a competitive environment, countries must continually improve their organization by importing "best practices" that prevail among their "competitors" (OECD & World Bank, 2008). Alain Supiot calls this the "dogmatics of universal competition" (Supiot, 2009). These indicators are the cornerstone of results-based management (Managing for Development Results, MfDR) and have become the key criteria for measuring the effectiveness of public policies, on the basis of quantitative targets. Governments are anxious to gain points in international rankings by any means, focusing especially on the Human Sustainable Development Index compiled by the United Nations Development Programme (UNDP). In this index, countries are ranked according to a series of indicators that implicitly set a norm, suggesting that the "lowest performing" countries need to better their scores. In this framework performance indicators have become a formidable tool in the

O. O. Sidibé (✉)
University of Bamako, Bamako, Mali

© The Author(s) 2022
A. Mennicken and R. Salais (eds.), *The New Politics of Numbers*,
Executive Politics and Governance,
https://doi.org/10.1007/978-3-030-78201-6_8

253

hands of a few international experts, who insidiously impose far-reaching public policy choices on states, in the absence of any real debate among citizens.

In Third World countries these indicators are in effect used by international funding bodies to orient development aid in the context of a change of strategy, shifting from "project aid" to budget support, in keeping with the Paris Declaration of 2 March 2005.[1] There is now an international consensus that management of project aid, largely controlled by funding bodies, undermines the responsibility of recipient countries, hampers mobilization of resources and raises transaction costs. In response to this criticism, the ministers in charge of development from developed and developing countries, and officials of bilateral and multilateral funding organizations convened in Paris on 2 March 2005. At this meeting they resolved to reform delivery and management of aid as stated in the Paris Declaration on Aid Effectiveness. To make aid more effective the Paris Declaration underscored the following five principles:

> *Ownership*: Recipient countries set their own public policy and strategy in the framework of Poverty Reduction Strategies (PRS), which are the reference documents of public policy;
> *Alignment:* Donor countries align their support with the recipient countries' national priorities, systems and procedures of public financial management, instead of each following its own system;
> *Harmonization*: Donor countries coordinate their choice of sectors to support (education, public health, etc.), control and reporting procedures to be put into place, and share information to avoid overlapping action;
> *Results*: Recipient countries work with partners to define results expected from funded programmes and on their evaluation;
> *Mutual accountability*: Donors and recipient countries are both accountable for results achieved in development action.

The general budget support recommended by the Paris Declaration is intended to improve mobilization of resources and foster greater mutual responsibility of donors and beneficiaries of aid.[2]

In the course of the 1990s, some African countries, among them Tanzania, Uganda, Mozambique, were able to shift from project aid to budget support. In West Africa, Mali was a pioneer of budget support aid, under a Framework Arrangement for general budget support agreed in March 2006 by the government of Mali and funding bodies. Agreements for sectoral budget support in the areas of education, public health and decentralization and institutional development were concluded respectively on 18 July 2006, 19 July 2006 and 11 September 2009. Under these agreements, the disbursement of aid in these sectors is contingent upon attaining quantitative figures for a number of indicators.[3] Advocates of this approach see in it a way to bring the authorities in recipient countries to report to funding institutions on the results of development programmes, in exchange for greater (although "under surveillance") responsibility in the use of the resources attributed. To link action undertaken to results obtained and more systematically monitor performance, the funding bodies outline results chains in loan agreements. These results chains form part of the logical framework that connects the objectives, the activities pursued to attain them, the resources deployed and the results obtained in relation to those expected, and the impact achieved.

This chapter illustrates how these performance indicators have contributed to undermine public policies in Africa. We look at several examples drawn from experience in Mali, particularly in education and public health. This research highlights the harmful effects of performance indicators (Sect. "The Harmful Effects of Performance Indicators"), queries the meaning of management by indicators (Sect. "Querying Performance-based Management in Third World Countries") and takes up the issue of the autonomy of aid recipient countries in management of their public policy (Sect. "Giving Aid Recipient Countries Greater Autonomy to Conduct Their Public Policy").

The Harmful Effects of Performance Indicators

Upon close examination it becomes clear that the use of performance indicators in managing public affairs often leads to outcomes that are contrary to the stated objectives (Desrosières, 2014). In the education sector in Mali, while indicators pertaining to the gross enrolment rate in the primary education fundamental cycle (71.0% in 2013) and to the completion rate for the fundamental cycle (51.0% in 2013) (Ministry of the Economy & Finances, 2014) attest to progress in universal schooling, they have nonetheless had undesirable effects on the education system.

It should be remembered that the widespread introduction of performance indicators for education in Africa was based on thinking that lent credence to the idea that education in Africa suffered more from poor governance than from inadequate funding (Boone, 1996; Mingat & Tan, 1998). This neoliberal theory, very much in vogue at the World Bank at the time, was supported by international comparisons of the internal efficiency of education systems, measuring the number of years in school against the share of spending for education in GDP (see Mingat & Suchaut, 2000; Olivier & Orivel, 1999).

International rankings do indeed reveal the low efficiency of African education systems in the 1990s, compared to Asian and Latin American countries where pupils stayed in school twice as long, on average, for the same proportion of GDP spent on education. This theory, positing that school performance is not fundamentally related to the volume of financial resources allocated to education, had earlier been advanced in other contexts in Brazil and in the United States, most notably in the Coleman Report in 1966 (Hanushek, 1997; Harbison & Hanushek, 1992). Subsequently, this theory served as the "scientific" basis for a number of decisions that can be seen as factors in the present crisis of education systems in Africa.

Under agreements with international aid donors, the national education ministers of African countries were pushed to apply brutal reforms with the sole aim of improving the internal yield of the systems, without regard for the quality of teaching. In this context various highly debatable measures were taken in order to maintain the largest body of teachers possible with the lowest possible payroll costs. Continuing to follow international comparisons, we recall that in the 1980s the Bretton Woods institutions worked hard to demonstrate that African teachers were overpaid, and that this hampered universal schooling. They strongly recommended that these countries recruit more teachers, for the same aggregate payroll expenditure. On the strength of this advice salaries were sometimes drastically cut, as in Cameroon, resulting in a demotivation of the teaching body, with foreseeable consequences for the quality of teaching. This theory was also the basis for the creation of different categories of teaching personnel in all African countries (non-civil-servant status, contract workers, volunteers, etc.). The first non-civil-servant teachers were hired in Mali in 1991, in general, technical and vocational secondary education. This practice was extended to primary education in 1992, and at an ever greater pace thereafter. In 2002, 95% of new teachers hired were contract workers.

The only justification for the introduction of this new type of staff was the desire to create new categories of teachers, without civil service status and trade union representation, and poorly paid, in order to attain the targets and improve the internal efficiency of the system as measured by indicators. The aim was to formally achieve universal school attendance, to improve ranking under the Human Sustainable Development Index (UNDP) as well as according to the Millennium Development Goals standard (MDG).

Today the undesirable effects of this policy cannot be denied. Impartial observers recognize that non-civil-servant teachers lack both the training and professional ethics required for this function. Corruption in schools is largely due to the status of this group of teachers. Poorly paid at the outset, poorly trained, receiving little consideration and without any real perspective for professional advancement, these teachers had few reasons to observe ethical standards, in which they had never been inculcated in any event. Having never felt included in this profession, they brought the educational system to its knees.[4] In hoping to economize on payroll costs, the country had disinvested in its essential resource, teachers.

Another unexpected effect of the massive introduction of non-professionals into the teaching body, in order to boost performance indicators, was to discourage civil-servant teachers. Many ultimately gave up in the face of the unorthodox practices of their young colleagues recruited from any number of different trades (accountants, plumbers, electricians, even individuals without degree or diploma) whose only motivation was to avoid unemployment. The most surprising aspect of this is that the Bretton Woods institutions, with all the brainpower within their ranks, did not foresee that once non-professional teachers had attained a level of critical mass within the system they too would seek to have the same advantages accorded to their civil-servant colleagues, which would wipe out the payroll savings purportedly achieved by hiring non-professionals. For quite a few years in Mali now, the issues pursued by teachers' unions have focused mainly on acquiring the same remuneration for non-civil servants and for civil servants.

Another paradox of this system is that the conditions for the application of structural adjustment programmes led countries like Mali to close most teacher training schools in the 1980s, and to implement voluntary early retirement schemes for civil servants in order to reduce overall remuneration costs, thus creating a shortage of qualified personnel in the education sector.[5] When in the 1990s social services were granted priority for hiring civil servants, there were no more qualified teachers in the job market. This led to a massive influx of insufficiently trained people in the categories of teachers created outside of the civil service, provoking serious disturbances in the education system (strikes, successive "blank" academic years, etc.). On top of this, in their desire to mechanically raise the schooling rate, education authorities were less rigorous in their oversight of ethnic-community-based schools and medersas, for instance, schooling dispensed in Arabic. In a country where French is still the official language this could only aggravate youth unemployment (Ministry of the Economy & Finances, 2014).

It was also with encouragement from the World Bank that African governments fostered the development of community schools, financed by village communities. These schools were housed in straw huts, with pupils seated on grass mats, without textbooks or school supplies. The teachers sometimes had not even completed primary school, and with salaries three to five times lower than remuneration in public schools, the quality of teaching was severely affected. Other highly debatable teaching reforms were undertaken in the pursuit of good indicator scores. In Mali, for example, one teacher might be assigned to teach two grade levels at the same time in the same room, the pupils doubling up on the benches (double load), or to teach two groups of pupils of the same grade level alternating morning and afternoon (double time).

The internal efficiency of a teaching cycle is also measured by the capacity to retain pupils from the beginning to the end of the cycle. To obtain a good score the system is likely to artificially limit the number of pupils repeating a year, even if the pupils have failed to improve, for the simple reason that repeating a year is considered to be a waste of public money (Mingat & Suchaut, 2000). Brutally applied, this way of thinking has without any doubt led to bargain-basement teaching in countries like Mali. To boost the completion rate for the fundamental cycle of primary education the National Education Ministry in Mali instituted in the 1990s a policy designed to pass as many pupils as possible to the next higher grade level, even with very poor results (as low as 2/10). This manifestly eroded the overall achievement level. Testing to measure basic skills (reading, writing, arithmetic) has shown a notable drop in achievement among pupils in Mali, and even a critical failure in subjects such as Mathematics and French (CONFEMEN, 2004). What is worse, and again with the notion of boosting indicator scores related to internal system efficiency, many African countries formed class groups of 100 or 200 pupils, assigned to harassed teachers, with disastrous consequences for the quality of teaching.

The consequences of these policies on teaching are such that today we question them, in pursuit of a more coherent system. Education systems are too complex to apply an analysis based on quantitative factors, as for factory production. If efficiency is defined as the ratio between a given result (duration of schooling) and financial resources consumed to attain it, this indicator is not in itself significant. We must look farther and measure the impact of the system as a whole. More broadly, one of the effects of mechanical application of performance indicators has been the creation of a two-tier education system in all African countries; public schools are open to all, with the prime mission of raising the schooling rate, while schools in the private sector are reserved for a rich minority of the population who seek a certain level of quality, and at the same time lessen pressure on the public purse.

The pernicious effects of performance-based management can be seen through indicators that measure budgetary expenditure in sectors deemed to be high priority. While indicators that measure the share of the national budget devoted to so-called high priority sectors (education, public health, agriculture) provide an incentive to concentrate resources in these sectors, they also have the drawback of pushing the government administrations in question to spend public money on measures that do not always improve the provision of services to the public.

In addition to questions of the quality of public expenditure and the capacity of these sectors to absorb this money, the use of these indicators drives ministries to analyse their budgets not in terms of needs, but in terms of percentage of the national budget, leading to sterile quarrels and competition throughout the budget process. In this respect it can be observed that priorities have shifted to infrastructure investment in the most recent generation of Poverty Reduction Strategies. As grain prices have been rising on world markets, agriculture, which was until recently neglected, is moving up in the list of priorities. While one can only express satisfaction with this particular evolution, as a general rule we observe regular changes in priority goals, an illustration of the fragility of objectives when they are not popularly supported.

Successive reforms, sometimes abandoned before completion, often for reasons related to the international agenda, without consultation of the people, is a source of real instability in public policy in developing countries. In this respect Africa appears to be a laboratory for all sorts of experiments.

In the area of justice, indicators pertaining to the number of judges per capita and to construction and renovation of judicial infrastructure (courthouses, prisons, housing for magistrates, etc.) give an idea of public budget support for this sector. These indicators do not, however, in themselves give information on the quality of justice, or on the independence of the judiciary, and even less on the level of satisfaction of those who have recourse to the courts. Quite the contrary, application of these two indicators combined pushes governments to concentrate their efforts on hiring judges and building courthouses, to the detriment of other crucial measures to institute fair remuneration of judicial personnel and to strengthen the ethics of the judicial apparatus as a whole.

One of the great weaknesses of "trigger indicators", i.e. those that determine the amounts of aid to be paid annually under funding agreements, lies precisely in the fact that recipient countries have incentives to focus on attaining the objectives that have been given, to the detriment of other aspects of public policy. The quest to reach the target figure set for the indicator becomes something of an obsession, detached from the policy it is meant to support. The finance ministry, anxious to cash in on the money promised by donors, puts pressure on other government administrations to attain the key indicator levels, and there is a great temptation to doctor the figures when objectives are not reached.

But a national indicator, by nature, does not reflect the conditions and the results of a public policy in distinct geographic areas. There are indeed strong disparities between regions. Looking at education in Mali, while the national rate of completion of the first cycle of schooling was 51.0% in 2013, this average encompassed significant differences across regions. The Bamako district registered the highest completion rate, with 77.9%, while in the districts of Kayes, Koulikoro, Ségou and Sikasso this rate stood at 44.4, 64.9, 49.5 and 53.5% respectively (Ministry of the Economy & Finances, 2014). The same disparities are found for public health indicators and access to drinking water. Globally speaking, the Bamako district is situated well above the national average, while interior regions, in particular in the north of the country, have lower scores. This underscores the dilemma of the northern part of the country, where repeated and ongoing rebellions are in part linked to differences in development between regions. These differences are not reflected in the national indicators published in international rankings.

Along the same lines, certain indicators must be carefully examined, and cross-analysed in conjunction with other data. While the pupil/teacher ratio for fundamental primary education (first cycle) is a relevant indicator, in that it describes the level of adult interaction with pupils, it must nonetheless be counterbalanced with the fact that in the schools with the most teachers (in cities) many of them are deployed in administrative positions, rather than in classrooms, or are simply extra staff on hand.

The public health indicator stating the percentage of the population that lives within five kilometres of a functional healthcare centre (56% in 2013) does not inform us as to the real accessibility of healthcare (Ministry of the Economy & Finances, 2014). We know that due to the paucity of medicines available in healthcare centres, and frequently to understaffing (as centres are located farther from large cities), the local population has little reason to go to these centres. People, especially the poorest, are likely to have recourse to traditional medicine, or worse to obtain remedies from "itinerant pharmacists" who are quacks and deliver counterfeit drugs. According to the 2013 report on implementation of PRS credits, barely one in three people in Mali visited a healthcare centre in the course of the year (Ministry of the Economy & Finances, 2014). It is rare that someone living in Mali would not have a health problem at least once a year, especially as malaria is endemic. The fact is that the economic accessibility of healthcare is the real challenge, in all African countries.

Furthermore, some indicators are difficult to quantify, contrary to what some may think. Two examples in the area of national health insurance ("sécurité sociale") coverage are the percentage of the population covered by secondary insurance ("mutuelles") and the percentage of the poor population that has medical coverage. This is a real preoccupation, as traditional social safety nets are becoming weaker, and are less and less able to cushion the social problems of the poor. However, it is difficult to compile reliable information on secondary health insurers, their subscribers and the services offered, as this is an informal sector that has sprung up in a legal and institutional framework that is inadequate.

It can also be noted that indicators are sometimes far from comparable between countries, contrary to what is intended. The 2013 report on implementation of PRS credits gives an unemployment rate of 7.9%, which is quite surprising when compared with that of developed countries (Ministry of the Economy & Finances, 2014). Knowing that this figure refers to persons of working age who meet both of the following conditions—without economic activity during the period, or without formal employment and availability to take a job–it is easy to understand that the indicator masks massive under-employment, to the order of 80% in most African countries. The vast majority of people considered to be unaffected by unemployment in fact work in subsistence agriculture or in the informal sector, and work only a few months in the year. The expansion of the informal economy is one of the key mutations of the labour market in Africa over the last two decades. The structure of employment has been considerably transformed by a shift from jobs in the formal sector to work in the informal sector, which has become a sponge absorbing the urban workforce. The informal sector adapts quickly to the real circumstances of the labour market, and the ease of taking and leaving work makes it attractive, in particular for vulnerable groups such as women. Research has shown that even within a single country, comparison of unemployment figures over the long term is problematic, because of the transformation of the phenomenon itself (on this issue see Salais, 1986; Thélot, 1985).

QUERYING PERFORMANCE-BASED MANAGEMENT IN THIRD WORLD COUNTRIES

In technocratic circles in the Third World, performance indicators are considered to be the miracle cure that will ensure proper use of public money. The most sceptical of technocrats see these tools as a practice imposed by international donors to force the leaders of recipient countries to finally be accountable for their management. But few people seek to understand what is really at stake. Indeed, these new tools of international cooperation based on quantitative objectives may appear comfortable for managers. A time frame and evaluation grid that are known in advance are components that can be reassuring for policy chiefs, who are looking for immediate results.

It is nonetheless true that a mechanical application of these tools leaves no room for in-depth thinking to accompany complex processes that involve fundamental changes, sometimes even a cultural revolution, setting off profound movements in society. One of the problems of performance-based management is to define relevant and legitimate targets. Reaching a consensus on targets is truly a challenge, as it is difficult to set priorities that are acceptable to a broad majority. While it may be easy to agree on objectives in a corporation (lowering costs, raising productivity and profits, etc.), finding a consensus is much more arduous in the public sector. The aims of public policy are multiple, and sometimes contradictory, so reaching consensus can be extremely laborious. It is true that it is the duty of those who govern to manage public resources properly and report on their use to citizens, and this makes the principle of Managing for Development Results (MfDR) legitimate. But first there must be agreement on the meaning and substance of this concept, for the results culture is not the same everywhere on the globe.

Let there be no mistake, the choice of indicators is not politically neutral. Since the beginning of the 1990s, in the wake of the fall of the Berlin Wall and with the growing move to democracy in Africa, "governance" has become the central theme of international aid and cooperation, the lens through which the relations between international donors and aid recipients are seen (James, 1998; Olivier de Sardan, 2007). A technocratic conception, inspired by the aims of reducing deficits and restricting the role of state government, now governs the relations between aid donors and recipients under cooperation agreements, with considerable room for sanctions against countries that deviate from prescribed practice in terms of good governance (on this issue see Meisel & Ould, 2008). In reality in Western countries, fervent advocates of MfDR (Canada, the United Kingdom, the United States, etc.), this reform inspired by neoliberal economic thought is intimately tied to systematic budget cuts as part of a move to clean up public finances, and these cuts become an end in and of themselves. Without a doubt the neoliberal free market movement of the 1980s that propounded the notion of "less government" and relegated social and cultural issues to the back burner had major repercussions on public policy in countries that receive aid.

What should we make of the fact that a country like Mali, heir to an ancient culture, home to world heritage sites in the old cities of Djenné, Tombouctou and Gao, guardian of thousands of ancient manuscripts, has no public policy indicator pertaining to the safeguarding of these treasures?

Likewise, social protection is analysed within the confines of the official outward-looking system, that posits the modern system as its model and refers only to the tools used in official regimes, which are meant to be extended more or less as is, to the rest of the population without taking into account peoples' needs, or existing mechanisms. It has to be recognized, however, that the state of healthcare coverage in Africa cannot be remedied using only existing tools associated with formal insurance regimens that cover between 5 and 10% of the population. There are other mechanisms of solidarity and redistribution, however. Research has described the faltering of traditional solidarity systems, under the effect of increasing focus on the individual, a process in African societies driven by monetization, urbanization and the consequences of neoliberal economics. This has led some authors, such as Axelle Kabou, to question the capacity of African society to stand up in the face of these social changes (Kabou, 1991). Others, including Vimard and Locoh, are willing to bet on the adaptive capacity of African solidarity mechanisms (Locoh, 1988, 1993a, b, 1995; Vimard & N'cho, 1997;). A number of solutions can be proposed as ways to integrate traditional solidarity into the global system of social protection. It is clear that indicators pertaining only to these covered by modern social security systems cannot be expected to reflect the reality of social protection in Africa.

In this context, how can a new social contract be devised, setting reciprocal obligations for families, communities, local authorities, civil society and the national government, in order to ensure that the poorest members of society have effective rights to social protection? What forms of consultation and cooperation between all the players are needed to renew thinking and means of action to address the real needs in the sector? These are the fundamental questions that must be answered to set up a complete and coherent social protection system.

In truth, the notions of ownership and responsibility touted by the Paris Declaration have not yielded "true" appropriation, but have resulted in national domestic policy piloted by donor countries. The Paris Declaration does not afford truly autonomous action, without which there can be no responsibility on the part of recipient countries. With a knife to their throats, the recipient states must sometimes accept to see their public policies guided by the key indicators that are imposed, with the hope that they may regain some degree of control later. Furthermore, while donor countries have made significant progress in coordinating their action, much remains to be done by the recipient countries to improve their internal coordination (among different government ministries) and set their own priorities for discussion with donors. This is due to weak national leadership, and to the weak institutional capacity of national administrations that do not yet possess the means to implement results-based management.

There is abundant literature that shows how institutional variables affect the performance of public policy (Burki & Perry, 1998; Fukuyama,

2004; Persson & Tabellini, 1997). The quality or weakness of institutions are factors that explain time and time again the success or failure of many aspects of reform in states that receive development aid (Burnside & Dollar, 1997; Dollar & Easterly, 1999; Sindzingre, 1998). On this point, it can only be regretted that under structural adjustment programmes in the 1980s these countries were led to drastically "shrink" their rosters of civil servants (which were in fact quite meagre already), and to create numerous programme management units. This was not conducive to a stable institutional environment. In Mali, for instance, all administrative reform measures in this period were undertaken in the name of "management of structures and staffing", a culture that continues to mark government administration in the country.

Likewise, and even if some donor countries have aligned practices with those of recipient countries, different types of aid may coexist, sometimes within the same programme, and with the same donor. This coexistence may impose considerable burdens on the national government in terms of record-keeping and reporting, taking time away from action to provide services to the people, which after all is the purpose of the aid. States deemed to be fragile, with little human resources, are forcibly given sophisticated tools that they are not well equipped to implement. In the same way, programme managers in aid recipient countries spend a good deal of their time collecting data on progress to attain "target values" for indicators, on reporting tasks and on discussion of these issues with representatives of donor countries. Clearly, this is time not devoted to work managing the programmes themselves, undermining the objectives of results-based management. In fact, performance-based management, invented in the private sector, is not readily transposed to the public sector, where it is less effective by reason of the very nature of public service. It would be pointless to try to measure with strict accuracy the cost of public action, precisely because these services are provided for all citizens.

In the Sahel countries, for example, government administration does not adequately cover the vast desert areas, and consequently the poor level of service to citizens (healthcare, education, safety and security of people and property, etc.) are factors of instability. In Mali, the sparse population in the north means low resources for local authorities as well as particular constraints in terms of economic, social and territorial organization (highly dispersed population, little infrastructure, open spaces, nomadic peoples). Low population density goes hand in hand with underprivileged areas that are poorly served in terms of basic amenities (water, schools, healthcare) and are remote and isolated. With little capacity to harness financial resources and hence low investment, low population and remote location, these territories are subject to additional costs that should not be borne by the local population alone. As equal access to public services is an imperative principle of republican government, it is not reasonable to apply across the board the same performance indicators–and most notably the ratio of public expenditure per capita–for public service throughout the country.

Referring to other contexts, Daniel Kaufman has given a very good description of the limits of management by indicators and this modelling of public action (Kaufman & Kraay, 2007). All this is to say that the real challenge facing developing countries, particularly in Africa, is to rebuild the capacities undermined by decades of structural adjustment and to regain a measure of negotiating power in their relations with international donors.

Giving Aid Recipient Countries Greater Autonomy to Conduct Their Public Policy

The dependence of developing countries with respect to donors is a notion that encompasses several different aspects (Naudet, 1999). In the most common sense, this dependence refers to the situation in which outside aid continues to weigh heavily in national budgets, while levels of development and internal financial capacity stagnate (see Azam et al., 1999). Dependence can also be apprehended through indicators, such as the share of aid in GDP, or in the conditions of aid allocation. Above all, dependence is characterized by little room to negotiate with donors. In this light the faltering capacity of recipient countries, the "fragile states", takes on new meaning, particularly in Africa. Chronic political instability, military coups and armed rebellion are among the ills that plague African states, described as oligarchic, neo-patrimonial, fragile, failed, among the long string of adjectives that flourish in the "post-colonial" approaches that predominate in political science literature since the 1980s (on this issue see Haut Conseil de la Coopération International, 2005; World Bank, 2006; but also Badie, 1992; Devarajan et al., 1999).

With very few exceptions (Cabo Verde, Botswana, South Africa, Rwanda), African countries struggle to establish state structures that are able to conduct strong public policy on their own, and to meet the needs of citizens, while resisting the demands of international donors (on this issue see Joseph, 1999; Mhonimpa, 1994; Prest et al., 2005). On this continent more than elsewhere, the relations between donors and recipient countries are skewed, and even coercive, with stringent conditions and ever more exacting timetables for implementation. Indeed, while in the spirit of the Paris Declaration indicators are to be chosen from the grids in PRS documents or from the logical frameworks of sectoral programmes, in truth it is the weight of the funding institutions that is the determining factor, in particular for setting the target values to be attained. Furthermore, this is the justification for the fact that negotiations on the budgetary arrangements that frame budgetary support last for months and absorb all the energy of programme managers. These arrangements set, among others, the "trigger indicators" that authorize disbursement of variable tranches of aid. The choice of "target values", i.e. the quantitative objectives to be reached each year as policy is implemented, are the focus of especially tough discussions. Be as it may, in the end it is always the donors who have their way, recalling the famous saying of the Malian philosopher Amadou Hampaté Bâ, who noted that "the hand that gives is always above the hand that receives". This situation is illustrated by the words of a representative of donor countries:

> The Malian government pilots the plane; we make sure there is a flight plan and that the altitude is properly measured; if the needle of the altimeter is not good, we will find better instruments for the pilot. (Bergamashi, 2009, p. 27)

Coordination between donors and the International Monetary Fund (IMF) imposes cross-linked conditions, increasing the pressure on the recipient countries. This type of condition–measures creating the obligation to link disbursement to progress markers set in advance under funding agreements–is in fact embedded in the mandate of international financial institutions, and is considered to be a sign of their credibility (Kanbur et al., 1999). It is hard to comply with these conditions, which explains the ambivalent attitude of recipient countries towards the reforms imposed by donors. Their formal acceptance often gives way to resistance in different forms when it comes to implementation (see Azam et al., 1999; Collier, 1997; Kahler, 1992). Faced with the difficulty of meeting some of the required conditions, recipient governments may turn to various subterfuges to postpone, circumvent or vitiate these rules. In some cases, conditions are suspended, knowing that from the outset recipient governments evaluate the costs of not complying with the conditions (see Bird, 1998; Bayart et al., 2002).

Paradoxically, the states with the least institutional capacity to manage the weighty apparatus of management by indicators are those on whom the donors impose the most stringent terms of conditionality, because they lack the capacity to negotiate. By setting ambitious goals the donors perhaps thought they would lift the recipients out of the danger zone; in fact, they have pushed them in deeper. In this respect, the annual joint review of aid by both donor and recipient countries are truly examinations, intended to assess the level attained for the indicators retained under the funding agreements. They are above all a humiliating experience for the recipient governments, when ordinary civil servants from aid agencies conduct themselves as judges, pronouncing sentences without appeal, setting the level of aid that will be disbursed, according to their appraisal of the results obtained.

It must also be said, however, that conditionality is sometimes exploited by the recipient governments themselves, for internal political purposes; donors can be held up as scapegoats, to justify unpopular (even if often essential) government decisions in the eyes of national public opinion (Collier, 1998; Rodrik, 1996). In reality, behind the pretext of granting greater autonomy to Third World countries to manage the aid they receive, the new approach based on budgetary support is a significant step towards a more political concept of aid, as part of a process to transform the political and social landscape of recipient countries, following pathways already defined by the donor countries. The donors' injunctions are based on a transfer of models, replacing the sociological foundation of states, as if an idealized and disembodied institutional reorganization could be decreed, instead of being worked out by interaction between public authorities and civil society. This forcible transfer of model aggravates the deficit of legitimacy of the state already found in African countries, and further undermines civic and citizen responsibility, by discrediting government action and widening the gap between citizens and the state.

Each donor country has its own world vision, its own framework of reference, that is vaunted in the country's relations with aid recipient countries. The desired transformations are imposed from the outside. This process can even go as far as direct supervision of entire sections of government, as in Sierra Leone, Iraq and Afghanistan. All this is to say that recipient countries should be able to construct their own public policies, in the context of their specific circumstances, without being forced to follow an outside model. Countries may be inspired by successful practices elsewhere, but freely and voluntarily, making the necessary adjustments. It is a matter of respecting the dignity of states, especially the most fragile among them. This movement must be advocated by democratic powers with real popular support and legitimacy from a fully involved civil society.

The fundamental issue is that donor countries must turn away from what Régis Debray, referring to Europe, has called a culture imbued with its own formulas when in dialogue with other countries (Debray, 2007). This is an enormous challenge, because it will oblige donors to change their intellectual processes and accept cultural diversity, something for which Western elites are not necessarily prepared. Experience teaches us that good governance is not achieved by decree. It is constructed by society in each country, in its specific circumstances. Political dialogue between donor and recipient countries naturally has a legitimate place, but the terms of this dialogue must be balanced. Each country should be allowed to pursue its own trajectory, according to its circumstances. The notions of the common good, the general interest and even of solidarity are not given the same meaning everywhere. Levy and Fukuyama have clearly shown how approaches diverge, depending on the situations found in each country (Levy & Fukuyama, 2008).

The concept of the effectiveness of public action is framed differently in the United States, in Europe and in Africa. Whereas Americans appraise the effectiveness of public policy through the lens of Schumpeterian "creative destruction" that enables a system to remain dynamic by eliminating its least effective agents, Europeans give preference to public funding based on a social contract. To promote economic and social development some Asian countries have emphasized the reinforcement of the state apparatus, while others, notably in Latin America, have preferred to bolster democracy in the political system.

CONCLUSION

The growing interest in using indicators to allocate development aid is closely connected to the needs of world capitalism. In this light, indicators are needed by investors, including international aid agencies that act as lenders, as a way to gather information on the quality of governance, and "country risk". The underlying idea for investors is to be assured of "good governance" and of a functioning market economy that will guarantee returns on investment. It is therefore hardly surprising to note that indicators are biased, even if the bias is hidden. This makes use of indicators problematic. In the absence of a universal common reference framework, countries do not necessarily share the same vision of development.

Despite these imperfections, the concept of performance indicators cannot be totally rejected. The problems arise when they are mechanically applied to rank countries, and even to sanction countries depending on their rank. This practice is less and less acceptable, even in schools for evaluating pupils. An indicator should be just that, literally furnishing an indication of the quality of governance, but not in itself a full appraisal of quality. From this point of view, indicators can send a warning on the state of public policy, and foster discussion, without issuing a judgement with no appeal, and are much less a guilty sentence.

Management by performance must in no case be allowed to be a new ideology, or a tool in the hands of a few technocrats who decide what is good for entire nations and impose profound changes on society. This approach can be legitimate only as an inclusive process intended to support discussion and joint efforts to seek the best use of public resources, according to criteria of intelligence, balance and fairness. Elected officials and citizens, through their organizations and associations, must take their full place in this process. To allow people to appropriate this discussion they must be properly informed as to the stakes and outcomes, with access to forums where views can be freely expressed, and the full range of opinion heard, in particular those voiced by the weakest groups in society.

NOTES

1. Project aid, whether managed by project unit or through institutional support more or less directly administered by funding bodies, is criticized for low disbursement rates, high overhead costs and weak appropriation by recipient countries. For these reasons a number of donors, including the European Union, are turning more and more often to budget support, i.e. aid paid directly into national budgets to fund all activities in the budget without specification (general budget support) or to support certain targeted sectors (sectoral budget support), most often public health and education (Feyzioglu et al., 1996).

2. On issues related to budget support see the study carried out by researchers at the University of Birmingham for the OECD Development Assistance

Committee (DAC) and other donors: IDD and Associates, Evaluation of General Budget Support: Synthesis Report. A joint Evaluation of General Budget Support 1994–2004, May 2006 (IDD & Associates, 2006).

3. On the choice of indicators for budget support, see: Introduction d'indicateurs de Résultat en Matière d'Appui aux Programmes d'Ajustement Structurel dans les Pays ACP, Etude réalisée à la demande de la Commission Européenne, Rapport de Synthèse, Vol.1, CERDI, June 2002.

4. For a discussion of teachers' capacities in Mali see: *La formation des enseignants dans la francophonie: diversités, défis et stratégies d'action* (2007). Montreal: AUF. *Profil de l'enseignant de qualité au Mali.* Oxfam, IE, OPT, SNEC, MEALN, undated. *Principes directeurs pour un profil de compétences national des enseignants du primaire.* Internationale de l'Education, Oxfam Novib, 2012.

5. In principle, teachers and public health personnel were not eligible for early retirement, so as to preserve these social services, but in fact they were the most severely affected, because workers in these sectors suffered from more difficult conditions than civil servants in other sectors.

REFERENCES

Azam, J.-P., Devarajan, S., & O'Connell, S. (1999). *Aid dependence reconsidered* (Working Paper WPS 99–5). Centre for the Study of African Economies.

Badie, B. (1992). *L'État importé. L'occidentalisation de l'ordre politique.* Fayard.

Bayart, J.-F., Hibou, B., & Khiari, S. (2002). Effets d'aubaine, les régimes autoritaires libérés des conditionnalités. *Critique Internationale, 14*(January), 7–11.

Bergamashi, I. (2009). Cadre stratégique de lutte contre la pauvreté. Aide budgétaire et dialogue politique au Mali. *Collection débats et controverses, N°2,* 27–38.

Bird, G. (1998). The effectiveness of conditionality and the political economy of policy reform: Is it simply a matter of political will? *Policy Reform, 2*(1), 89–113.

Boone, P. (1996). Politics and the effectiveness of foreign aid. *European Economic Review, 40*(2), 289–329.

Burki, S. J., & Perry, G. (1998). *Beyond the Washington Consensus: Institutions matter.* World Bank.

Burnside, C., & Dollar, D. (1997). *Aid, policies and growth* (Policy Research Working Paper No. 1777). World Bank.

Collier, P. (1997). The failure of conditionality. In C. Gwin & J. M. Nelson (Eds.), *Perspectives on aid and development.* Overseas Development Council, Policy Essay No. 22.

Collier, P. (1998). The role of the state in economic development: Cross-regional experiences. *Journal of African Economies, 7*(2), 38–76.

CONFEMEN (2004). Enseignants contractuels et qualité de l'enseignement au Mali: quels enseignements?

Debray, R. (2007). *Un mythe contemporain: le dialogue des civilisations*: CNRS Editions.

Desrosières, A. (2014). *Prouver et gouverner: Une analyse politique des statistiques publiques*. La Découverte.

Devarajan, S., Rajkumar, A. A., & Swaroop, V. (1999). *What does aid to Africa finance?* (Policy Research Working Paper No. 2092). World Bank.

Dollar, D., & Easterly, W. (1999). *The search for the key: Aid, investment, and policies in Africa* (Policy Research Working Paper No 2070). World Bank.

Feyzioglu, T., Swaroop, V., & Zhu, M. (1996). *Foreign aid's impact on public spending* (Policy Research Working Paper No. 1610). World Bank.

Fukuyama, F. (2004). *State-building, governance and the world order in the 21st century*. Cornell University Press.

Hanushek, E. A. (1997). Assessing the effects of school resources on student performance: An update. *Educational Evaluation and Policy Analysis, 19*(2), 141–164.

Harbison, R. W., & Hanushek, E. A. (1992). *Education performance of the poor: Lessons from rural northeast Brazil (A World Bank Book)*. Oxford University Press.

Haut Conseil de la Coopération International. (2005). Les acteurs français dans le post-conflit. *Rapport de la Commission "Crises, prévention des crises et reconstruction"*. HCCI.

IDD and Associates. (2006). Evaluation of general budget support: Synthesis report. *A joint Evaluation of General Budget Support 1994–2004*.

James, H. (1998). From grandmotherliness to governance: The evolution of IMF conditionality. *Finance and Development, 35*(4), 44–47.

Joseph, A. (1999). *L'Aide française au développement*. DIAL et Ministère de la Coopération.

Kabou, A. (1991). *Et si l'Afrique refusait le développement ?* L'Harmattan.

Kahler, M. (1992). External influence, conditionality, and the politics of adjustment. In S. Haggard & R.R. Kaufman (Eds.), *The politics of economic adjustment*. Princeton University Press.

Kanbur, R., Sandler, T., & with Morrison, K. M. (1999). *The future of development assistance: Common pools and international public goods* (Overseas Development Council, Policy Essay No. 25).

Kaufman, D., & Kraay, A. (2007). *On measuring governance: Framing issues for debate*. Draft issues prepared for the round table on measuring governance. The World Bank Institute.

Levy, B., & Fukuyama, F. (2008). *Development strategies: Integrating governance and growth* (World Bank Discussion Paper). World Bank.

Locoh, T. (1988). Structures familiales et changements sociaux. In D. Tabutin (Ed.), *Population et sociétés en Afrique au Sud du Sahara* (pp. 441–478). L'Harmattan.

Locoh, T. (1993a). Familles africaines face à la crise. *Afrique contemporaine, 166* (April–June), 3–14.

Locoh, T. (1993b). La solidarité familiale est-elle un amortisseur de la crise? *Pop Sahel, 19*(August), 20–25.

Locoh, T. (1995). Familles africaines, population et qualité de vie (Les dossiers du CEPED, No. 31 [p. 48]). CEPED.

Meisel, N., & Ould, A. J. (2008). *La "bonne gouvernance" est-elle une bonne stratégie de développement?* (Vol. Parocument de travail n° 58). Agence française de développement.

Mhonimpa, M. (1994). *Ethnicité et démocratie en Afrique. L'Homme tribal contre l'homme citoyen.* L'Harmattan.

Mingat, A., & Suchaut, B. (2000). *Les systèmes éducatifs africains. Une analyse économique comparative.* Editions De Boeck Université.

Mingat, A., & Tan, J. P. (1998). *The mechanics of progress in education: Evidence from cross-country data* (Policy Research Workig Paper [Vol. 2015]). World Bank.

Ministry of the Economy and Finances. (2014). *Rapport 2013 de mise en oeuvre du Cadre stratégique pour la croissance et la réduction de la pauvreté (CSCRP) 2012–2017.* Ministry of the Economy and Finances, Mali.

Naudet, J.-D. (1999). *Réflexion sur la notion de dépendance à l'aide extérieure.* DIAL.

OECD, & World Bank. (2008). *Sourcebook: Emerging good practice in managing for development results* (3rd ed.). World Bank.

Olivier de Sardan, J.-P. (2007, January 23–25). Gouvernance despotique, Gouvernance chefferiale, et Gouvernance postcoloniale, in *Entre tradition et modernité: quelle gouvernance pour l'Afrique? Actes du Colloque de Bamako* (pp. 109–131). IRG.

Olivier, E., & Orivel, F. (1999). Les comparaisons internationales de l'efficience interne des systèmes éducatifs. In J.-J. Paul (Ed.), *Administrer, gérer, évaluer les systèmes éducatifs.* ESF.

Persson, T., & Tabellini, G. (1997). *Political economics and macroeconomic policy* (Discussion Paper No. 1759). Centre for Economic Policy Research.

Prest, S., Gazo, J., & Carment, D. (2005). Conference on Canada's policy towards fragile, failed and dangerous states. Working out strategies for strengthening fragile states—The British, American and German experience. CIFP.

Rodrik, D. (1996). Why is there multilateral lending? In M. Bruno & B. Pleskovic (Eds.), *Annual World Bank Conference on development economics 1995* (pp. 167–193). World Bank.

Salais, R. (1986). L'émergence de la catégorie moderne de chômeur: les années 1930. In R. Salais, N. Baverez, & B. Reynaud (Eds.), *L'invention du chômage. Histoire et transformations d'une catégorie en France des années 1980* (pp. 77–123). PUF.

Sindzingre, A. (1998). Crédibilité des États et économie politique des réformes en Afrique. *Economies et sociétés, 4*, 117–147.

Supiot, A. (2009, February). Justice sociale et libéralisation du commerce international. *Droit social*, No 2, 131–141.

Thélot, C. (1985). Les traits majeurs du chômage depuis 20 ans. *Economie et statistique, 183*(December), 37–59.

Vimard, P., & N'cho, S. (1997). Évolution de la structure des ménages et différenciation des modèles familiaux en Côte d'Ivoire 1975–1993. In M. Pilon, T. Locoh, E. Vignikin, & P. Vimard (Eds.), *Ménages et familles en Afrique: approches des dynamiques contemporaines* (pp. 101–123). Les Études du CEPED n° 15.

World Bank. (2006). *Fragile States—Good practice in country assistance strategies*. World Bank.

CHAPTER 9

Archaeology of a Quantification Device: Quantification, Policies and Politics in French Higher Education

Corine Eyraud

Over the last fifteen years, I have aimed to enter into the analysis of broader phenomena and processes by decoding the genesis and uses of quantification devices.[1] The accounting reform of Chinese state enterprises, for instance, makes it possible to grasp the essence of the Chinese economic reforms of the 1990s (Eyraud, 1999, 2003). The construction and implementation of a system of performance-based management in French universities informs us about the profound transformations these organizations have undergone in the last two decades (Eyraud, 2014; Eyraud et al., 2011). As Alain Desrosières points out, these various studies suggest that "it is possible to look at the same time at social or political philosophies and seemingly technical tools, considering them as a totality" (Desrosières, 2000, p. 84).

C. Eyraud (✉)
Department of Sociology, Aix-Marseille University, CNRS, LEST,
Aix-en-Provence, France
e-mail: corine.eyraud@univ-amu.fr

© The Author(s) 2022   275
A. Mennicken and R. Salais (eds.), *The New Politics of Numbers*,
Executive Politics and Governance,
https://doi.org/10.1007/978-3-030-78201-6_9

However, my recent research on the transformation of French state accounting[2] (Eyraud, 2013) has shown that an accounting system can be employed in relation to different, sometimes conflicting logics and objectives, for example, in attempts to show that the state is heavily indebted or not so much, to improve public management, or to encourage the outsourcing of public activities. The link between a specific type of quantification device and a particular social or political philosophy does not seem to be univocal; both do perhaps not constitute "a totality". Hence, the nature and conditions of their linkage have to be questioned. I did this first through a renewed analysis of the French state accounting reform (Eyraud, 2016). This led me to put forward a grid for the analysis of quantification devices which seems capable, first, of casting light on the very nature of this link and, second, of making visible the possible choices involved.

The purpose of this chapter is to test this grid by revisiting our work on performance indicators for French universities. I choose thus to start from the analysis of a concrete quantification device—i.e. performance indicators—and conduct a kind of "archaeologic" analysis, using it as a lens for investigating and understanding the changes French universities have undergone since the mid-2000s. Drawing on the works on quantification done by French social scientists, such as Alain Desrosières (1988 with Thévenot; 1998 [1993], 2003, 2008a, b, 2014), Robert Salais (1986, 2004, 2010, 2016), Alain Supiot (2010, 2015) and Laurent Thévenot (1979, 1983, 1990, 2016), this paper seeks ultimately to enhance our understanding of reactivity.[3]

The analytical grid I propose distinguishes between three different levels that exist within a quantification device, each of which is examined in a separate section of this chapter. First, there is what might be called the bedrock level: a quantification device is grounded in a founding vision that is generally congruent with a particular form of state or economic system. Second, there is what might be called the intermediate level: a quantification device contains a conception of the objectives and "*raisons d'être*" of the entity that is quantified. Third, there is the level that relates to the micro-conventions of calculation: philosophies can be hidden at this microscopic level and give a particular orientation to the device. However, the analysis of these three levels does not tell us everything about the orientation of the device and the effects it can produce; the device is part of a larger configuration, the context of its deployment and its uses have thus to be examined. That constitutes the fourth dimension of the analytical grid, which will be examined in the fourth section of this paper. Returning to the question of the device as a totality, the conclusion will show that this specific device is not a very integrated assemblage, which explains its real but at the same time also limited effects.

The analysis is based, first, on in-depth interviews with government and university officials, second, on the study of official documents, reports and archival materials from parliament, central government, the Ministry of Higher Education and several universities, and, third, on participant observation. I conducted participant observations in my role as a member of the Governing Board and Finance Committee at my own university. Further, I was a special adviser to the President of my university on performance management systems. I attended the training on performance indicators and performance management for universities provided by the Ministry, and I participated in the implementation of the performance indicators in my own university. This variety of materials allowed me to follow the numbers from their birth through their very detailed construction process to their concrete uses.

THE BEDROCK: NPM, LOLF

Performance indicators are the latest form of public statistics. The development of their current form and uses began in some European countries, such as Sweden or the United Kingdom, in the late 1980s, and was a part of much a broader phenomenon: the rise of New Public Management (NPM). This term is quite ambiguous, even stretchy: it is used to speak about government reforms implemented in Great Britain during the Thatcher government, in the United States during the Reagan and Clinton administrations, in the Netherlands under a Christian-Democrat government, or in Sweden and New Zealand under Labour governments. All these reforms have a number of common features but also many differences (Hood, 1995). The shift in doctrines of public accountability is part of their common ground. Before NPM, democratic accountability depended on limiting corruption, waste and incompetence in public administration. To this end, the public sector was kept sharply distinct from the private sector in terms of ethos, methods of management, organizational design, people and career structure. An elaborate system of procedural rules was designed to prevent favouritism and corruption. In contrast NPM involved a very different conception of public account-ability, with different patterns of trust and distrust. As Hood (1995, p. 94) writes:

> The basis of NPM lies in reversing the two cardinal doctrines of public administration; that is, lessening or removing differences between the public and the private sector and shifting the emphasis from process accountability towards a greater element of accountability in terms of results. Accounting was to be a key element in this new conception of accountability, since it reflected high trust in the market and private business methods and low trust in public servants and professionals.

In this process, the introduction of the "*Loi organique relative aux lois de finances*" (henceforth referred to as "LOLF") was in France a very important step. This law, passed in 2001 and taking effect in 2006, introduced performance management and private accounting to the state and public services. The first objective of the LOLF was to make the government and the public services accountable to parliament for the results of their actions, and to give more power to parliament over budgetary policies and choices. Since 2006, French MPs have two new documents for the budget debate: an annual performance plan and an annual performance report for each public policy. The first one deter-mines the objectives for the following year; quantified indicators (known as performance indicators) are used to quantify these objectives and set the targets which have to be reached. The second one gives an account of the results achieved (relative success or relative failure) over the last few years. National performance indicators were set at each level of govern-ment and for all public bodies, so that a performance-oriented form of management was introduced throughout the public services and public administration, which was the second objective of the law.

A performance measurement system for public policies is based on the idea that the state is accountable for the results of its actions, which is based on, as Hood highlighted, a specific conception of public account-ability which emerged from the 1980s, and which now prevails.[4] This is what might be called the bedrock level of quantification: a quantification device, such as a performance indicator, is grounded in a founding vision, an ontology.[5] It would be possible to go deeper and view this conception of public accountability as being rooted in a specific way of governing that might be called "government by objectives" (Thévenot, 2015).[6]

From these specific conceptions of governing and accountability, the French government made several choices. First, in probably a typically French way, a law (here the LOLF) was used to establish the system of performance management, using a standardized, centralized, top-down approach. Second, the LOLF speaks of "performance" and not of "results". At the beginning of the 2000s, this was the choice made by many governments (such as the UK and the US), but not all (see Canada, for example). Whereas "results" can be seen as a quite a neutral word, the notion of "performance" carries many connotations (Jany-Catrice, 2016): ideas of outstanding qualities, of achievement (specifically in sports), of excellence, of winning; the notion of competition being not so far from them. Third, the LOLF chose to assess public performance only by means of quantitative indicators, relying on the belief that only numbers are able to report on public action and social reality, and, hence, demonstrating a "trust in numbers" (Porter, 1995).

A quantification device is thus the bearer of some great fundamentals: here, it implies a specific way of governing (government by objectives), a particular conception of public accountability, a standardized, centralized and top-down approach, a focus on performance, and a specific way of assessing public performance (via quantitative indicators) which is based on a belief in numbers.

Performance Indicators for French Universities: What Are Their *Raisons D'être*?

The formulation of the first annual performance plans introduced by the LOLF involved designing a set of performance indicators. However, what is the "performance" of a public policy? What do we expect from a prison or from a school? What does a "well-performing university" do? Do we expect it to produce graduates adapted to the labour market and fitting the needs of private companies? Do we expect it to allow women and men from different social origins to attain the same levels of education? Do we expect it produce a lot of patents that enhance the competitiveness of national companies? Do we expect it to broaden human knowledge? Do we expect it to provide a fulfilling and motivating working environment for its staff and students? Measuring the performance of a public institution is clearly built on a system of values, and it involves the making of fundamental, societal choices on what is important to measure.

In 2006, the first annual performance plan for higher education and research specified 33 indicators for universities with targets to be attained within five years. Where did these indicators, which French universities and the French Ministry of Education were accountable to Parliament for, come from? The LOLF could have been an opportunity for large democratic debate on the outputs and outcomes we expect from our public policies and public services. However, in my view, the LOLF was a missed opportunity for democracy, the choices of indicators were made in a completely technocratic way: they were the result of discussions and negotiations (from 2003 to 2005) between the ministry in charge of the public policy in question (here the Ministry of Education, Higher Education and Research, which I will refer to as the Ministry of Education), on the one hand, and the Treasury, on the other. Surprisingly, the discussions between the two ministries were not about the expected outcomes of higher education, but directly focussed on the indicators themselves: one ministry proposing certain indicators, which were then very often refused by the other.

To prevent the application of a "realist epistemology" to numbers, Desrosières suggested to talk not about "measurement" but about the "quantifying process". As he put it:

> The use of the verb 'to measure' is misleading because it overshadows the conventions at the foundation of quantification. The verb 'quantify', in its transitive form (make into a number, put a figure on, numericize), presupposes that a series of prior equivalence conventions has been developed and made explicit [...]. Measurement, strictly understood, comes afterwards [...]. From this viewpoint, quantification splits into two moments: convention and measurement. (Desrosières 2008b, pp. 10–11)

In our case, that means that the ministries argued about measurements before agreeing on what the performance of higher education is. We can analyse these controversies through the interviews I conducted with officials of both ministries and try to understand the rationales of each.

For the Ministry of Education, the indicators served various objectives. Firstly, the Ministry conceived them as incentives for universities to act in a certain way. As a senior official in this Ministry put it:

There is a political will behind each indicator; we would like each indicator to be an incentive for universities to be better in a certain field or to develop in a certain way.

The indicators, then, are seen as signals.

Secondly, the Ministry thought about how the indicators could be used during its negotiations of the education budget with the Treasury. The Ministry proposed for example the following indicators: cost per student, student–teacher and student–administrative staff ratios, all of which are quite low in France compared to other OECD countries with similar GDP per inhabitant. These indicators would allow the Ministry to justify an increase of its budget, in particular in relation to staff numbers. Thirdly, the Ministry wanted to present a positive image of higher education. This can be seen in several interviews, where interviewees for example said:

We're not masochists. You try to have indicators which can only improve; you do not want to be shot for it.
We wanted to show what was working well.

In contrast, the Treasury's main concern was about its uses of the indicators during budgetary negotiations with ministries. It was very aware of their potential use as devices to justify the need of enhanced income and staff numbers. See, for example, the following comment from a senior Treasury official:

We absolutely did not want indicators which allow our counterparts to say 'to get better results on this indicator, we need more money, more teachers, more premises, more computers or anything else'.

The Treasury refused, for this reason, nearly all the indicators that were put forward by the Ministry of Education. Contrary to the third objective of the Ministry of Education, indicators which presented a negative image of the Ministry to the public, which depreciated or undermined the value of its activities, would turn into a useful weapon for the Treasury. They allowed it to be in a stronger position to negotiate the education budget. Several quotations from the Ministry of Education showed that the Treasury was pressing for "negative" indicators:

The Treasury wanted to impose figures that you can find in tabloid head-lines, such as failure rate, dropout rate and so on, a negative picture in fact.

The next section will show that the very limitations of the indicators were exploited in this struggle.

By analysing not only the controversies around the indicators but also the final indicators themselves, we can now try to make visible the value systems on which they are based. The works of Boltanski and Thévenot (2006 [1991]) and Boltanski and Chiapello (2007) are useful for that. To analyse disputes and controversies, Boltanski and Thévenot (2006) identified seven "orders of worth" which imply systematic and coherent principles of evaluation, justification and legitimacy. Each of these "orders" (or "worlds") gives importance to different values:

- The *inspired world* values imagination and creation.
- The *domestic world* values tradition, long-term relationships and the respect of hierarchies.
- The *fame world* values celebrity and public opinion.
- The *civic world* values collective interest, solidarity, equality and democracy.
- The *market world* values competition and the exchange of goods and services on a profit basis.
- The *industrial world* values efficiency, productivity and technical competences.
- The *projective world*, which features prominently in the *New Spirit of Capitalism* (Boltanski & Chiapello, 2007), values flexibility, mobility, attractiveness and networks.

It is possible to link each of the indicators (some refer to teaching, some refer to research activities) of the annual performance plan for higher education to a specific world; only four of the above listed worlds are relevant for the purposes of our analysis. For each of these worlds, several topics and associated indicators can be identified. For the indus-trial world, for example: production volume (percentage of people with a university degree); production "failure" (non-completion rates); lead time (rate of PhD students defending their PhD thesis within three years); and efficiency (percentage of university building capacity in use). For the market world: revenues (percentage of revenues coming from intellectual property rights); competitiveness (share of world scientific publications, percentage of patents deposited by universities); and inclusion in the labour market (employment rate of graduates three years after graduating). For the projective world: attractiveness (percentage of foreigners among masters, doctoral and postdoctoral students, and academics); visibility (two-year citation impact); and networks (rate of participation in European Framework Programmes). For the civic world: fairness (ratio between foreign and home students' success rates).[7]

In summary, as far as teaching is concerned, the industrial world (with eleven indicators) is dominant, whereas the projective (four indicators), market and civic worlds (three indicators each) are present but in much weaker form. Especially the indicators related to the civic world are quantitatively weak but also qualitatively poor: the indicators chosen do not send out a strong political signal, for example towards widening participation and democratizing higher education. With regards to the indicators for research, six of them can be linked to the market world and five to the projective world. It is ultimately a widely economics-based idea of performance that emerges from the chosen indicators, focussing on revenues, efficiency, competitiveness and insertion in the labour market. Some of the indicators, especially those related to research, also focus on attractiveness and networking, being thus closely akin to "knowledge economy" theories. On the other hand, the political and civic dimensions are not very pronounced at all.

THE UPPER STRATUM: THE MICRO-CONVENTIONS OF CALCULATION

There is a large number of possible choices available at the most granular, even microscopic, level of each calculation. These choices have also been debated by the ministries. The Ministry of Education and the Conférence des Présidents d'Universités (CPU) [Association of University Presidents] often preferred indicators expressed in absolute terms, which they consider better for showing the high activity levels and social usefulness of higher education. The following excerpt from a letter from the CPU to the Secretary of Higher Education clearly illustrates this:

> Our general analysis of the indicators put forward [by the Treasury] is that several of them are disadvantageous, are negative for universities. While

French universities have to accept all the students who have passed the *baccalauréat*, they will be judged on qualitative criteria, such as success rates within a certain period of time (three years for undergraduates and PhD students for example). It is absolutely necessary that universities can also show quantitative results, such as the number (that is in absolute terms) of graduates they have trained.

By contrast, the Treasury systematically refused indicators expressed in absolute terms, and even refused their relative expression for a temporal analysis, such as growth rate. Let us consider, for instance, the indicator aiming to measure the objective of "Producing scientific knowledge at the best international level". The two ministries struggled fiercely over it. The Ministry of Education promoted the number or the growth rate of French publications in internationally recognized journals, whereas the Treasury promoted the percentage of French publications in internationally recognized journals.[8] The latter was finally chosen, so the Treasury won the battle on this indicator.

If we go back to the "orders of worth" analysis presented above, it is possible to regard the absolute measure as related to the industrial world, underlining production volume and the increase of this production volume; it is rooted in a productivity-based perspective. The relative measure can be seen as linked to the market world and being rooted in a market-based perspective, all the more so because the term "market share" is used in several speeches of Treasury officials, as well as many of the interviews I conducted with them. Furthermore, the latter choice allows comparisons between countries, and between universities or research centres. This choice, as Desrosières stated, "creates a new world in relation to which everyone has to position himself" (Desrosières, 2008b, p. 15); it makes benchmarking and ranking possible, and it makes it possible to put under pressure the universities or research centres which are at the bottom of the league (see also Dixon & Hood, 2016). As Ozga wrote: "Comparison defines the new mode of governance [...]. Comparison is war by other means" (Ozga, 2008, p. 268).

Even the limitations of the indicators can be used for this war, as a member of the Ministry of Education put it:

> The Treasury really put pressure on us to calculate some of the indicators in a certain way. For example, we had to fight really hard to make sure that the rate of PhD students defending their thesis within three years

should take into account students who have their viva before the 31st of December and not only before the beginning of the new academic year in September. This would change the result by more than 20 percentage points.[9]

The determination of the targets to be reached in the mid-term constitutes further evidence of this struggle between positive and negative pictures. The Ministry of Education wanted them to be set at an achievable level, the Treasury wanted them to be as high as possible. Furthermore, this negative picture tarnishes the image and perception of the public service in question in the eyes of MPs and in the eyes of the public. Indicators can then be used to justify reforms: this sector is doing badly, hence new policies are needed. For instance, the percentage of French publications in internationally recognized journals has been the most widely publicized indicator when talking about the quality of French research, fuelling alarmist discourses. Nicolas Sarkozy used this indicator in his speech of 22 January 2009 to legitimate the implementation of the Law on Liberty and Responsibility of Universities (the LRU) which was passed in 2007, but denounced by a great part of the French academia.

To sum up, the analysis of the construction process of the indicators picked up three elements: the indicators conceived as signals, the development of comparability and an economics-based idea of performance. This confirms Desrosières's (2008a) analysis. "Markets, incentives, benchmarks and rankings" have been, since the 1980s, "new and increasing features of public statistics" (Desrosières, 2008a, p. 112). But these indicators are only "loosely linked to each other" (Miller, 1992, p. 84). This way of developing quite an inconsistent set of indicators seems a specificity of NPM, considering that each field of social reality has its own dynamic separated from the others. In contrast, macroeconomic or national accounting aggregates are a very different kind of statistics, highly interconnected and based on a conception of the economy as a whole entity.

The analysis has also shown that a quantification device, such as a performance indicator, is the result, down to its smallest detail, of power struggles between the actors involved. Hence, "the moment of indicator design is a defining moment which will shape the future" (Desrosières, 2014, p. 47), and it is therefore a moment particularly important to analyse.

THE LIFE OF THE DEVICE: CONTEXT, USES AND DEVELOPMENTS

To understand the orientation of a quantification device and the effects it can produce, the context of its deployment and its uses have also to be examined.

2006–2012

The development of performance indicators is still at a relatively early stage in France. We must remember that one of the main objectives of the LOLF was to make the government and the public services more accountable to parliament for the outcomes of their actions. The performance indicators were supposed to be used by the MPs during the budget debate. However, all the reports and speeches during the national budget debates [10] were based on budget figures, describing the evolution of each policy measure; and the discussions among MPs turned to the relevance of these budget decisions. Performance indicators were not used at all in these debates. Two reasons at least can explain that. Firstly, the LOLF did not link performance to funding. As Lambert and Migaud, two MPs and fathers of the LOLF, pointed out:

> Managing a public entity is not the same as managing a company. For the state, there is no direct link between the level of budget funds and the objectives to be achieved. To decide the level of appropriations, the notion of needs will remain the most important. It is thus possible that achievements will have no budgetary impact. (Lambert & Migaud, 2006, pp. 13–14)

Secondly, the indicators are rather meaningless for the MPs. They are quite technical. Because of the objective of consistency, the indicators have changed little since 2006, and the MPs, as was mentioned, did not participate in their construction. MPs do not find the indicators meaningful in relation to public policy making. Since they discuss policy options and consider to be at the heart of political choices budgetary decisions, they use budget figures. Furthermore, the budget debate is conceived to be more of a debate between MPs from different political parties, rather than an exchange between parliament and public administration.

As shown above, within the government, the Treasury and the Ministry of Education anticipated, while negotiating the indicators, that these might be used during the budgetary negotiations. But the performance indicators were not used at that moment either; as a senior official from the Treasury put it:

> We were afraid that the different ministries would use the performance indicators to ask for an increased budget—that is the reason why we were so tough during the discussions about these indicators. But in fact, nobody uses them during the discussions of the ministries' budgets; we speak about money, about staff, about policies, not about performance, neither them nor us.

The performance indicators were, in the end, not used in the two situations where they were expected to be used. However, they were going to be very powerful in a different, unexpected way. A French state reform, called the "General Revision of Public Policies" (RGPP), launched by Nicolas Sarkozy in 2007 immediately after his election, initiated a new usage of them: performance indicators became tools for resource allocation from the Ministry of Education to universities. To make this understandable, we must briefly explain the history of the French university funding system as summarized in Table 9.1.

The new system of resource allocation, called "*Sympa*" (which can be translated as "cool") resulted from discussions between the Department of Higher Education of the Ministry of Education, the parliament and the Conférence des Présidents d'Universités (CPU) [Association of University Presidents]. The system had two parts: one depending on activities (80%) and one depending on performance achievements (20%). Although, this does not seem to be a big change compared to the previous system, it actually was, as the newly introduced activities criterion had also a performance dimension. It is now no longer merely the number of students registered at the beginning of the year, but the number of students who sit the exams, and it is no longer the number of staff, but the number of "publishing academics" that counts (and non-publishing academics hamper the performance of their own university). If one incorporates this last criterion into the "performance share", the share makes up more than 50% of the budget of most universities,[11] although it is the 80–20 ratio which has been taken up in the parliamentary reports and documents and speeches produced by the ministries and CPU.

Table 9.1 French university funding system (from mid-1980s to 2012) (compiled by the author)

Up to mid-1980s	From mid-1980s to 2008	From 2009 to 2012 (Sympa)
100% block grant[a] based on: – Number of students – Number of administrative and academic Staff – Surface area	70–80% block grant based on: – Number of students – Number of administrative and academic staff – Surface area 20–30% contractual resources negotiated and based on projects	80% 'activity-based share' based on: - Number of students sitting the exams[b] – Number of publishing academics[c] 20% 'Performance Share' based on: *For Teaching:* – Undergraduate success rate[d] – Number of Master degrees delivered[e] *For Research:* – Grades of university research centres (A+, A, B or C) – Number of PhD degrees delivered

[a]The tuition fees were very low (they are in fact more comparable to registration fees than to tuition fees); they were set by the Ministry of Education and were the same for all universities. They were taken off the block grant. The system remains the same today: annual tuition fees were around 300 euros for BA and MA degrees in 2020–2021

[b]Subject to a weighting based on the field (exact sciences, natural sciences, social sciences and humanities) and the level (undergraduate or graduate) of the degree

[c]Subject to a weighting based on the field: exact sciences (COEFF 2.5), natural sciences (COEFF 2.6) and social sciences and humanities (COEFF 2.0), and multiplied by the grade of the research centre: a publishing academic who works in an A+ unit is weighted 2.0, in an A unit 1.5, in a B unit 1.0 and in a C unit 0.5

[d]Weighted by the grant holders' ratio

[e]Subject to a weighting based on the field

Although I have not examined the negotiations and controversies behind the production of the new resource allocation system, we can analyse its indicators. Some of these are derived from the LOLF, however, at least four different rationales can be noticed: first, a pure performance logic awarding good and bad marks (counting the number of publishing academics, awarding grades to research centres); second, an attempt to consider social and cultural inequalities (for example, by weighting the success rate by the grant holders' ratio)[12] driven by equity concerns; third, the acknowledgement of a university's activity (for instance the number of Master degrees and PhD degrees delivered, so an indicator in absolute terms as the CPU asked for the LOLF indicators); fourth, a weighting by fields, which is a remainder from the first resource allocation model (see also Table 9.1).

Quite technical and poorly publicized, the *Sympa* system received very little attention beyond the circles of the CPU, Ministry of Education and some well-informed MPs. These parties welcomed the new resource allocation system for several reasons. First, it was the result of negotiations in which these parties had been involved, and different rationales had been taken into account. Second, it was deemed to be an objective system based on clear criteria which replaced the 20–30% of contractual resources which previously had to be negotiated between a university's management team and the Ministry. Third, besides its performance dimension, it was supposed to make visible inequalities existing between universities and, hence, could help address and reduce these. Fourth, parallel to the introduction of the new system, the government committed to a general increase of the higher education budget; so each university was supposed to benefit from *Sympa*, and under-resourced ones were supposed to benefit more.

However, it quickly became apparent that things should turn out very differently. The Ministry decided to increase the budget by far less than originally promised, and it also decided not to put the increase into *Sympa*'s envelope, but a separate "Undergraduate Success Programme" (*Plan Licence*), which should become a key measure for the government. In fact, the *Plan Licence* is still attached to the name of Valérie Pécresse, then Minister of Higher Education. The *Plan Licence* was allocated to each university without using the *Sympa*'s criteria, and through negotiations with each university. As a staff member of the Ministry of Education put it:

> Actually, the Ministry wanted to keep the power in its hands, at least what it thought was power; it wanted to have something on the table. And to be able to use it to encourage universities to apply voluntarily some new regulations (for example encourage them to merge). The result is: it was the tougher university president, the one with political support, the one who was president of an already well known university, etc. who earned the most.

Since the *Sympa*'s envelope remained steady, gains for some meant losses for others. The CPU, on the other hand, refused the notion of an "over-resourced" university and rejected any redeployment of resources between universities. Because of the decision of the Ministry and the position of the CPU,[13] the possibility of reducing inequalities faded away, while the effects of *Sympa's* performance dimension became quickly evident.

Performance is now financially rewarded, while non-performance is financially punished. In this context, having indicators which take into account the inputs or the conditions of teaching and research is very important. Performance can be linked with the academic level of the students, or with good working conditions, such as the number of administrative staff, which allows academics to do less administrative work and to have more time for research. In the same way, non-performance can be linked to "low" inputs and difficult working conditions, something which *Sympa*'s weighting tried to take into account, but now the punishment is only going to make worse.

Merton already revealed this phenomenon in the 1960s, and he called it the "Matthew effect" (Merton, 1968) referring to the following passage in the gospel of Matthew: "For everyone who has will be given more, and he will have an abundance. Whoever does not have, even what he has will be taken from him" (*New Testament*, Matthew 25:29). We can also speak of "cumulative advantages": a favourable relative position becomes a resource that produces further relative gains, so the rich get richer at a rate that makes the poor become relatively poorer. And the richest universities are generally the ones where most of the students come from privileged social backgrounds. So, performance-based financing, which is a frequent component of New Public Management, leads to a concentration of resources around those who already have the most,[14] which has significant implications for our conceptions of equality and justice. Suleiman already noted in 2003 that a lot of the proposals coming from New Public Management theories "have little to do with bureaucracy in itself and much to do with the distribution of public resources" (Suleiman, 2003, p. 20). But often this political dimension is hidden: "The allocation of resources seems to result instead from the dynamism and the quality of individuals and institutions" (Le Galès & Scott, 2010, p. 132).

Generally speaking, this way of funding introduces competition between universities: once the total budget for higher education is decided, the fact that some universities have financial rewards, and so more money, naturally means that other universities will have less. Competition takes place at the heart of performance-based financing systems. Furthermore, these "*Sympa* indicators" have impacted universities' policies. *Sympa*'s weighting pattern has indeed had an incentive effect: an increase of five hundred students who pass their degree (for the same number of students passing the exams) would generate an additional budget of €80,000, while a 2.5% increase of the number of publishing academics would mean an additional budget of €474,000. In this context, a lot of universities chose to redirect financial and human resources from teaching, not towards marketing, as Espeland and Sauder (2007) observed for American Law Schools, but towards research. Furthermore, it is advantageous for universities to recruit new academics for research centres ranked A + or A, and not for the ones graded B or C. Lastly, a lot of universities chose not to hire research officers, but to recruit only university lecturers and professors, because the activity of the former is made invisible by *Sympa*, although the potential lack of research officers does not bode well for the development of science.

In addition, more and more universities decided to introduce a performance-based resource allocation system internally, especially for their research centres. They introduced a variable part for their budget, for example 15%, based on performance indicators, such as the ratio of publishing academics and the grade of the research centre (which is partly based on the same ratio). But most often the management teams of the research centres do not know anything about the criteria on which the allocation of the remaining 85% is based. They are rather committed to the variable component, arguing that at least here the criteria are known and clear, and that it is a more transparent and fairer way of allocating resources, not depending on personal relationships and lobbying. With this system, they know what they have to improve and so they have the feeling of being able to contribute to the sound management of public money.

But, this link between performance and funding also produced what can be called "punitive practices": some research centres excluded the non-publishing academics (to increase the ratio of publishing ones); some decided not to pay for the costs when a non-publishing academic gave a presentation at a conference. The presence of such practices depends largely on the disciplines; they are quite rare in the social sciences but quite common in economics and management,[15] all the more so because the criteria used to decide whether one is publishing or non-publishing are, as will be discussed later, more strictly defined in economics and management than in the social sciences and the humanities.

This way of governing public services by financial incentives is really new in France. The relationships within the state and public services (and between them) are thought of in terms of microeconomic theory, specifically agency theory: "Society is viewed as a system of essentially self-interested 'elementary particles'" (Supiot, 2015, p. 216); institutional and individual actors are thus thought of in terms of *homo economicus*, whose actions can be driven and controlled through a system of punishments and rewards. This is often the basis of management by objectives in the private sector and in the context of New Public Management. Vinokur (2008) summarized these changes when talking about a shift from the model of "obligation of means + trust" (the obligation of means resulting in a bureaucratic *ex ante* form of control, and the trust in job security for civil servants) to a model of "obligations of results + distrust". We already highlighted the "low trust in public servants and professionals" when referring to Hood (1995, p. 94), which accompanied the rise of NPM and its new conception of accountability.

Finally, the performance indicators have widely replaced, within universities, the previous statistics they produced internally for a better understanding of their students. Performance indicators, and specifically those used by *Sympa*, became the dominant metrics compared to the statistics produced for acquiring knowledge about students' characteristics. This trend, which can also be found in health care and the social services,[16] is part of "the shift away from the social welfare state as guarantor of basic solidarities and rights, access and treatment for all, to the state as a provider of services" (Jany-Catrice, 2016, p. 129).

Since 2012

In 2012, the resource allocation model of *Sympa* was abandoned by the new French government set up under the presidency of François Hollande. This demise has to be linked to several protests from students and academia from 2007 onwards. As mentioned before, the government, set up in May 2007 under the presidency of Nicolas Sarkozy, had immediately passed a new law on Liberty and Responsibility of Universities (the LRU Law).[17] This law led to great protests, strikes and demonstrations, from 2007 to 2009. Initially, the protests were against the new law as a whole; later (in 2009), they partly focused on the new status introduced for university lecturers and professors by the new law, under the so-called "*modulation des services*". Here, those who were classified as non-publishing might be given, by the president of their university, more teaching. Opponents to the reforms widely criticized the quantitative evaluation of the research centres and individuals, which led to centres being ranked and academics categorized as publishing or non-publishing.

The AERES (*Agence d'Évaluation de la Recherche et de l'Enseignement supérieur*), whose setting up in 2006 was already controversial, was in charge of these assessments. It developed, during the summer of 2008, the criteria that defined an academic as publishing or non-publishing, and it designed a ranking of scientific journals for each discipline: an academic was considered as publishing, if s/he had published two articles in a journal ranked A or B during the past four years. In the fall of 2008, many petitions emerged signed by academics, academic professional organizations and trade unions, scientific committees and editorial boards of academic journals; some of these petitions were against the ranking system, some against the priority given to publication in journals at the expense of other scientific activities (including the publishing of books), and some were against forms of quantitative evaluation all together. In response to these, in October 2008, the AERES allowed the academic disciplines themselves to identify a list of scientific journals without ranking them, and it made it possible to count books and book chapters as publications (at the discretion of the respective assessors). In 2009, AERES published new lists of journals, and while nearly all the social sciences and humanities decided not to rank the journals listed by them, economics and management studies did and still do so.

To bring the strike against the new status for lecturers and professors to an end, the government added that non-publishing academics might be given more teaching by the university president only if the academics concerned agreed. Critics of the ranking of research centres arose again a few years later, leading to new petitions in 2011. In response to all these protests, the Socialist Party committed during the presidential campaign (2012) to organize a National Conference on Higher Education and Research. François Hollande was elected, and the National Conference was held in November 2012. The previous system of assessment was, among other things, widely criticized during the Conference, and the participants proposed that research centres should no longer be graded and ranked. A working group, comprising representatives of the Ministry and the CPU, was set up in April 2013 to re-examine the *Sympa* model. At the beginning of 2014, the AERES replaced the grades with a "textual appreciation" of research centres, which made *Sympa* obsolete. The working group was supposed to produce a new resource allocation model by the end of 2014, but the Ministry and the CPU did not manage to agree, and the working group stopped working in 2015.

From the above, four sticking points can be at least identified. First, the Ministry conceived *Sympa* as a decision-support tool, providing it with some room for manoeuvre and negotiations. In contrast, the CPU wanted the system to automatically calculate the budget of each university. Second, the Ministry would have liked to include the payroll in *Sympa*, but the CPU refused. Third, there was disagreement about the indicators themselves. The negotiating bodies seemed to agree on replacing the number of publishing academics by the number of academics, but they disagreed on the performance indicator for research that would replace the grades of research centres. The Ministry proposed two indicators, one measuring the participation in European Framework Programmes, and one based on the number or the percentage of *Institut Universitaire de France* laureates. The CPU questioned the method of calculation of the first and refused also the second arguing that it measured individual performance but did not evaluate the collective performance of a research centre or a university as a whole. Finally, a controversy about the weighting factors by field led to the conduct of a cost analysis of teaching and research in order to base the factors on objective information. As a result, the decision-making process on the budget allocated by the Ministry to each university became, contrary to what was initially intended, even less transparent.

CONCLUSION

This chapter has traced the life of performance indicators in French higher education from their birth in the beginning of the 2000s to the end of the 2010s. Analytically, it distinguished between three levels that make up a quantification device, such as performance indicators: (a) the bedrock or the ontology of the device; (b) the intermediate level made up of the conceptions about the *"raisons d'être"* of the quantified entity; and (c) the upper stratum comprising the micro-conventions of calculation, and it analyse the context in which the indicators were deployed. This study has shown how relevant it is to enter into the analysis of a specific field by decoding the genesis and uses of its quantification devices.[18] As Salais has highlighted: "The choice of the indicators, the construction of data and their uses reveal the normative assumption of the policies" (Salais, 2004, p. 298). The three levels and the context are the result of socio-historical processes in which different social actors participate, bearing different philosophies or value systems. These processes can lead to great coherence between levels and context. But they can also lead to a weakly integrated device that exhibits many "gaps". As Kurunmäki, Mennicken and Miller, drawing on Deleuze and Guattari's works, put it: "The unity of such assemblages derives only from the co-functioning of their components; the relations that are formed among them" (Kurunmäki et al., 2016, p. 399). We can try now to capture the salient points of this assemblage.

This assemblage is based on the idea that the state is accountable for the results of its action, which is a specific conception of public accountability which emerged from the 1980s and which now prevails. The state chose to assess results by, and only by, quantitative indicators, which reveals a belief in numbers and a "realist epistemology". From these starting points, this assemblage became economics-based and competition-oriented. Firstly, by choosing the vocabulary of "performance" (rather than the more neutral term "results"). "Performance" carries many connotations and introduces the notion of competition. Furthermore, the various negotiations between different parts of public administration led to a widely economics-based idea of performance focussed on revenues, efficiency, competitiveness, insertion into the labour market, attractiveness and networking, being thus closely akin to "knowledge economy" theories. On the other hand, political and civic dimensions of governing were almost absent. The chosen indicators and the way they are calculated made comparisons, benchmarking and ranking possible; they came to be conceived as signals towards universities. A performance-based financing system was implemented that introduced financial rewards and punishments, which was strengthened by some of the *Sympa* indicators, such as the number of publishing academics and the grades of the research centres. This system led to the production of "Matthew effects" and it introduced competition between universities. Because of *Sympa*'s weighting pattern, it encouraged universities to redirect financial and human resources from teaching towards research.

Together with the *"modulation des services"*, it made the development of punitive practices against individuals possible. This was a new way of governing public services in France: the relationships within the state and public services (and between them) became thought of in terms of microeconomic theory, and specifically within the framework of the agency theory; collective and individual actors were thought of in terms of *homo economicus*. All of these elements are quite coherent.

However, the LOLF did not, at its beginning, link performance to funding, thus there was no performance-based funding system at the national or ministerial levels. The indicators resulted from discussions and negotiations; they resulted from power dynamics and different strategies which introduced different logics. The present analysis has tried "to disentangle such multiplicities" (Kurunmäki et al., 2016, p. 397) of rationales. Aside from the competitive logic, there was also strong support for a transparent and automatic funding system that would be able to counteract nepotism and arbitrariness (and at the same time a refusal from the Ministry of such a transparent and automatic system). Porter (1995) and Supiot (2015) showed that quantification devices are also an essential part of "government by rules" and of democracy. There was a strong demand for a funding system based on the needs of the universities, more than on their results; a well-informed MP, for example, welcomed in 2014 the change of indicator from the number of publishing academics to the number of academics. There was finally a will to take into account the social and cultural inequalities in order to reduce them.

Furthermore, several protest movements within academia have strongly criticized the rationale of competition, the quantitative mode of evaluation, the definition and use (*modulation des services*) of the status of "publishing academic", the rankings of academic journals, the grading of research centres and the punishing practices. These movements were victorious in some respect: the *modulation des services* is now only possible if the non-publishing academic agrees with the ruling; the grades for research centres have been abolished; the publication of books and book chapters can be taken into account for the social sciences and humanities; and journals are no longer ranked in these disciplines. These alterations have reduced the impact of the competitive and punitive dimensions of the device and explain why, together with the tenure and status of *"fonctionnaire"* for the great majority of academics,[19] it has become, at least in the social sciences and humanities, a very limited "engine of anxiety" (Espeland & Sauder, 2016).

In summary, the reforms studied here, and the quantification devices implied, do not form a very integrated assemblage. This explains why the changes that the performance indicators were supposed to help bring about were at the same time significant and limited, in comparison with what may have occurred in the UK higher education system for example (see also Eyraud, 2016).

The three-part analytical distinction introduced in this chapter, combined with an analysis of the context in which a quantification device, such as a performance indicator, is deployed, makes visible the very broad range of possible options. One can agree with the idea that the state and the public services must be accountable to citizens, regarding it as a significant democratic progress. But this does not mean that only numbers are able to report on public action and social reality; being accountable is not just about reporting numbers. As Supiot pointed out: "To confuse measurement and assessment inevitably dooms us to lose our sense of proportion; assessment is not only measurement; assessment requires that the measurement is referred to a value-based judgement which gives it meaning" (Supiot, 2010, p. 82). De Gaulejac went even further: "We should abandon the economist, objectivist and mathematical conception of assessment and adopt a qualitative, democratic and dynamic one" (De Gaulejac, 2012, p. 77). Indeed, the very idea of measuring performance by indicators should be questioned, since there may exist other preferable ways of assessing public services.

However, even if quantitative evaluation should not be the only way, quantitative indicators can be useful in the process. In that case, the starting point, from which the indicators emerge, should be a wide-ranging public debate on what is expected from public policies. As Gadrey (1996)proposed, Boltanski and Thévenot's (2006) "orders of worth" can help to specify the different expectations placed on indicators. Furthermore, detailed definitions of the indicators should not be a way of surreptitiously encapsulating values and hierarchies. These micro-conventions of calculation should be drawn up with an aim of impartiality, and when a choice is needed it should be made democratically in order to construct what could be called a "shared objectivity". Finally, the uses of the indicators should also be carefully scrutinized and debated. The decision to base funding on performance indicators has powerful effects, such as an increase in inequality. The analytical grid proposed here provides a blueprint for building "what should be, in our view, a satisfying process of quantification" (Salais, 2016, p. 133).[20]

NOTES

1. I use the term "device" referring to Foucault's concept of "dispositif". Foucault defines dispositif as "a thoroughly heterogeneous ensemble consisting of discourses, institutions, architectural forms, regulatory decisions, laws, administrative measures, scientific statements, philosophical, moral and philanthropic propositions—in short, the said as much as the unsaid (…). The dispositif itself is the system of relations that can be established between these elements" (Foucault, 1994, p. 299). The term thus emphasizes the complex and varied nature and the systemic dimension of these "ensembles". I have chosen to use the notion of "device", even if the English translation of Foucault's "dispositive" has given rise to extensive discussions. Some of the published translations retain the term in French; others opt for various solutions such as "apparatus", "device", "arrangement", etc.

2. In 2006, the French state moved from a specific public accounting system to a business accounting system.

3. This concept refers to Espeland and Sauder's work. They define "reactivity" as "the idea that people change their behaviour in reaction to being evaluated, observed or measured. [...] Because people are reflexive beings who continually monitor and interpret the world and adjust their actions accordingly, measures are reactive" (Espeland & Sauder, 2007, pp. 1–2). Desrosières has also insisted on this aspect of quantification throughout his work (see for example, Desrosières, 2008b, p. 12).

4. In the same vein, Miller (1990) demonstrated the interrelation between accounting and the state, and Desrosières showed that there is a degree of congruence between modes of governance, conceptions of the state and statistical tools (Desrosières, 2014, pp. 33–58).

5. Miller and Rose (1990; 1992) used the notion of "programme".

6. It would be possible to go even deeper into the roots of Western civilization and the way it conceived government (Supiot, 2015).

7. A table which groups the various indicators into the different orders of worth can be found in Eyraud (2014, p. 81).

8. We can note here that even if the two ministries did not agree on the way of calculation, they implicitly agreed on the principle that scientific production had to be measured by, and only by, publications in academic journals.

9. In France, the great majority of PhD vivas take place from October to December. This quotation also shows the absolute need, if one wants to understand statistical figures, to go into the details of definitions, delimitations and methods of calculation (Eyraud, 2008). It is one of the reasons why international comparisons using statistical data are so difficult to handle properly.

10. I have followed the budget debates for higher education on the parliamentary channel each year since 2007, and, on that matter, there is no change even after 2012.
11. The calculation can be made using the universities' *Sympa* data sheets, which I managed to get hold of for two universities.
12. Grants are allocated based on means-testing parental incomes. The idea behind is that scholarship students are the ones with low cultural capital, thus, with a lowest probability of success which has to be taken into account in the performance measurement of a specific university.
13. Neither the Ministry of Higher Education nor the CPU are homogeneous organizations. These were decisions and positions that had to be won inside these organizations.
14. This process is reinforced by the different "policies of excellence" launched since 2010 and their competitive funding arrangements.
15. I did not conduct any interviews within natural science research centres.
16. But not in all domains. Dubet showed, for example, that quite a similar way of governing French secondary schools pushed them to produce social data on their pupils to justify their choices and to obtain additional resources (Dubet, 2016, p. 387). Statistics produced for acquiring knowledge about pupils existed before, but only at the national level, not at the school level.
17. To understand more about the law and the protests, one can read in French Vinokur (2008) and in English Briggs (2009). The English page of Wikipedia is also quite informative: https://en.wikipedia.org/wiki/2007%E2%80%9309_university_protests_in_France, accessed 15 September 2019.
18. Even if it does not provide a complete analysis of the field, because some of its characteristics and transformations are beyond the scope of the quantification devices. For a comprehensive understanding of the recent transformations of French higher education, one should integrate at least the severe budgetary constraints with which the universities are confronted (Henry & Sinigaglia, 2014; Sinigaglia, 2018), the different "policies of excellence", the policy of university grouping and merging, and the change of universities legal status (Eyraud, 2020).
19. The situation is different for casual workers, especially for the young generation. The LRU introduced the possibility to recruit "casual lecturers", but a lot of academics, academic organizations and trade-unions opposed it, so few governing boards of universities decided to hire people under this new status. Things may change quickly as a result of a new law, the "*Loi de Programmation de la Recherche*" (LPR) passed in December 2020.
20. I develop this idea further with several concrete examples in Eyraud (2019).

REFERENCES

Boltanski, L., & Chiapello, E. (2007). *The new spirit of capitalism*. Verso (French edition, 1999).

Boltanski, l., & Thévenot, l. (2006 [1991]). *On justification: Economies of worth*. Princeton University Press (French edition, 1991).

Briggs, R. (2009). President Sarkozy, La Princesse de Clèves, and the crisis in the French higher education system. *Oxford Magazine, Second Week* (Trinity Term), 4–6.

Eyraud, C. (2020). Université française: mort sur ordonnance? *Droit et société*, N°105(2), 361. https://doi.org/10.3917/drs1.105.0361.

De Gaulejac, V. (2012). *Manifeste pour sortir du mal-être au travail*. Desclée de Brouwer.

Desrosières, A. (1998 [1993]). *The politics of large numbers: A history of statistical reasoning*. Harvard University Press.

Desrosières, A. (2000). L'usage des statistiques dans l'étude des inégalités sociales. In *Définir les inégalités, des principes de justice à leur représentation sociale* (pp. 74–92). Ministère de l'Emploi et de la Solidarité (DREES/MIRE).

Desrosières, A. (2003). Managing the economy: The state, the market and statistics. In T. M. Porter & D. Ross (Eds.), *The Cambridge history of science* (pp. 553–564). Cambridge University Press.

Desrosières, A. (2008a). *Gouverner par les nombres*. Presses de l'Ecole des Mines de Paris.

Desrosières, A. (2008b). *Pour une sociologie historique de la quantification*. Presses de l'Ecole des Mines de Paris.

Desrosières, A. (2014). *Prouver et gouverner: Une analyse politique des statistiques publiques*. La Découverte.

Desrosières, A., & Thévenot, L. (1988). *Les catégories socioprofessionnelles*. La Découverte.

Dixon, R., & Hood, C. (2016). Ranking academic research performance: A recipe for success? *Sociologie Du Travail, 58*, 403–411.

Dubet, F. (2016). Les instruments et l'institution. *Sociologie Du Travail, 58*, 381–389.

Espeland, W. N., & Sauder, M. (2007). Rankings and reactivity: How public measures recreate social worlds. *American Journal of Sociology, 113*(1), 1–40.

Espeland, W. N., & Sauder, M. (2016). *Engines of anxiety: Academic rankings, reputation, and accountability*. Sage.

Eyraud, C. (1999). *L'entreprise d'état chinoise. De l'institution sociale totale vers l'entité économique?* L'Harmattan.

Eyraud, C. (2003). Pour une approche sociologique de la comptabilité. Réflexions à partir de la réforme comptable chinoise. *Sociologie du Travail, 45*(4), 491–508.

Eyraud, C. (2008). *Les données chiffrées en sciences sociales. Du matériau brut à la connaissance des phénomènes sociaux.* Armand Colin.

Eyraud, C. (2013). *Le capitalisme au coeur de l'état: comptabilité privée et action publique* Bellecombe-en-Bauges Ed. du Croquant.

Eyraud, C. (2014). Reforming under pressure: Governing and funding French higher education by performance indicators (2006–2012). In P. Mattéi (Ed.), *University adaptation in difficult economic times* (pp. 75–88). Oxford University Press.

Eyraud, C. (2016). Quantification devices and political or social philosophy: Thoughts inspired by the French state accounting reform. *Historical Social Research, 41*(2), 178–195.

Eyraud, C. (2019). La sociologie et le chiffre. Ou existe-t-il des chiffres objectifs? In A. Pariente (Ed.), *Les chiffres en finances publiques* (pp. 247–265). Le Kremlin-Bicêtre: Mare & Martin (Collection Droit & Gestions publiques).

Eyraud, C., El Miri, M., & Perez, P. (2011). Les enjeux de quantification dans la LOLF. Le cas de l'enseignement supérieur. *Revue Française de Socio-Économie*(7), 149–170.

Foucault, M. (1994). *Dits et écrits 2, 1972–1975* (Vol. 1). Editions Gallimard.

Gadrey, J. (1996). *Services: La productivité en question.* Desclée de Brouwer.

Henry, O., & Sinigaglia, J. (2014). De l'autonomie à la mise sous tutelle. *Savoir/agir, 29*, 15–24.

Hood, C. (1995). The "New public management" in the 1980s: Variations on a theme. *Accounting, Organizations and Society, 20*(2/3), 93–109.

Jany-Catrice, F. (2016). Evaluating public policies or measuring the performance of public services? In I. Bruno, F. Jany-Catrice, & B. Touchelay (Eds.), *The Social Sciences of Quantification* (pp. 123–135). Springer.

Sinigaglia, J. (2018). Mes enfants l'heure est grave: il va falloir faire des économies. *Actes de la recherche en sciences sociales, 221–222*(1), 20. https://doi.org/10.3917/arss.221.0020.

Kurunmäki, L., Mennicken, A., & Miller, P. (2016). Quantifying, economising, and marketising: Democratising the social sphere? *Sociologie Du Travail, 58*, 390–402.

Lambert, A., & Migaud, D. (2006). La LOLF: Levier de la réforme de l'État. *Revue Française D'administration Publique, 117*, 11–14.

Le Galès, P., & Scott, A. (2010). A British bureaucratic revolution? Autonomy without control or 'Freer actors more rules'. *Revue française de sociologie, 51*(English annual selection), 119–146.

Merton, R. K. (1968). The Matthew Effect in science. *Science, 159*, 52–64.

Miller, P. (1990). On the interrelation between accounting and the state. *Accounting, Organizations and Society, 15*(4), 315–338.

Miller, P. (1992). Accounting and objectivity: The invention of calculating selves and calculable spaces. *Annals of Scholarship, 9*(1–2), 61–86.

Miller, P., & Rose, N. (1990). Governing economic life. *Economy and Society, 19*(1), 1–31.

Ozga, J. (2008). Governing knowledge: Research steering and research quality. *European Educational Research Journal, 7*(3), 261–272.

Porter, T. M. (1995). *Trust in numbers: The pursuit of objectivity in science and public life.* Princeton University Press.

Rose, N., & Miller, P. (1992). Political power beyond the state: Problematics of government. *British Journal of Sociology, 43*(2), 172–205.

Salais, R. (1986). L'émergence de la catégorie moderne de chômeur: les années 1930. In R. Salais, N. Baverez, & B. Reynaud (Eds.), *L'invention du chômage. Histoire et transformations d'une catégorie en France des années 1980* (pp. 77–123). PUF.

Salais, R. (2004). La politique des indicateurs. Du taux de chômage au taux d'emploi dans la stratégie européenne pour l'emploi (SEE). In B. Zimmermann (Ed.), *Les sciences sociales à l'épreuve de l'action: Le savant, le politique et l'Europe* (pp. 287–331). Éditions de la Maison des Sciences de l'Homme.

Salais, R. (2010). Usages et mésusages de l'argument statistique: le pilotage des politiques publiques par la performance. *Revue Française des affaires sociales*(1–2), 129–147.

Salais, R. (2016). Quantification and objectivity: From statistical conventions to social conventions. *Historical Social Research, 41*(2), 118–134.

Suleiman, E. (2003). *Dismantling democratic states.* Princeton University Press.

Supiot, A. (2010). *L'Esprit de Philadelphie. La justice sociale face au marché total.* Seuil (English translation: Verso, 2012).

Supiot, A. (2015). *La gouvernance par les nombres.* Fayard (English edition, Verso 2016).

Thévenot, L. (1979). Une jeunesse difficile: les fonctions sociales du flou et de la rigueur dans les classements. *Actes de la recherche en sciences sociales* (26–27), 3–18.

Thévenot, L. (1983). L'économie du codage social. *Critiques De L'economie Politique, 23–24,* 188–222.

Thévenot, L. (1990). La politique des statistiques: les origines sociales des enquêtes de mobilité sociale. *Annales E.S.C., 45*(6), 1275–1300.

Thévenot, L. (2015). Autorités à l'épreuve de la critique. Jusqu'aux oppressions du 'gouvernement par l'objectif'. In B. Frère (Ed.), *Le tournant de la théorie critique* (pp. 216–235).Desclée de Brouwer.

Thévenot, L. (2016). From codage social to economie des conventions: A thirty years perspective on the analysis of qualification and quantification investments. *Historical Social Research, 41*(2), 96–117.

Vinokur, A. (2008). La loi relative aux libertés et responsabilités des universités: essai de mise en perspective", Revue de la régulation. Capitalisme, institutions, pouvoirs, n°2, http://regulation.revues.org/document1783.html. *Revue de la régulation. Capitalisme, institutions, pouvoirs, 2,* http://journals.opened ition.org/regulation/1783.

PART III

Voicing for Democracy

Quantification = Economization? Numbers, Ratings and Rankings in the Prison Service of England and Wales

Andrei Guter-Sandu and Andrea Mennicken

Since the late 1980s, the HM Prison Service in England and Wales has undertaken a series of steps to transform its prison establishments into calculating, economically minded, performance-oriented institutions. This happened in the broader context of wider New Public Management reforms (Hood 1991, 1995). Of particular importance, in this context, was the government's engagement in prison privatization and the introduction of private sector accounting and management consulting expertise (Mennicken, 2013). The government hoped that the introduction of private prisons would help to "provide an alternative standard against which to measure the performance of public prisons, thereby bringing about improvements in the public sector" (James et al., 1997,

A. Guter-Sandu (✉) · A. Mennicken
Department of Accounting and Centre for Analysis of Risk and Regulation,
London School of Economics and Political Science, London, UK
e-mail: a.guter-sandu@lse.ac.uk

A. Mennicken
e-mail: a.m.mennicken@lse.ac.uk

A. Mennicken and R. Salais (eds.), *The New Politics of Numbers*,
Executive Politics and Governance,
https://doi.org/10.1007/978-3-030-78201-6_10

p. 9). Privatization was promoted to transform prison establishments from inflexible, inefficient, rules-based bureaucratic organizations to cost-conscious, performance-oriented entities. The rise and spread of standardized performance measures paralleled these developments. Quantified performance targets were introduced to make public and private prison performance outcomes more visible and transparent and to enhance competitiveness amongst individual establishments (Ministry of Justice, 2009, but see also Home Office, 1988, 2000).

This chapter uses the case of prison privatization in England and Wales to scrutinize what it means to "economize the social" through numbers. By economizing we refer to the processes through which individuals, activities and organizations are constituted or framed as economic actors and entities (Çalışkan & Callon, 2009; Miller & Power 2013). Emphasis is placed on the process by which a supremacy of the economic over society, including politics and domestic life, is articulated and established (Miller & Power, 2013). Economizing defined in such broad terms has many components: First and foremost, "it implies a concern with the idea of efficiency – governing aimed at enhancing individual or collective performance, the reduction of wastefulness, and the imposition of rationing through calculation" (Kurunmäki et al., 2016, p. 396). Economizing also encompasses the creation and expansion of markets, and an enhanced focus on competition to improve performance (Çalışkan & Callon, 2010; Davies, 2014). Lastly, financialization, understood as the rise and expansion of financial markets, financial expertise and (capital) investment rationales, can be seen as a variant of economization (Mennicken & Espeland 2019, p. 234). In this context, French scholars in particular have drawn attention to the processes whereby things (e.g. higher education) or human beings (e.g. prisoners) are turned into assets evaluated in their capacity to create value from the perspective of an investor who expects calculable future returns (Chiapello, 2015; Muniesa et al., 2017).

Quantification undergirds all these processes of economization. Quantification and commensuration are key conditions for economic calculation and action (Mennicken & Espeland 2019). Cost accounting, such as the National Reference Cost Index for NHS hospitals or the calculation of prison unit costs, for instance, helps instantiate ideas of efficiency and frugality at the heart of hospital care and the management of prison establishments. Ratings and rankings (e.g. of organizational entities such as universities or prisons) enable benchmarking and the stimulation of competition (e.g. between universities), and are often closely linked with the establishment of (quasi)markets in the public services. Furthermore, especially over the past fifteen years or so, we have seen an increase in quantified social impact assessments, such as social return on investment figures, aimed at making the value of charity and public sector work knowable and visible from an investor's perspective (Barman, 2016; Hall et al., 2015).

In the following, we scrutinize the multiplicity of such quantification practices and their implication in different processes of economization in the Prison Service of England and Wales. First, we show how prison privatization in the 1990s was accompanied by a rise of prison performance metrics, rankings and ratings aimed at facilitating a shift from "governing by rules" to "governance by numbers" (Miller 2001; Supiot 2015). We observe a market-oriented utilization of quantification, aimed at stimulating competition amongst and between public and private prison providers. Next, we turn to attempts aimed at undoing such economization. We trace efforts of "moralizing" prison performance metrics through the development of measures of the quality of prison life (Liebling, 2004). We show how such moralizing quantification eventually came to be undermined by austerity policies and related economizing practices of curtailment and frugality. Lastly, we discuss the implication of prison quantification in processes of financialization and (capital) investment rationales, focusing on the introduction (and abolishment) of the world's first social impact bond in Peterborough prison.

We argue that to uncover the multiple effects of economization and quantification brought about by market-oriented new public management reforms and prison privatization, one needs to set presumed dichotomies between the public and the private aside and turn instead to the multiplicity of economizing practices (here: curtailing, marketizing, financializing) and their implication in different forms of quantification. Ironically, in the case of England and Wales, numbers and state contracts governing privately managed prisons also shielded these establishments from economization (e.g. budgetary savings requests); and it is the public prisons that have been exposed the most to measures of government austerity.

Furthermore, we need to be careful not to equate quantification with economization. Although quantification is an important condition *for* economization, it is only recently that quantification has been largely annexed by the phenomenon dubbed neoliberalism (Kurunmäki et al., 2016). We call for greater attention to the complex interplays unfolding between different practices of quantifying and economizing. In so doing, we also argue for closer scrutiny of the conditionality of the performativity of quantification (Butler, 2010; Kurunmäki et al., 2016). We need to pay closer attention to the conditions under which numbers produce (economizing and other) effects, and we need to better understand the varying nature and extent of those effects.

Quantifying and Marketizing: Prison Privatization, Quantification and the Ethos of Contestability

Since the 1980s, public services in the UK have undergone a series of far-reaching reforms. Instead of state coordination, market competition came to be seen as an effective lever for driving efficiency and innovation (Hood 1991; Miller & Rose 1990; Pollitt, 1993).[1] The reforms, labelled "New Public Management" by scholars (see, e.g., Hood, 1991), were based on a conception of accountability that, as Hood (1995, p. 94) writes, reflected high trust in the market and private business methods, and low trust in public servants and professionals. New Public Management ideas also took hold in the Prison Service (Bennett et al., 2008; Bryans, 2007; Coyle, 2005; Coyle et al., 2003; Liebling, 2004; Liebling et al., 2011), providing fertile ground for plans of prison privatization. It was assumed that selective privatizations and the threat of market testing public prisons, where the HM Prison Service had to bid against private prison providers for the running of its own prison establishments, would stimulate the development of a "business culture" (Black, 1993) and "yield cross-fertilization benefits" (Harding, 2001). "The market" was called upon to discipline public prison administrators, financially and operationally (see, e.g., Home Office, 1988 and the introductory quote). Competition between and amongst public and private prison providers should help drive down costs (e.g. costs connected to the running of prison establishments), increase quality (e.g. the quality of prison life) and effectiveness (e.g. the effectiveness of operational procedures).

The first private prison establishment was opened in England and Wales in 1992 (Wolds Remand Prison, which was a newly built prison) (James et al., 1997). In the same year, the first public prison (Strangeways Prison) was also market tested (Prison Reform Trust, 1994). Market testing permitted the private sector to compete directly with the public sector for the management of prisons that were considered to be "failing", that were not meeting performance targets, for instance with respect to cost management or security standards, evidenced for example by prisoner escapes or riots. As Black notes:

> Market testing within the Prison Service will be twofold: First, under the provisions of the Criminal Justice Act 1991, the Prison Service will have to compete with the total privatization of certain prison establishments. [...] Internal services currently run by the Prison Service are also to be contracted out, both to the public and private sectors. This is the second tier of market testing. [...] In the open market, an agency status Prison Service will have to bid with other competitive tenders in order to run its own existing establishments, under certain proscribed conditions. (Black, 1993, pp. 27–30)

The introduction of market testing and privatization coincided with an "institutional crisis" in the Prison Service (King & McDermott, 1989; Resodihardjo, 2009). Between 1980 and 1987, prison expenditure had increased in real terms by 72% (James et al., 1997, p. 48). The over-crowding of prisons had become a serious issue. By 1990, England alone was imprisoning more people than any other Western European country (Pozen, 2003, p. 263). In the early 1980s, a third of the offenders in custody were sharing with one or two others cells designed for only one person and, in June 1989, Wadsworth prison in London had only eight cells with access to sanitation at night versus 1149 without access (ibid.). Furthermore, in 1990 the Prison Service experienced a series of severe prison riots. The riots had started in Strangeways Prison in April 1990, and spread thereafter to more than 20 prisons throughout the country. As Resodihardjo (2009, p. 93) writes, for Great Britain, these were "the most serious series of riots ever experienced... When the quiet returned, three people had died, 133 inmates and 282 prison staff had been injured and there [sic] the cost of the damage ran into millions of pounds".

In view of these issues, and the lobbying efforts of the security industry, neoliberal think tanks and the general preference of the Thatcher government for free enterprise, prison privatization and market testing were put forward as solutions to the various problems facing the Prison Service, despite fierce opposition from the entire British penal lobby and the Labour Party. From 1992 onwards, the Prison Service began to be reorganized in terms of a "mixed economy", with a mix of public and private prison service providers (King & McDermott, 1995). And by the late 1990s the debate had shifted from whether the government should principally allow private prisons or not to how it could best govern them, including the assessment of the efficacy and effects of privatization schemes (National Audit Office, 2003; Pozen, 2003).

In these reforms, a key role came to be assigned to quantification and "auditability" in terms of numbers (Power 1997). Standardized performance metrics should help make prison performance visible and governable across public and private prison establishments. These quantification tools were supposed to turn prisons into competitive, market-oriented "accounting entities" (Kurunmäki, 1999). Whereas in the 1970s in the Prison Service, and in the public services more generally, corporate planning had been the main mechanism through which central oversight was exercised, in the late 1980s, this began to change, and policy-makers turned to ideas of market coordination.

Market-oriented quantification, via accrual-based accounting, performance measurement and prison ratings, should facilitate an extension of the rationality of the market to the Prison Service, and public administration more generally, domains previously viewed as non-market and non-economic (Davies, 2014; Kurunmäki et al., 2016). Quantification thus became implicated, as Foucault put it, in a much broader process of governmental reform, characterized by the transformation "from a market supervised by the state to a state under the supervision of the market" (Foucault, 2008, p. 116) (see also Bruno & Didier, 2013; Miller & Rose, 2008; Supiot 2012, 2015; Rose & Miller, 1992; Davies, 2014).

The first set of standardized performance metrics was introduced into the Prison Service in 1992–93 by Derek Lewis, the Prison Service's first chief executive who was recruited from the private sector. These metrics included, amongst other things, the number of prisoner escapes, the number of assaults (on staff, prisoners and others), the number of hours spend in purposeful activity, the proportion of prisoners held in unit of accommodation intended for fewer numbers, the proportion of prisoners held in prisons where prisoners are unlocked on weekdays for a total of at least 12 hours and information about the average cost per prisoner place (Prison Reform Trust, 1996). In subsequent years, the number of prison Key Performance Indicators (KPIs) steadily increased from 8 to 18 in 2000–01 (Liebling, 2004, pp. 58–63), to 28 for the newly created National Offender Management Service Agency (NOMS) in 2008–09 (joining up the prison and probation services).

These KPIs included public protection measures (measured in number of escapes), Offending Behaviour Program (OBP) completions, OBP starts, number of completed drug rehabilitation programmes, drug testing results, employment of prisoners upon release, accommodation of prisoners upon release, number of serious assaults, overcrowding data (target: number of prisoners held in accommodation units intended for fewer prisoners does not exceed 26% of the population), staff sickness, race equality data (target: at least 6.3% of prison staff should be from ethnic minority groups), costs per prisoner place and audit compliance (NOMS, 2009, pp. 18–19).

In 2003, such quantifications were given a further boost by the introduction of benchmarking and composite performance ratings. The benchmarking and rating programmes were introduced following Lord Carter's correctional review in which he called for the establishment of the principle of "contestability" within the HM Prison Service (Carter, 2003; Home Office, 2004). According to Nellis (2006, p. 53), Carter's review exemplified "the messianic managerialism"—the re-engineering of existing structures and functions to produce "guaranteed", quantifiable and externally verifiable behavioural outcomes—that had come to characterize New Labour's approach to modernization (see also Pollitt, 1993; Power, 1997). From 2003, prison performance measures of safety, security, rehabilitation and economic efficiency came to be put together in a weighted scorecard, drawn up in the fashion of Kaplan and Norton's Balanced Scorecard (Kaplan & Norton, 1992), on the basis of which composite performance ratings are drawn up, similar to the "star ratings" of the NHS. Public and private prisons are rated and compared on a 1 to 4 performance scale. Level 4 is awarded to excellent establishments that are delivering "exceptionally high performance". Level 1 indicates a "poor performer". Since the introduction of the ratings, according to a study by Bryans (2007), prison governors find themselves operating in a "more competitive" and "less collegiate" world; "more than ever before, their focus is on how prisons are performing relative to other similar prisons" (Bryans, 2007, p. 74) (see also Bennett et al., 2008).

Underlying the introduction of the prison ratings was a belief in the power of market incentives, and the aspiration to govern through competition (Mehrpouya & Samiolo, 2016). As Bryans (2007, p. 73) writes, with the help of the ratings poorly performing prisons were publicly identified and given six months in which to improve their performance. A failure to improve meant that the prison faced the threat of closure or being contracted out to the private sector. The performance measurements thus redefined the prison as a separate performance-oriented unit—a calculating accounting entity, responsible for its own success and failure. The prison ratings enabled comparisons between public and private sector prison performance and they helped render ideas about competition and competitiveness operable. The government also used (and still uses) quantified performance targets in the definition, detailing and monitoring of private prison contracts and the operation of a financial penalty system if such contracts are non-fulfilled.

Before quantified performance measures were introduced, prison values and objectives had been articulated in Circular Instructions, which had been criticized for being "uncoordinated and uncosted" and "lacking a mechanism of ensuring the initiatives they contain are implemented" (Lygo, 1991; Prison Reform Trust, 1992). Of course, the created "market" for prison services is highly imperfect. It is a market that is highly regulated and not characterized by free trade and exchange. The goods traded are not private goods: they are public services aimed at the delivery of public security, punishment and rehabilitation at a reasonable cost. Private prisons act on behalf of the government. They are supervised by the state and accountable to parliament. Private prison providers have to obey the same rules and regulations as public prison establishments. The state is the sole "buyer" of prison services, acting in the interest of a third party—the public. Prison services do not represent a good that is consumed. Prisoners do not have "a choice" and, also, prison establishments cannot freely choose which prisoners they want to house.

One could argue that ideas of marketization and privatization did not lead to the creation of an actual market. First and foremost, we observe the emergence of calculative bureaucracies, created in the name of the market, and a corresponding rise in managerialism (Bennett, 2019; Liebling & Crewe, 2013), akin to what Boltanski & Thévenot (2006 [1991]) have labelled in terms of an "industrial world", characterized by attempts aimed at "asserting greater managerial control within establishments over both staff and prisoners, making them more ordered and more legitimate inside and out" (Bennett, 2016, p. 8; but see also Bryans, 2007; Coyle, 2005). Furthermore, expanding quantification and calculation did not automatically contribute to an enhancement of administrative capacities. First of all, the prison performance measures and ratings placed new demands *on* administrative capacities, for example with regard to the expertise required to make quantification operable, to work with and make sense of the numbers.

LIMITS OF MARKETIZING QUANTIFICATION

Operational and financial performance measures, summarized in published prison ratings, were aimed at stimulating organizational intro-spection and inter-organizational competitiveness. They were enrolled in attempts aimed at "exacting responsibility" (Miller 2001, p. 380) following ideals of total, finely calibrated control. Yet, at the same time, we see also the creation of new zones of opacity, invisibility and non-accountability (Bennett, 2016; Liebling, 2004; Mennicken, 2013).

Research has shown that accounting entities are fictional and network-effacing superimpositions on complex organizations (Mennicken & Power 2015, p. 213). Underlying the concept of an accounting entity is the fiction of separable economic units (Hines, 1988) and aspi-rations to use accounting, such as performance measurement, as a tool for the mapping and managing of social and economic relations (Hopwood 1984; Kurunmäki, 1999; Miller 2001; Miller & Rose, 1990). The boundaries that delineate an organization, such as a prison, as a performance-oriented, economic unit separate from other organiza-tions are not clear-cut, natural or fixed. The making of an accounting entity is not so much an economic than a political process (Kurunmäki, 1999): "The actors who identify entities and define their limits are many and varied, and may speak on behalf of legal, economic, social, polit-ical, aesthetic and professional interests" (Kurunmäki, 1999, p. 220). Also in the Prison Service quantification and "accountingization" were, and still are, contested and commensuration, "the transformation of different qualities into a common metric" (Espeland & Stevens, 1998), a "congenitally failing" (Miller & Rose, 1990) undertaking. Even seem-ingly straightforward measures, like measures of cost, are far from being unproblematic. Should prisons costs, for example, be expressed as a per diem rate for prisoners, or as a fixed sum assuming 100 per cent occupancy?

Inconsistencies exist between the performance measures and targets for public and private prison entities (National Audit Office, 2003). Attempts aimed at establishing comparability are further undermined by variation in the characteristics of different prison entities (size, location, design, function and age). The contracts negotiated for each private prison differ, making it difficult to draw comparisons between different public and private prison establishments. It is also difficult to establish an exact correlation between financial penalties that private prisons occur and their operational performance (National Audit Office, 2003). Financial deductions can be reduced following negotiations between the Prison Service and the private prison contractors. According to the National Audit Office (2003, p. 17), these negotiations are not solely concerned with the prisons' operational performance, but take also account of problems of "inflexible contract monitoring". In other words, negotiating penalty deductions allows for the flexing of otherwise inflexible 15-year or 25-year contracts, which normally are based upon fixed key performance indicators (KPIs) and targets. See for example the dialogue below between Martin Narey (Director General of the HM Prison Service in England and Wales, 1998–2003) and Alan Williams (Member of Parliament) in a Public Accounts Committee hearing:

> *Mr Williams (Committee of Public Accounts)* Can you just clarify a point which is genuine misunderstanding? In Figure 9 on page 15, cost per penalty point, Ashfield is £94, Altcourse is £293. Why is the cost per point three times higher in one than in the other?
>
> *Mr Narey (Commissioner for Correctional Services)* It is because each penalty point regime— and I confess to having had to have a tutorial on this just this morning—is unique to that particular prison. If you want to compare how a particular prison has performed against another one, you cannot just look at the penalty points incurred, but you will see that prisons have very different base lines, that is, the number of penalty points which are tolerated before a financial penalty is enforced. That again reflects the fact that we have different schemes for different prisons.
>
> *Mr Williams* Why was Ashfield set at £94? Why is it so different, particularly in view of subsequent events? The worst offender of the lot, yet it has the least disadvantageous penalty point system?
>
> *Mr Narey* It has been significantly at a disadvantage in terms of the money we have taken from it, a total of £4.2 million, as Mr Beeston said. Whatever the penalty point regime, the fact is that we have been able to use sanctions against Ashfield going way beyond the use of penalty points, in this case in closing places and saying this is not a safe enough place in which to put young prisoners so as to make a very significant financial sanction.
>
> *Mr Williams* Was the £4.2 million based just on penalty points or on other factors?

Mr Narey No, the £4.2 million was primarily based on closing places and saying we do not think this is a place which is safe enough or good enough to meet our standard requirements so we are not going to put young people into there. (House of Commons, 2003, Ev18)

Further it is difficult to attribute re-offending rates to individual prison performance, as prisoners are regularly transferred between prisons, transcending prison boundaries and the accounting for them (Bastow, 2013; Mennicken, 2013). The Prison Service struggled, and still struggles, to identify the most appropriate KPIs. As the Prison Reform Trust (1996, p. 3) observes:

First, people may not agree as to the most important goals. Second, measuring performance in a quantitative way gives no indication of quality. Third, goals which cover general areas of work can be measured using a number of different KPIs. For example, helping prisoners to return to the community is not achieved solely by providing more than the minimum visiting entitlements; it is also achieved by enabling prisoners to gain educational qualifications, or by helping them find accommodation after release. Fourth, the date may not be accurate or objective. Finally, even if there is agreement as to the KPIs, the target performance may be set too high or too low.

Following these and other criticisms the KPIs were reviewed, repaired and extended (see also Liebling, 2004). They constitute a moving target, and a platform for an ever-expanding apparatus of quantification and calculation, which itself has been described as unwieldy and uneconomical (Bennett, 2016; Coyle, 2005, 2008). Prison governors have to meet increasingly detailed reporting demands, facing "constant oversight from internal auditors and external inspecting bodies" (Coyle, 2005, p. 97; but see also Bennett, 2016). Prison governors also find it difficult to prioritize amongst the various performance measures. They are often overwhelmed by the reporting demands, which take them away from "where the action is", the "floor" and day-to-day interactions with prisoners and prison officers (Bennett, 2016; Bryans, 2007; Coyle, 2005, 2008).

Moralizing Versus Economizing Numbers

Quantification involves a transformation of quality into quantity, of subjective experience into objectified knowledge (Espeland & Stevens 2008; Kurunmäki et al., 2016). In so doing, it prepares the ground for new possibilities of governing (Kurunmäki et al., 2016; Miller 2001; Supiot 2015). But, as Kurunmäki et al. (2016, p. 395) highlight, we need to be careful to separate quantification from economization, for not all quantification implies economization. Although the rise of prison performance measurement in England and Wales was, in large part, animated by market-oriented reforms and ideals of competition, the introduction of the performance measures gave also rise to the creation of an unwieldy calculative bureaucracy and new information systems that needed to be managed and fed.

Furthermore, we should not forget that the performance measures, at least at their onset, also had a "democratizing" ambition, an aspiration to hold managers, public administrators, and civil servants to account, so as to counteract nepotism and arbitrariness (Kurunmäki et al., 2016; Lewis, 1997; Prison Reform Trust, 1996). Prison interest groups, such as Prison Reform Trust, for instance, had welcomed the introduction of the performance measures, as they allowed for insight into areas of prison activity which had not been publicly accounted for previously (Prison Reform Trust, 1996). Of course, the performance measures were also criticized by the very same groups. The Prison Reform Trust, for instance, questioned the purpose of the prison league tables arguing that they were useful as an internal management tool but did not provide a clear picture of prison life (Solomon, 2004, p. 2). According to criminologist Alison Liebling (2004, p. 26), the market-oriented prison performance ratings and the regime aspirations arising out of them "left crucial questions of moral responsibility and individual transformation untouched".

Following such criticisms, the measurement system came to be reformed from within. In 2000, the Home Office commissioned Professor Alison Liebling, Director of the Prisons Research Centre at Cambridge University, with the task to develop "quantitative measures of qualitative dimensions of prison life" (Liebling, 2004; Liebling & Arnold, 2002). Liebling and her team developed new performance measures along two dimensions: relationships (respect, humanity, trust, staff-prisoner relationships and support) and regimes (fairness, order, safety, well-being, personal development, family contact and decency) (Liebling,

2004). These measures were aimed at "moralizing prison management" (Liebling, 2004) and at counter-acting the economization and managerialization of prison life. The newly developed "Measures of the Quality of Prison Life" (MQPL) were fully rolled out for both public and private sector prisons in 2012, when they became part of the weighted scorecard of prison performance and were fully incorporated into the 1–4 prison performance ratings under the heading of "decency" (making up 28.6% of the overall prison performance rating). Liebling and her team perceived the performance measures as a mediating, rather than economizing, instrument (Miller & O'Leary, 2007)—as a mechanism that could be utilized to link up and mediate between conflicting concerns and prison values, such as those of security, economy and decency. They used quantification as a way to bring prison values relating to questions of rehabilitation, care and decency back in, and to give prisoners "a voice" through the introduction of a standardized survey aimed at capturing their day-to-day experiences (Liebling, 2004).

It is beyond the scope of this chapter to provide a detailed assessment of the success of this quantification project in reshaping and rebalancing prison value configurations. Further, as Robert Salais' contribution in this volume reminds us, we need to be cautious when labelling this as "democratization". Questions of democracy do not only concern issues of consultation and participation (e.g. participation in processes of performance measurement, or the capturing of multiple voices through KPIs). As Salais argues in this volume, democratizing quantification would also entail a critical engagement with the very logic of "managing by numbers", with the very process of capturing and transforming a prisoner's voice via KPIs, and the extent to which such quantification is still able to take account (or not) of individual capabilities, hopes and desires. Nonetheless, we ought to acknowledge that the Prison Service's KPIs contributed not only to an infusion of the Prison Service with market-oriented ideals of efficient, economic management. The performance measures also served as a platform for debate about prison values and reform, not least because of the public attention and criticism they attracted.

At the same time, such measures can themselves become challenged and changed by economization. Put differently, quantification is not only an instrument of economization; it can also be subjected to economization and destabilized as a result. Liebling's Measures of the Quality of

Prison Life (MQPL), for instance, soon after their system-wide introduction and establishment, came to be undermined by austerity policies and related economizing practices of curtailment and frugality. In the wake of the 2008 financial crisis and government austerity measures, concerns with cost and "economies of scale" came to overrule the Prison Service's "balanced" performance measurement system. Definitions of failure were narrowed to definitions of failure in economic (i.e. cost management) terms. In a report by the HM Chief Inspector of Prisons we read (HM Chief Inspector of Prisons for England & Wales, 2013, p. 7):

> The National Offender Management Service (NOMS) as a whole (that is, prison, probation and headquarters functions) had to make savings of £246 million on top of the £228 million savings delivered in 2011–12. This represented a further reduction of seven per cent of NOMS' resource budget against the spending review baseline. Public sector prisons alone had to find savings of around £80 million. NOMS overall savings were delivered by a combination of workforce restructuring; market testing and privatisation of entire establishments and specific services; standardising costs and services; and reconfiguring the prison estate by closing some smaller, older prisons and increasing the size and number of very large establishments.

The bulk of savings that the Prison Service had to deliver was achieved through a changed estate management strategy. As is highlighted in the quote above, this included not only land sales, but also the closing of smaller, older prisons (with approx. 150–250 certified places of normal accommodation on average) that, in comparison, were costlier to run than new, large establishments with more than 1,500 places where economies of scale can be realized (see, e.g., Oakwood Prison). According to the National Audit Office, by the end of 2013–14, the changed estate management strategy contributed £71 million of savings since 2010 (National Audit Office, 2013). The National Audit Office notes that the estate strategy's explicit focus was on cost measurement and reduction (i.e. economizing understood in terms of curtailment) and that this limited how far it could address concerns with quality and performance (National Audit Office, 2013, p. 5). The annual prison performance ratings and HM Inspectorate of Prisons' regular reports were not considered in the estates management strategy. Only cost calculations were taken into account. Of the 18 prisons closed or identified for closure by December 2013, eight were considered to be "high performers" on dimensions related to Liebling's Measures of the Quality of Prison life (National Audit Office, 2013, p. 30). Different forms of quantification thus came to be hierarchized; leading to a prioritization of issues of security and cost over objectives of individual rehabilitation, measures of decency and the quality of prison life (Bennett, 2019; Liebling & Crewe, 2013).

Ironically, contracted-out prisons were largely spared from these cuts, due to the inflexible nature of the 15- or 25-year contracts under which they are operating and "because of the cost and difficulty of terminating contracts early" (National Audit Office, 2013, p. 6). In other words, state contracts governing privately managed prisons shielded these establishments *from* economization in the form of budgetary savings requests, and public prisons were exposed the most to measures of government austerity. This does not imply that private prison establishments are not economized, but different mechanisms of economization (and quantification) are at work here, and it is important to differentiate between these. Private prisons, for example, might be more financialized than public prisons, and it is to financialization as one specific modality of economization, and the implication of quantification in such processes of financialization, that we turn next.

Quantifying and Financializing: Accrual Accounting and Social Impact Bonds

Financialization can be seen as one distinctive mode of economizing that has taken hold in the Prison Service, as well as in other parts of the public services in England and Wales, over the past 38 years, first through the government's Financial Management Initiative (FMI), which was launched in 1982, later through the Private Finance Initiative (PFI), which was launched in 1992, and more recently through the experimentation with Social Impact Bonds (for instance in Peterborough Prison) (Anders & Dorsett, 2017; Disley et al., 2015; Joliffe & Hedderman, 2014). According to van der Zwan (2014), studies of financialization interrogate how an increasingly autonomous realm of global finance has altered the underlying logics of the industrial economy and the inner workings of democratic society. Financialization encompasses a range of different developments connected to the rise and spread of finance: the emergence of new accumulation regimes (including the increasing importance of financial services and financial assets on a company's balance sheet); the increasing importance of capital markets that is paralleled by an ascendency of the shareholder orientation; the financialization of everyday life; and the rise and spread of financial economics, for example into accounting.

Also the public services, including the Prison Service of England and Wales, have been affected by such processes of financialization. With prison privatization, new stakeholders entered the picture: financial investors, such as banks and shareholders, who seek financial returns on their investments. When this chapter was written, 13 prisons out of 117 were private and contractually managed by G4S Justice Services, Serco Custodial Services and Sodexo Justice Services. In 2019, these private companies housed nearly 20% of the prisoners in England and Wales (20% of approx. 82,000 prisoners).[2] All three security firms are globally operating corporations listed on multiple stock exchanges. G4S Justice Services, for instance, is listed on the London and Copenhagen stock exchanges. Serco Custodial Services is listed on the London Stock Exchange. Sodexo Justice Services is part of the Sodexo Group that is traded on the Paris Bourse and the New York Stock Exchange.

Private prison managers are not only subject to government oversight and scrutiny. They are also wary of stock price reactions. *Inter alia*, this brings to the fore concerns with reputation management—a prison establishment's appearance to the (financial) market. Globally operating security corporations are interested in economic gains. They are concerned with winning new prison contracts and renewing old ones, and with avoiding both adverse publicity and drops in stock price (Volokh, 2002, p. 1870). Quantification is here not only implicated in the determination of operational prison performance against nationally stipulated KPIs, but also in financial profit and loss calculations, and a logic of (financial) "capitalization" (Muniesa et al., 2017), where prison establishments and prisoners come to be viewed as (financial) assets, as vehicles for the generation of future (financial) returns (see also Birch & Muniesa, 2020).

Such a financialization of the prison organization is intertwined with a shift in the conventions underpinning (economic) quantification (Boltanski & Thévenot, 2006 [1991]; but see also Chiapello, 2015; Chiapello & Walter, 2016). Chiapello (2015) highlights the progressive diffusion of financialized conventions of quantification. Such conventions have been developed and spread by accounting and finance professionals, for example in the form of net present value calculations, probability-based estimates of financial value, and market prices as true value benchmarks (see also Muniesa et al., 2017). Such financialized conventions of quantification have not only been utilized in private prisons. They have also travelled into public prisons. As Liebling (2004, p. 71) writes, with privatization a "new rationality of governance" (Miller & Rose, 1990, 2008) was introduced into the (public) Prison Service, where auditing and accounting practices originating from "the world of finance" came to be applied to the non-financial practices and systems of the Prison Service.

Private Sector Accrual Accounting

In particular, private sector budgeting and accrual accounting were promoted as "a necessary precondition for identifying inefficiencies" and "managing capital budgets properly". As O'Quinn and Ashford (1996, p. 31) put it: "While adopting accrual accounting and budgeting is technical, it is a necessary precondition for identifying inefficiencies, improving services and saving taxpayers money".

Unlike cash accounting, private sector accrual accounting records the changes in value of assets and liabilities and distinguishes between operating and capital flows (Ellwood & Newberry, 2007; O'Quinn & Ashford, 1996). Expenses are recorded as incurred and revenues as earned (rather than when cash changes hands). Fixed assets, such as buildings or equipment, are capitalized, i.e. they appear as a separable item on an organization's balance sheet. The introduction of business-like accrual accounting was supposed to allow for better measurement of costs and revenues and more efficient and effective use of resources, for example, through charges for fixed assets, or calculations of full costs of providing a public service, which could then be compared with the prices charged by outside suppliers (Ellwood & Newberry, 2007).

In the Prison Service, the introduction of private sector accrual accounting was largely driven by the desire to establish financial comparability between public and private prison providers, and to stimulate and govern private financial investment (see also Ellwood & Newberry, 2007). Private sector accrual accounting was supposed to enable the development of benchmarks for inter-organizational financial comparisons. It should help establish a common business language, making financial reports readable across the public and private sectors. Thereby, it should increase inter-organizational competition and strengthen "financial transparency" and "fiscal responsibility" (O'Quinn & Ashford, 1996).

Yet, numerous studies have also challenged the relevance of accrual accounting to governments and have highlighted difficulties of its practical implementation (see, e.g., Ezzamel et al., 2014; Ellwood & Newberry, 2007; A. Bruno & Lapsley, 2018; Barton, 2004). Such difficulties were also experienced in the Prison Service. The assumed benefits proved difficult to realize. The valuation and recognition of assets, such as property and equipment, was far from straightforward, given that no market exists to determine their "fair value", including financial losses due to impairment. Furthermore, assets, such as prison buildings or the equipment they contain, do not generate any cash revenues for the government, rather they represent a future stream of expenses.

Moreover, private prisons and the costs of running those remained off-balance sheet. This applies also to the costs connected to the preparing and monitoring of private prison contracts. Private prison establishments are governed by contract. These contracts fall into two main categories. Design, Construct, Manage and Finance (DCMF) contracts, also known as Private Finance Initiatives (PFI) that are typically 25 years long and Manage and Maintain Contracts that are typically 15 years long. These contracts are negotiated behind closed doors and they are not available to the public, as they fall under the commercial confidentiality clause. "Fiscal transparency" is thus punctuated, the "representational faithfulness" of the Prison Service's accrual-based accounts highly limited, and the objective of financial comparability and contestability undermined.

Finally, financial and operational prison performance, and their respective quantification, are largely de-coupled. It is not easy to assess their interplay from the Prison Service's financial and management reports, and there is no systematic mapping of the impact of financial management on a prison's operational effectiveness, including measures concerning the quality of prison life, and vice versa.

Despite these limitations, the accounting reforms introduced new ways of thinking about what the state consists of. The state does not "pay" or "fund" any longer. Instead, it "invests" in an accountable manner (Mennicken & Muniesa, 2017; Muniesa et al., 2017). The reforms also led to the rise of new experts, the emergence and empowerment of financial experts and expertise. The transformations shifted the locus and focus of governing. They redefined relations between the state, public and private prison providers, "making up" prison providers in financial business terms, introducing private-sector oriented forms of calculation and financial responsibility. Such reforms also paved the way for experimentations with new forms of financing, supporting the rise of "social finance" initiatives, aimed at stimulating investments that generate financial returns for the (private) investor while including measurable positive social impact, e.g. in the form of reduced recidivism rates. It is to such forms of investments, and the implication of quantification in these, that we turn last.

Social Impact Bonds

By way of concluding this section, we would like to draw attention to a relatively new class of financial instrument, so-called Social Impact Bonds (SIBs). Social Impact Bonds are aimed at "socializing finance"—"rethinking finance for social outcomes" (Social Finance, 2009). Social Impact Bonds are based on a commitment from government to use a proportion of the savings that result from improved social outcomes to reward non-government investors that fund the intervention activities. They are based on a contract negotiated with government that includes definitions of a success metric (e.g. 1-year reoffending rate for short-sentence offenders in a specified geographic area); a specific target population (e.g. offenders aged over 18 leaving prison after a sentence of less than 12 months and returning to a specified geographic area); and the value of success (i.e. the amount returned to investors for a given improvement in the social outcome; generally a proportion of the related savings to government) (Social Finance, 2009). The private investment is used to finance a range of interventions to improve the target social outcome over the contract period (often around 5 years). If the interventions are successful and the social outcomes improve, government pays investors a reward based on the pre-agreed payment schedule (this scheme is also referred to as "payment by results"). SIBs seek to "align government policy priorities with the interests of non-government investors and social service providers" (Social Finance, 2009). They are aimed at creating "a rational investment market" and aligning "the financial and social return on investment" (Social Finance, 2009).

In the Prison Service, a pilot with a Social Impact Bond was started in 2010 in Peterborough Prison. This Social Impact Bond was aimed at reducing the reoffending of approximately 2000 male prisoners who were discharged from HMP Peterborough after serving a sentence of less than 12 months between 2010 and 2014. Social Finance, a not-for-profit financial intermediary, developed the model, raised the finance and was performance managing the bond. The Ministry of Justice commissioned the service, with part of the outcome payments contributed by the Big Lottery Fund. Organizations involved in delivering the service were: St Giles Trust; Ormiston Children and Families Trust (Ormiston); SOVA; YMCA; Peterborough and Fenland Mind (Mind) (Disley et al., 2015; Civil Society Media, 2014).

The investment pool totalled £5 million from 17 social investors including: the Barrow Cadbury Charitable Trust; the Esmée Fairbairn Foundation; the Friends Provident Foundation; the Henry Smith Charity; the Johansson Family Foundation; the Lankelly Chase Foundation; the Monument Trust; the Panahpur Charitable Trust; the Paul Hamlyn Foundation; and the Tudor Trust. If re-offending was reduced overall by at least 7.5%, investors received a minimum repayment of 2.5%. The greater the drop in re-offending beyond this threshold, the more the investors would receive. The total payments by Government were capped at £8 m (or £7 m in real terms) and return to investors was capped at 13% annual IRR (Civil Society Media, 2014).

The Peterborough Social Impact Bond was part of a broader government strategy to outsource the financing of government services and to stimulate more third sector involvement in the delivery of these. An interim report of results (Joliffe & Hedderman, 2014) showed that the frequency of re-conviction events for the Peterborough SIB cohort 1 was 8.4% lower compared to a matched national control group (142 re-conviction events per 100 offenders in Peterborough's cohort 1 compared to 155 re-conviction events per 100 offenders nationally). This means that the provider (One Service) was on track to achieve the 7.5% reduction target for the final payment based on an aggregate of both cohorts, but that the pilot had not achieved the 10% reduction target for cohort 1 (Joliffe & Hedderman, 2014).

The final cohort impact evaluation (Anders & Dorsett, 2017) found that the provider had managed to reduce the number of reconviction events amongst those discharged from HMP Peterborough by 9.7% for cohort 2. Reduction across both cohorts was estimated to be 9.0%, which reached the minimum threshold of 7.5% across all cohorts, and was sufficient to trigger payment under the terms of the SIB contract (Anders & Dorsett, 2017). The evaluation of the Peterborough SIB was far from straightforward. In the evaluation of the first cohort, it took the evaluators 11 months alone to agree the sample and obtain all the data needed to begin the analysis (Joliffe & Hedderman, 2014). And, unlike random control allocation, the applied method of propensity score matching (PSM), although regarded as one of the best ways of matching quasi-experimentally, could not take account of unmeasured differences which may account for variation in reconviction rates aside from "treatment received" (Joliffe & Hedderman, 2014). It was also not possible to precisely replicate the approach adopted in the dry run because of data quality issues, including missing data pertaining to the type of offence (Joliffe & Hedderman, 2014, p. 4).

Apart from such measurement challenges and the costliness of the evaluation, SIBs like the Peterborough SIB have also been criticized for their short-term orientation, high transaction costs, and their focus on the fulfilment of quantifiable targets that dictate the rate and amount of payment received by contracted service providers (Cooper et al., 2016; Edmiston & Nicholls, 2018; Jeamet & Salais 2019). As Edmiston and Nicholls (2018) highlight, this can compromise service quality and integrity due to "gaming" and perverse incentives (see also Cooper et al., 2016; Jeamet & Salais, 2019). Social investors are likely to have different and potential conflicting motivations. Some may prioritize financial return on investment, whereas others may focus on the social impact rather than financial rates of return (Edmiston & Nicholls, 2018).

In the case of the HMP Peterborough SIB, according to the final process evaluation report produced by RAND Europe (Disley et al., 2015), the private provider (One Service) made extensive efforts to engage offenders and ex-offenders through the provisioning of through-the-gate and post-release support (addressing practical problems, such as housing, benefits, training and education). Yet, often the contact with One Service was ended after a few months and a longer-term engagement with ex-offenders was difficult to achieve (Disley et al., 2015). It was also highlighted that different organizations hired by One Service shouldered unequal burdens in delivering the service, and interviewees perceived contractual relationships behind the Peterborough SIB to be complex (Disley et al., 2015).

In addition, not all reductions in re-offending (more precisely re-conviction) bring realizable savings, where costs of existing services are largely fixed and prison numbers are not a direct product of re-offending (Fox & Albertson, 2011). The Peterborough SIB's focus on reduced recidivism rates is itself highly questionable. It is based on what Robert Salais has termed in this volume "governance-driven quantification", where meeting contractual requirements is prioritized over the (long-term) needs of subjects (see also Morley, 2021; Supiot 2015), and root causes underlying the commitment of crime remain largely unaddressed. As a recent OECD report on SIBs stated: "There is a fear that this strong focus on results can change the public service ethos or lead to a narrow mechanical determinism in service delivery" (OECD, 2016, p. 16, cited in Morley, 2021). SIBs reframe public service users, in this case prisoners, as potential revenue sources rather than conscious agents and citizens, and they often lack engagement with subjective experiences and preferences (Cooper et al., 2016; Jeamet & Salais, 2019; Sinclair et al., 2021).

However, the actors involved in the Peterborough SIB differed from those engaged in the prison privatization schemes described earlier. Most of these actors are located in the not-for-profit voluntary sector, and boundaries between the public and private have become increasingly blurred. As Edmiston and Nicholls (2018) remind us, we need to be careful not to dismiss the heterogeneity of these actors and their varied motives. We need to be careful not to overlook the heterogeneous kinds of private capital present within public services and their dynamic influence on service operations and delivery.

On the one hand, SIBs foster the penetration of financial criteria and issues into what were previously non-economic areas (Sinclair et al., 2021). Yet, on the other hand, they can also come to be utilized for the "socializing" of finance, as some investors might forego a higher rate of return for the sake of social impacts, because of their charitable orientation (Disley et al., 2015; Edmiston & Nicholls, 2018). We need to be mindful of the multiple ways in which quantification and financial concerns come to be interlinked—reinforced, mitigated or undermined—in the day-to-day realization of such schemes. Of course, this does not mean to say that (financial) "governance-driven quantification" (Salais, in this volume) is suspended. But we need to attend to the conditionality of its performativity (Butler, 2010) and scrutinize the circumstances that enable its scope and depth to increase and intensify (or not).

CONCLUSION

This chapter has explored different dynamics of economization and quantification in the Prison Service of England and Wales. In so doing, it concentrated on neoliberal, Anglo-American dynamics of quantification and economization, characterized by attempts aimed at the remaking of everything and everyone in the image of *homo oeconomicus* (Brown, 2015;

Davies, 2014; Supiot 2012). In the UK, such developments were kick-started under the Thatcher government (1979–1990) and later continued under the rule of Labour. This chapter has scrutinized what it means to "economize the social" drawing attention to three different modalities of economizing and the varied implication of quantification in these: ideas and practices of marketizing, politics of austerity (curtailment) and processes of financializing.

First, we examined the rise of market-oriented quantification in the form of prison ratings and rankings aimed at stimulating competition amongst public and private prison providers. We then attended to the limits of such quantification and showed how in particular criminologists undertook efforts to undo such economization through the development of "alternative" forms of quantification measuring the quality of prison life. These efforts, in turn, were undermined by the government's austerity policies, and related economizing and quantifying practices which brought concerns with economies of scale and cost management to the fore. Lastly, we explored the implication of prison quantification in processes of financialization, drawing attention to practices of private-sector accrual accounting and experimentations with Social Impact Bonds (SIBs).

We underscored the importance of unpacking the multiplicity of economization and quantification. Quantification is an important condition for economization, but should not be equated with it. To develop a better understanding of the multiple roles that quantification can come to fulfil we need to investigate interactions between different quantification regimes (e.g. cost accounting regimes, prison performance ratings and quantifications of the quality of prison life) and their implication in different "orders of worth", as Boltanski and Thévenot (2006) would put it. We need to follow the numbers across the different sites of their production and circulation (see also Kurunmäki et al., 2016, and the other chapters in this volume). Further, we need to be mindful of the different facets of economization. This chapter focused on neoliberal forms of economizing and quantifying—reforms that were undertaken in the name of the market, involving a logic of competition and capital investment rationales. Of course, we should not forget about other forms of economizing and quantifying—involving, for instance, state planning, public statistics, redistribution and public welfare rationales (see here also the contributions by Mespoulet and Amossé in this volume).

Finally, we have shown how economizing quantification can be punctuated, undermined and undone. State contracts governing privately managed prisons also shielded these from economization, such as government austerity measures. The introduction of prison performance ratings, aimed at enhancing transparency, comparability and inter-organizational competitiveness, was accompanied by the creation of new zones of opacity, invisibility and unaccountability. Private prison contracts as well as SIBs, for instance, are negotiated behind closed doors, and they can further the production of mutual lock-ins that, in many respects, suspend the logic of market discipline. Furthermore, the market-oriented reforms led to a creation of an unwieldy calculative bureaucracy, new quantification systems which needed to be fed and maintained, which could be described as uneconomical.

For those interested in the powers of quantification in all its forms, the challenge is to get to grips with its multiple modalities and intertwinement in different programmes and processes of reform. As Kurunmäki et al. (2016, p. 400) remind us, the performance of calculating selves, prisoners, prison officers and prison governors, may be evaluated by others without their knowledge, or against their wishes. Prisoners and prison governors may seek to influence prison ratings in their favour and tamper with the numbers. On the other hand, quantification can be turned against programmes of marketization, financialization and austerity. This, as also the chapters by Salais and Thévenot in this volume show, happens when ruling mechanisms of quantification and programmes of governing (e.g. governing in the name of the market or efficiency) are subjected to scrutiny, debate and critique, when forms of disruption are sought that go beyond "gaming the numbers", when numbers become attached to dreams and schemes of doing things differently (Kurunmäki et al., 2016, p. 400).

Acknowledgements This work was supported by the Economic and Social Research Council (grant number ES/N018869/1) under the Open Research Area Scheme (Project Title: QUAD—Quantification, Administrative Capacity and Democracy). The QUAD project is an international project co-funded by the Agence Nationale de la Recherche (ANR, France), Deutsche Forschungsgemeinschaft (DFG, Germany), Economic and Social Research Council (ESRC, UK), and the Nederlands Organisatie voor Wetenschappelijk Onderzoek (NWO, Netherlands). Both authors thank the other contributors to this volume, and in particular Robert Salais, for comments on earlier versions of this chapter.

Mennicken thanks the Wissenschaftskolleg zu Berlin for generous intellectual and financial support, and the Quantification Focus Group at the Wissenschaftskolleg (2013-14) for stimulating discussions.

Notes

1. This and the next two sections are based in part on Mennicken (2013, 2014).
2. See   https://www.justice.gov.uk/about/hmps/contracted-out   and https://www.gov.uk/government/collections/prison-population-statis tics, accessed 19 June 2020.

References

Anders, J., & Dorsett, R. (2017). *HMP Peterborough social impact bond: Cohort 2 and final cohort impact evaluation*. National Institute of Economic and Social Research.

Barman, E. (2016). *Caring capitalism: The meaning and measure of social value in the market*. Cambridge University Press.

Barton, A. D. (2004). How to profit from defence: A study in the misapplication of business accounting to the public sector in Australia. *Financial Accountability and Management, 20*(3), 281–304.

Bastow, S. (2013). *Governance, performance, and capacity stress: The chronic case of prison crowding*. Palgrave Macmillan.

Bennett, J. (2016). *The working lives of prison managers: Global change, local culture and individual agency in the late modern prison*. Palgrave Macmillan.

Bennett, J. (2019). Reform, resistance and managerial clawback: The evolution of 'reform prisons' in England. *Howard Journal, 58*(1), 45–64.

Bennett, J., Crewe, B., & Wahidin, A. (Eds.). (2008). *Understanding prison staff*. Willan.

Birch, K., & Muniesa, F. (Eds.). (2020). *Assetization: Turning things into assets in technoscientific capitalism*. MIT Press.

Black, J. (1993). The prison service and executive agency status—HM prisons plc? *International Journal of Public Sector Management, 6*(6), 27–41.

Boltanski, L., & Thévenot, L. (2006 [1991]). *On justification: Economies of worth*. Princeton University Press (French edition, 1991).

Brown, W. (2015). *Undoing the demos*. Zone Books.

Bruno, A., & Lapsley, I. (2018). The emergence of an accounting practice: The fabrication of a government accrual accounting system. *Accounting, Auditing and Accountability Journal, 31*(4), 1045–1066.

Bruno, I., & Didier, E. (2013). *Benchmarking: L'Etat sous pression statistique.* Zones.

Bryans, S. (2007). *Prison governors: Managing prisons in a time of change.* Willan.

Butler, J. (2010). Performative agency. *Journal of cultural economy, 3*(2), 147–161.

Çalışkan, K., & Callon, M. (2009). Economization, part 1: Shifting attention from the economy towards processes of economization. *Economy and Society, 38*(3), 369–398.

Çalışkan, K., & Callon, M. (2010). Economization, part 2: A research programme for the study of markets. *Economy and Society, 39*(1), 1–32.

Carter, P. (2003). *Managing offenders: Reducing crime—A new approach.* Prime Minister's Strategy Unit.

Civil Society Media. (2014). *Charity finance yearbook 2014.* Civil Society Media.

Chiapello, E. (2015). Financialisation of valuation. *Human Studies, 38*(1), 13–35.

Chiapello, E., & Walter, C. (2016). The three ages of financial quantification: A conventionalist approach to the financiers' metrology. *Historical Social Research, 41*(2), 155–177.

Cooper, C., Graham, C., & Himick, D. (2016). Social impact bonds: The securitization of the homeless. *Accounting, Organizations and Society, 55*, 63–82.

Coyle, A. (2005). *Understanding Prisons: Key issues in policy and practice.* Open University Press.

Coyle, A. (2008). Change management in prisons. In J. Bennett, B. Crewe, & A. Wahidin (Eds.), *Understanding prison staff* (pp. 231–246). Willan.

Coyle, A., Campbell, A., & Neufeld, R. (Eds.). (2003). *Capitalist punishment: Prison privatization and human rights.* Zed Books.

Davies, W. (2014). *The limits of neoliberalism: Authority, sovereignty and the logic of competition.* Sage.

Disley, E., Giacomantonio, C., Kruithof, K., & Sim, M. (2015). *The payment by results social impact bond pilot at HMP Peterborough: Final process evaluation report.* RAND Europe: Ministry of Justice Analytical Series.

Edmiston, D., & Nicholls, A. (2018). Social impact bonds: The role of private capital in outcome-based commissioning. *Journal of Social Policy, 47*(1), 57–76.

Ellwood, S., & Newberry, S. (2007). Public sector accrual accounting: Institutionalising neo-liberal principles? *Accounting, Auditing and Accountability Journal, 20*(4), 549–573.

Espeland, W. N., & Stevens, M. L. (1998). Commensuration as a social process. *Annual Review of Sociology, 24*, 313–343.

Espeland, W. N., & Stevens, M. L. (2008). A sociology of quantification. *European Journal of Sociology, 49*(3), 401–436.

Ezzamel, M., Hyndman, N., Johnsen, A., & Lapsley, I. (2014). Reforming central government: An evaluation of an accounting innovation. *Critical Perspectives on Accounting, 25*(4/5), 409–422.

Foucault, M. (2008). *The birth of biopolitics: Lectures at the College de France, 1978–1979.* Palgrave Macmillan.

Fox, C., & Albertson, K. (2011). Payment by results and social impact bonds in the criminal justice sector: New challenges for the concept of evidence-based policy? *Criminology and Criminal Justice, 11*(5), 395–413.

Hall, M., Millo, Y., & Barman, E. (2015). Who and what really counts? Stakeholder prioritization and accounting for social value. *Journal of Management Studies, 52*(7), 907–934.

Harding, R. W. (2001). Private prisons. In M. Tonry, & J. Petersilia (Eds.), *Crime and justice: An annual review of research* (Vol. 28). University of Chicago Press.

Hines, R. D. (1988). Financial accounting: In communicating reality, we construct reality. *Accounting, Organizations and Society, 13*(2), 251–262.

HM Chief Inspector of Prisons for England and Wales. (2013). *Annual Report 2012–13.* TSO.

Home Office (1988). *Private Sector Involvement in the Remand System.* HMSO.

Home Office (2004). *Reducing crime, changing lives: The government's plans for transforming the management of offenders.* Home Office.

Hood, C. (1991). A public management for all seasons? *Public Administration, 69*(1), 3–19.

Hood, C. (1995). The "new public management" in the 1980s: Variations on a theme. *Accounting, Organizations and Society, 20*(2/3), 93–109.

Hopwood, A. G. (1984). Accounting and the pursuit of efficiency. In A. G. Hopwood & C. Tomkins (Eds.), *Issues in public sector accounting* (pp. 167–187). Philip Allan.

House of Commons. (2003). *Committee of Public Accounts. Forty-ninth report of session 2002–03: The operational performance of PFI prisons.* House of Commons.

James, A. L., Bottomley, A. K., Liebling, A., & Clare, E. (1997). *Privatizing prisons: Rhetoric and reality.* Sage.

Jeamet, A., & Salais, R. (2019). Revising European governance toward capability and human rights-based benchmarking and allowing voices of stakeholders to be heard in social investment. *Horizon 2020 No 649447 Report WP7.3–4.* Leuven.

Joliffe, D., & Hedderman, C. (2014). *Peterborough social impact bond: Final report on Cohort 1 analysis.* QinetiQ University of Leicester.

Kaplan, R. S., & Norton, D. P. (1992). The balanced scorecard: Measures that drive performance. *Harvard Business Review, 70*(1), 71–79.

King, R. D., & McDermott, K. (1989). British prisons 1970–1987: The ever-deepening crisis. *The British Journal of Criminology, 29*(2), 107–128.

King, R. D., & McDermott, K. (1995). *The state of our prisons.* Clarendon Press.

Kurunmäki, L. (1999). Making an accounting entity: The case of the hospital in Finnish health care reforms. *European Accounting Review, 8*(2), 219–237.

Kurunmäki, L., Mennicken, A., & Miller, P. (2016). Quantifying, economising, and marketising: Democratising the social sphere? *Sociologie du travail, 58,* 390–402.

Lewis, D. (1997). *Hidden agendas: Politics, law and disorder.* Hamish Hamilton.

Liebling, A. (2004). *Prisons and their moral performance: A study of values, quality, and prison life.* Oxford University Press.

Liebling, A., & Arnold, H. (2002). Measuring the quality of prison life. *Home Office Research Findings No. 174.* Home Office.

Liebling, A., & Crewe, B. (2013). Prisons beyond the new penology: The shifiting moral foundations of prison management. In J. Simon & R. Sparks (Eds.), *The SAGE handbook of punishment and society* (pp. 283–307). Sage.

Liebling, A., Price, D., & Shefer, G. (2011). *The prison officer* (2nd ed.). Willan.

Lygo, A. S. R. (1991). *Management of the prison service.* Home Office.

Mehrpouya, A., & Samiolo, R. (2016). Performance measurement in global governance: Ranking and the politics of variability. *Accounting, Organizations and Society, 55,* 12–31.

Mennicken, A. (2013). "Too big to fail and too big to succeed": Accounting and privatisation in the Prison Service of England and Wales. *Financial Accountability and Management, 29*(2), 206–226.

Mennicken, A. (2014). Accounting for values in prison privatization. In S. Alexius, & K. Tamm Hallström (Eds.), *Configuring value conflicts in markets* (pp. 22–42). Edward Elgar.

Mennicken, A., & Espeland, W. N. (2019). What's new with numbers? Sociological approaches to the study of quantification. *Annual Review of Sociology, 45,* 223–245.

Mennicken, A., & Muniesa, F. (2017). Governing through value: Public service and the asset rationale. *Risk and Regulation, Winter, 2017,* 9–11.

Mennicken, A., & Power, M. (2015). Accounting and the plasticity of valuation. In A. Berthoin Antal, M. Hutter, & D. Stark (Eds.), *Moments of valuation: Exploring sites of dissonance* (pp. 208–228). Oxford University Press.

Miller, P. (2001). Governing by numbers: Why calculative practices matter. *Social Research, 68*(2), 379–396.

Miller, P., & O'Leary, T. (2007). Mediating instruments and making markets: Capital budgeting, science and the economy. *Accounting, Organizations and Society, 32*(7–8), 701–734.

Miller, P., & Power, M. (2013). Accounting, organizing, and economizing: Connecting accounting research and organization theory. *Academy of Management Annals, 7*(1), 557–605.

Miller, P., & Rose, N. (1990). Governing economic life. *Economy and Society, 19*(1), 1–31.

Miller, P., & Rose, N. (2008). *Governing the present: Administering economic, social and personal life.* Polity Press.

Ministry of Justice. (2009). *Capacity and competition policy for prisons and probation.* Ministry of Justice.

Morley, J. (2021). The ethical status of social impact bonds. *Journal of Economic Policy Reform, 24*(1), 44–60.

Muniesa, F., Doganova, L., Ortiz, H., Pina-Stranger, A., Paterson, F., Bourgoin, A., et al. (2017). *Capitalization: A cultural guide.* Ecole des Mines.

National Audit Office. (2003). *Report by the Comptroller and Auditor General: The operational performance of PFI prisons.* HMSO.

National Audit Office (2013). *Managing the prison estate.* TSO.

Nellis, M. (2006). NOMS, contestability and the process of technocorrectional innovation. In M. Hough, R. Allen, & U. Padel (Eds.), *Reshaping probation and prisons: The new offender management framework* (pp. 49–68). Policy Press.

NOMS. (2009). *NOMS Annual Report and Accounts 2008–09.* NOMS.

O'Quinn, R., & Ashford, N. (1996). *The Kiwi effect: What Britain can learn from New Zealand.* Adam Smith Institute.

Pollitt, C. (1993). *Managerialism and the public services* (2nd ed.). Blackwell.

Power, M. (1997). *The audit society: Rituals of verification.* Oxford University Press.

Pozen, D. E. (2003). Managing a correctional marketplace: Prison privatization in the United States and the United Kingdom. *Journal of Law and Politics, 19*, 253–282.

Prison Reform Trust. (1992). *Comments on the report by Admiral Sir Raymond Lygo "Management of the Prison Service".* PRT.

Prison Reform Trust (1994). *Privatisation and market testing in the Prison Service.* London Prison Reform Trust.

Prison Reform Trust. (1996). *The prisons league table: Performance against key performance indicators.* PRT.

Resodihardjo, S. L. (2009). *Crisis and change in the British and Dutch prison services.* Ashgate.

Rose, N., & Miller, P. (1992). Political power beyond the state: Problematics of government. *British Journal of Sociology, 43*(2), 172–205.

Sinclair, S., McHugh, N., & Roy, M. J. (2021). Social innovation, financialisation and commodification: A critique of social impact bonds. *Journal of Economic Policy Reform, 24*(1), 11–27.

Finance, S. (2009). *Social impact bonds: Rethinking finance for social outcomes.* Social Finance.

Solomon, E. (2004). *A measure of success: An analysis of the Prison Service's performance against its key performance indicators 2003–2004.* Prison Reform Trust.

Supiot, A. (2012). *The spirit of Philadelphia: Social justice vs. the total market.* Verso.

Supiot, A. (2015). *La gouvernance par les nombres.* Paris: Fayard (English edition, Verso 2016).

van der Zwan, N. (2014). Making sense of financialization. *Socio-Economic Review, 12*(1), 99–129.

Volokh, A. (2002). A tale of two systems: Cost, quality, and accountability in private prisons. *Harvard Law Review, 115*(7), 1868–1891.

CHAPTER 11

The Shifting Legitimacies of Price Measurements: Official Statistics and the Quantification of *Pwofitasyon* in the 2009 Social Struggle in Guadeloupe

Boris Samuel

The historic mobilization experienced by Guadeloupe in early 2009 resulted in a 44 day-long strike, whose watchword was the struggle against the high costs of living and *pwofitasyon*. The social movement denounced the opacity of the state's regulation methods and of the management of the main economic sectors, particularly in large-scale retail. The strike was led by a group called LKP (*Lyannaj Kont' Pwofitasyon*—l'alliance contre la *pwofitasyon*), which used numbers as a weapon to analyse, claim and negotiate. *Pwofitasyon* is a creole word which

B. Samuel (✉)
French National Research Institute for Sustainable Development (IRD), Centre d'études en sciences sociales sur les mondes africains, américains, et asiatiques (CESSMA), University of Paris, Paris, France
e-mail: boris.samuel@ird.fr

© The Author(s) 2022
A. Mennicken and R. Salais (eds.), *The New Politics of Numbers*, Executive Politics and Governance,
https://doi.org/10.1007/978-3-030-78201-6_11

gained prominence at the time of the 2009 conflict. The LKP collective translated it into French by the expression *outrageous exploitation* ("exploitation outrancière"). It means abusive economic exploitation, with the connotation that this exploitation is rooted in both colonial and capitalist relations. The *Union générale des travailleurs guadeloupéens*, a leftist union had already used the term in social conflicts before, since 1997 at least (Ruffin, 2009). The identification of *pwofitasyon* always rests on the same idea: players holding a dominant position in a given market or in an economic activity capture an undue profit, resulting from the existence of high sales prices. The denunciation of the *pwofitasyon* thus makes the quantified (re)evaluation of the profit or the abusive margins a passage not only possible, but also necessary.

In the case studied here, the fight against *pwofitasyon* resulted in a multiplication of calculation work, which was at the very heart of the 2009 movement. The essential role of figures in the 2009 struggle was not limited to the issue of price formation. The platform of protest put forward by the LKP carried a broad set of measures aimed at the revaluation of purchasing power. One of its central demands, for instance, was the introduction of a wage bonus of 200 Euros for those on low incomes. The negotiations consisted of a series of number battles around this and other demands. The parties involved in these standoffs were public institutions—the French National Institute for Statistics and Economic Studies (INSEE); the General Directorate for Competition, Consumption and Fraud Prevention (DGCCRF)[1]; the Inspectorate General of Finances (IGF); the Court of Audit; and others—as well as non-state actors urging the administrations to produce new quantitative analyses, audits and figures, such as the unions and the LKP. This ability to propose new frameworks for thinking about the economy, as well as new quantification methods was one of the strongest points of the mobilization.

How far has the "statactivist" (Bruno et al., 2014) momentum of the LKP and other non-state actors been capable of shifting the legitimate price measurement methods and the social construction of the price debate, and by what means? This is the question that this study addresses. Based on empirical observations of the calculations used during and in the wake of the movement, and analysing the conditions of their implementation as well as the discussions they triggered, this article will attempt to assess the processes by which the new measures became legitimate.

The text is organized as follows: The first part will show that the 2009 social conflict involved a variety of actors, differing in their relations to

calculation and also in the quantification methods they used. The struggles with numbers lead to unequal outcomes, in which technical but also
relational resources determined the balance of power between the actors.
The second part will question the legitimacy of the new quantifications of
high costs of living appearing in the wake of the social crisis, at a time
when official price analysis and measurement studies were particularly
poor. It will compare the different quantification methods under discussion during and after the negotiations: some estimates of abusive margins
proposed by the LKP, which, although effective to impose a public debate
on the *pwofitasyon*, were considered too simplistic to become legitimate;
a measurement of price differences between Guadeloupe and mainland
France undertaken by the INSEE, which did not sufficiently highlight
abusive profits to become visible and legitimate, although it was meant to
be a reference; and, finally, a practice widely used in official reports, the
press, or political speeches, which selected and displayed individual prices
in order to report, and denounce, the existence of abusive pricing practices. It is this latter utilization of figures which proved to be the most
efficient in the public space under consideration here.

The article illustrates the shifting legitimacies of price measurements
after the 2009 social conflict: the common and rigorous statistical
methods used by the INSEE created controversy, because they did not
display abusive prices with sufficient strength, while the innovative but
clumsy quantification practices of the LKP led public actors to adopt new
ways to account for pricing practices and for the price level gaps with
mainland France.

The Quantification of *Pwofitasyon* in the 2009 Battles for Power

Quantification as a Mode of Action for a Variety of Players

The first protests leading to the January and February 2009 general
strikes were triggered by an unexplained rise in fuel prices. In the French
overseas *départements*, unlike as in the rest of France, fuel prices were
administered and they were fixed via a "formula",[2] which was periodically
revised, taking into account a variety of parameters, such as freight and
employment costs, profit margins granted to the operators, international
prices, or the USD exchange rate (according to Decree 2003–1241 of 23

December 2003) (see Bolliet et al., 2009, pp. 8–9) (Autorité de la concurrence, 2009a, pp. 4–5). The "formula" used for the price calculation was deemed to guarantee an equitable regulation of the sector, but in practice criticisms of the industry's lack of transparency and its abusive pricing mechanisms had become stronger and stronger in the years preceding the conflict.

In 2005, an association of fishermen—the Association of Sea Fishermen of North Basse-Terre—complained that the tax free price paid by their profession had unexplainably increased by 70% between 2003 and 2005, while the all-inclusive prices had progressed much less (Gircour & Rey, 2010, p. 86). Facing the impossibility of understanding the price determination mechanisms (Les pêcheurs exigent de la transparence, newspaper article in France-Antilles Guadeloupe, 18 August 2006), their criticism targeted the regulation techniques employed by the DGCCRF. Here, the "formula" was debated for the first time outside of the administration. Then, in 2007, one of the major wholesale importers of the island, Didier Payen,[3] went on a crusade against fuel supply policies and price-setting mechanisms in Guadeloupe. In a detailed study, not dissimilar to what an audit firm could have supplied, he showed that cost differentials were, at least to a large extent, due to the supply policy in place for over 40 years—a policy favouring local refining and granting a monopoly for importing and refining to a private company (the Société Anonyme de Raffinerie des Antilles, SARA)[4] (Payen, 2009, p. 33). Furthermore, he showed that the calculation of pump prices contained various obvious and unjustifiable irregularities, such as the double-counting of certain taxes (as for instance in the case of the accounting for the tax on used oils) (Payen, 2009, p. 29). Although his work gave rise to discussions among the island's economic and administrative actors in 2007 and 2008—the report was even endorsed by the *Regional Economic and Social Council* of which Didier Payen was a member—his intervention did not alter the methods used by the DDCCRF (regional outpost of the French General Directorate for Competition, Consumption and Fraud Prevention, the DGCCRF, in Guadeloupe), which continued to apply the same "formula" to set prices.

The social movement against *pwofityason* examined here arose in this context. In the first half of 2008, international oil prices rose sky high, before starting to decline in July. Yet, as the months went by, pump prices continued to rise in the various French Overseas departments. The peak was reached on 1 October 2008.[5] A collective of entrepreneurs whose activities were hard hit by the increase, was the first to protest against this unbearable situation: it called for a strike in November 2008 and set up the first roadblocks in December. From the outset, Yves Jégo, Secretary of State for Overseas territories, showed himself to be receptive to the protest movement's messages. He, too, was somewhat suspicious of the administration's way of regulating the oil sector, which he considered opaque and potentially collusive (Jégo, 2009, p. 89).

In December 2008, the Inspection Générale des Finances (Inspectorate General of Finances, IGF) was asked to investigate the situation, and the Competition Authority was seized in February 2009. Pending the conclusions of these audits, the State adopted transitory measures, applying an immediate 31 centimes reduction per litre for lead-free petrol, and a 22 centimes reduction for diesel (Guadeloupe: An agreement to reduce oil prices triggers the removal of the roadblocks, 2008). To compensate SARA's loss of income, the agreement reached with the collective of entrepreneurs also provided for temporary State transfers to the company. Far from appeasing the social situation, the agreement actually fuelled the conflict: The *Lyannaj Kont' Pwofitasyon* (LKP), the alliance against *pwofitasyon,* was set up on 5 December 2008 upon the call of the *Union Générale des Travailleurs de la Guadeloupe*,[6] the main Guadeloupian trade union, which was completely opposed to the State's compensatory transfer to SARA. The very object of their anger were precisely the profits made by the company, which they considered to be illegitimate, i.e. *pwofitasyon.* Hence, it called for a general strike on 16 December, the day after the agreement was signed with the entrepreneurs (Gircour & Rey, 2010, p. 97).

LKP's accusations were incomprehensible to the administration, which denied any fault. The "formula" may of course have been clumsy, since price revisions were not frequent enough to guarantee a good matching of prices at the pump with international market fluctuations. However, according to the administration, no collusive or irregular practices occurred,[7] and despite the existence of certain dysfunctionalities, price regulations had mostly suffered from a lack of adequate consumer information. The March 2009 General Inspectorate's report confirmed these assertions. But it also acknowledged that the complexity of the "formula", and its outdated character, had made fuel prices opaque and vulnerable to calculation errors. The impact remained low, however: only 8 centimes were due to these errors, that is 5% of the price, and not 40%, as asserted by Payen.

LKP's call for a strike was, however, the starting point of a wider movement. The radical trade unions, UGTG (Union générale des travailleurs guadeloupéens), CGTG (Centrale des travailleurs unis) and CTU (Confédération générale des travailleurs de Guadeloupe), aspired to start a general strike on a broad set of claims going well beyond oil prices. Between December 2008 and January 2009, the collective prepared a broad platform of protest comprising 165 points.[8] The denunciation of *pwofitasyon* and the issue of purchasing power formed the platform's base (Ruffin, 2009).[9] The reasoning justifying the denunciation of fuel prices was replicated in numerous sectors, considering that prices were seen to hide abusive margins more generally: the LKP thus drew attention to the possibility of abuses on the markets for "basic necessity items", such as transportation, water, rents, electricity, communications, etc. To address this situation, the collective demanded the adoption of a variety of measures: the promotion of transparency, both in the private sector and public services, for example through the conduct of a programme of audits, or the creation of a "workers' research office" (Bureau d'études ouvrières, BEO) intended to help trade unions monitor prices; the bolstering of purchasing power via a series of social transfers to households (the symbolic claim being a bonus of €200 for all employees below a certain level of salary, and other claims pertaining to an increase of the social minima); interventions on price formation (the LKP demanded for example a *"significant reduction of all taxes and margins on basic necessity products and on transportation"* as well as the freeze of certain prices, such as rents and fuel); the fight against the pre-eminence of the importing companies and the promotion of Guadeloupian products.

What does the above teach us about the role of calculations and figures in the formation of social and political relations in Guadeloupe? Economic calculation played several roles here. The mobilization was aimed at fighting against the social and economic relations which calculation had helped establish (for example, the determination of fuel prices through the "formula" had immediate and daily consequences for the entire region). Besides, various actors turned calculation into a weapon, mobilizing their analytical capacity to denounce or even accuse the private firms and public administration of abusive pricing practices. Lastly, calculation was used by administrative and political players who carried out audits and controls in order to promote transparency and arbitrate the conflict, making calculation also a mediation tool.

Moreover, social actors stood out by their plural use of calculation and differed in their position with respect to the handling of figures. For the administration in charge of regulating the fuel sector, the handling of the "formula" echoed a routine task of "government at a distance". The interviews I conducted showed that the executives in charge were concerned with professionalism and accuracy, while being subjected to strong pressure by the economic actors. Since the existence of calculation errors was somehow part of the routine in their eyes, they also demonstrated the administration's relative indifference towards citizens (Herzfeld, 1992; Hibou, 2012, pp. 128–129). For some executives also, the handling of the formula could possibly reflect the collusion with the operators of the oil sector, who were seeking to draw maximum profits from the framework established by the "formula", but this could not be proven.[10] Here, different players used calculations and economic analysis to denounce the arbitrariness and opacity of price management. Didier Payen decided to undertake his own investigation of fuel price formation, making large use of quantification and collecting information via his personal network. His task was difficult: his report underlined how it had not been easy to obtain relevant information, the administrations hardly being open to his interrogations. Nevertheless, he belonged to an economic elite that had some access to information and power circles and was also able to spread his message. His report was even published by the regional Economic and Social Council.

This was not the case for the fishermen, who a priori turned out to be the victims of the calculation's arbitrariness. By initiating an inquiry into the "formula", they seemed to fight David's fight against Goliath, but while facing the opacity of price determination, they finally obtained some pieces of the puzzle of the fuel sector's management. Central political and administrative authorities could also act as counter-powers to the local authorities: The State Secretary, and the supervisory bodies questioned the way the figures were used, claimed their right to inspect and audit the data, and reaffirmed their capacity to impose sanctions in cases of proven circumvention of the rules. Finally, by demanding the end of *pwofita-syon* and claiming the existence of abuses, the LKP's relation to data was two-sided: they expressed doubt and uncertainty as to the integrity of the calculation methods and considered it was a sufficient reason to challenge the legitimacy of power practices, and to enter into a struggle with the administrative authorities. Furthermore, later events of the movement (see also the next section below) show that the collective used calculation as a weapon and an accusatory tool, even when it could not prove the existence of abuses.

Thus, there are different styles of calculation, which are characterized by varying capacities of actors to master calculation techniques, varying access to information, varying relationships to the political authority and different motivations in making the calculations.

These varying styles of calculation also reflect different historical trajectories. Didier Payen, in particular, was not only interested in social dialogue and defending his own entrepreneurial interests: he was a member of MEDEF and acted in his capacity as a representative of heads of businesses; he was also a notorious supporter of free market ideas. He highlighted the virtues of free trade continuously, and disputed the regulatory measures taken by the French authorities. His calculation techniques expose this multi-layered social and political position, for example he mixes the writing of pamphlets with the work of an audit. His approach was not that of an auditing firm, as can be observed immediately from the style of presentation: large characters, flashy colours to highlight the most important findings and underline the denunciative tone. His approach was reminiscent of the correspondence between chambers of commerce and the administrative authorities during imperial times, when merchants and settlers from the islands challenged state decisions in order to obtain free marketing rights (Lemercier, 2008; Tarrade, 1972, pp. 224–285).

The LKP's and unions' calculation techniques also deserve to be questioned from a historical standpoint. The use of calculation and technicity for purposes of activism must be considered in the light of the specific history of unions in Guadeloupe, and of their relations towards the administrations.[11]

The Role of Technicity and Expertise in the Negotiations

The struggle breaking out in January 2009 with the general strike showed that LKP's actions related in a variety of ways to state administrations and to the logics of expertise. On the one hand, the collective's skill and capacity enabled it to negotiate with the State on its own ground, in particular, because some of its members stemmed from the administration. The collective also had close links with most technical administrative bodies, such as the INSEE, which could support the activists during the negotiations with their expertise. Nevertheless, the negotiations were also an unequal process, in which highly skilled negotiators from the

State's Overseas department cabinet in Paris succeeded in getting the upper hand over LKP's leaders, who did not have the same access to economic information, and who did not enjoy the same calculative skills when it came to the design and discussion of new public policy instruments. Such observations ask for further investigation of the links that these "statactivist" (Bruno et al., 2014) mobilizations maintain with state administrations and expertise. Historically, both unions and employers' organizations fought to impose their conceptions of quantification, and they employed quantitative skill in their struggles; yet, they were also often backed by statistical administrations (Stapleford, 2009; Touchelay, 2014; Volle, 1982). What kind of situation is reflected by the Guadeloupian case? What kinds of links between the LKP's activist use of numbers and administrative expertise can be uncovered in this instance?

Thanks to the protests, LKP quickly met with resounding success. After the indefinite general strike which was launched on 20 January 2009, a series of important demonstrations began (Calimia-Dinane, 2009). On Saturday, 24 January, and Sunday, 25 January, over 10% of the island's population are said to have marched in the streets of Pointe-à-Pitre. The Prefecture, impressed by these successes, agreed to enter negotiations. It also took a decision for which it would later on be much blamed by the State's Overseas department: the Prefect agreed to a live television broadcast of the negotiations (Jégo, 2009, p. 54). The live broadcasting by Canal 10, from 24 to 28 January, was an unprecedented event. Thanks to their skills and mastery of the economic and social issues under discussion, the LKP members stood up well to the Prefect and his administrative directors. On each of the points up for negotiation, the administrative directors—often coming from mainland France like most in the state administration hierarchy, and in Guadeloupe for just a few years—faced union members who had managed to build solid knowledge of the files over many years. At that time in Guadeloupe, both among activists, the administration, the negotiators and the press, the LKP was praised for its skills, and placed on a pedestal, whereas the administration was allegedly found to have been incapable and at fault. The course of the negotiations confirmed LKP's victory. The televised discussion was interrupted early, on an order from Paris. Prefect Desforges read a message from the State Secretary Yves Jégo denouncing the way the negotiation had been turned into a "tribunal" (Jégo, 2009, p. 54). And Yves Jégo decided to come to Guadeloupe in person to settle the matter.

At the beginning of the strike, the movement thus succeeded in gaining the upper hand over the state players by demonstrating its ability to use "government tools", such as administrative files and techniques of economic calculation (Desrosières, 2008, p. 59). This evidences a profound transformation of the modes of political action in the Département. After the violent struggles that had occurred during the 1970s, which included armed and terrorist action,[12] since the 1990s, the left-wing anti-establishment and separatist movements moved onto different institutional ground (Daniel, 1997; Réno, 2001). Its leaders, many of whom were born after Guadeloupe's "Departmentalization", got to know (and to challenge) the state apparatus from the inside. The separatist parties became very successful in local elections by asserting their management abilities, while facing a political class seen as corrupt and unreliable. This shift, however, remained limited to political parties only. The main unions within LKP, such as UGTG and CGTG, continued to present themselves as the legatees of the radicalism of past struggles. Continuing to refer to the traumatic memory of the great repressions of the 1960s and 1970s, such as the May 1967 episode, when police fired at crowds, resulting in a number of victims, still kept secret to date by the French State,[13] these unions continued to use force and inflexibility as weapons, sometimes even advocating resort to violence (Braflan-Trobo, 2007).

Yet, the use of such radical methods generated a deep division among trade unions in present-day Guadeloupe. Major strikes had often resulted in a divided society. In this respect, LKP's approach marked a break. The trade unions united with political parties and a number of associations to form an unprecedented alliance (Bonilla, 2010; Bonniol, 2011, p. 92; Chivallon, 2009; Gircour & Rey, 2010, p. 101; Larcher, 2009). The LKP could take advantage of a generational renewal. By grouping officials from different economic and social sectors, customs officers, company executives, political leaders, academics, representatives of consumer organizations, etc., it benefited from the arrival of union leaders who came from the very heart of the bureaucratic and political system. Elie Domota, head of LKP at the time, and originally from UGTG, the main Guadeloupian autonomist union, was for example deputy director of the ANPE in Guadeloupe. Alain Plaisir, the collective's economist at the time, was a customs officer with a thorough knowledge of economic policies and the tax system. Thus, LKP was in a position to initiate the struggle via the administrative field itself. It was comfortable with the handling of figures and administrative data, presenting itself as the institutions' interlocutor. In this respect, the movement can be seen as a "XXIst century movement".[14]

Nevertheless, only a few months after the mobilization, the initial impressions of success started to fade. The idea that the LKP could play on an equal footing with the public administration thanks to its skills was contradicted by the observation that the movement—with its limited means—was facing a dominating State apparatus. In many respects, the fight was unequal. A closer look at one of the negotiations helps to get a sense of the multiple factors shaping the power relations that developed between by the LKP and the other parties around economic policies. The negotiation of the wage agreement resulting in a €200 bonus for workers earning less than 1.4 times of the SMIC (guaranteed minimum wage) affords in particular a better understanding of the movement's relation to economic and statistical expertise.

The negotiation involved the State, the local authorities and the social partners (trade unions and heads of businesses). It took place shortly after the adoption of a new social system in France, called the "active solidarity income" (*Revenu de Solidarité Active*, RSA), which provided a bonus to all persons whose income was below the minimum wage. But this system had not yet been applied outside mainland France, in spite of repeated appeals from the overseas departments' elected representatives (Le RSTA moins avantageux que le RSA? Newspaper article, France-Antilles Guadeloupe, 15 May 2009). The negotiation's aim was to determine the overall financial effort that could be made by each of the parties (region, regional council, State and companies) in order to pay a bonus. The total amount obtained would determine the salary threshold below which the bonus could be paid, and therefore the number of beneficiaries. The first phase in the negotiation had made it possible to find an agreement close to the wishes of the LKP. Under the auspices of Yves Jégo, the MEDEF accepted that the bonus would apply to all employees earning less than 1.6 of the SMIC (guaranteed minimum wage). But the State secretary was disavowed by the government and the agreement was adjourned before it could be sealed and signed, probably due to pressure from the lobbies of heads of businesses in Paris (Jégo, 2009, p. 11). Because of this U-turn by the government, the unions' position became more extreme, and tough, even clearly violent, methods propagated by certain radical fringes of the LKP, such as Alex Lollia's "GTL" (*Groupe d'intervention des travailleurs en lutte*) (Gircour & Rey, 2010, p. 14), emerged at this point. Shops were closed by force to comply with the general strike order; extremely tough road blocks took place, night-time violence, lootings of shopping centres, clashes with police forces erupted, even causing the death of a trade unionist, Jacques Bino, who was shot dead near a roadblock during the night of 18–19 February (Gircour & Rey, 2010, p. 123).

In the midst of this tense situation, a team of negotiators was dispatched from Paris by Matignon. The discussions turned into a tug of war. The LKP refused to go beneath the threshold of 1.6 of the SMIC (guaranteed minimum wage), which it considered had already been obtained during the first negotiating phase. The other parties could not, or did not wish to, finance the total sum. At this point, the discussions placed calculations at the centre of an open conflict, and INSEE's (National Institute for Statistics and Economic Studies) regional office played an important part in the mediation. It lay the foundations for the conduct of the discussions, by supplying all parties with the figures required for the calculation of various scenarios, in particular figures concerning employment and the distribution of income by branches.[15]

INSEE's regional office was in direct contact with the negotiators, often quite informally. On LKP's side, Alain Plaisir, the collective's "economist" (and also Secretary General of the Centrale des Travailleurs Unis, CTU) communicated with INSEE's regional office, sometimes directly from the negotiation's backstage. The delegation of negotiators could present its requests to the INSEE internally through the administrative channel, either directly or via the Prefecture's services. The negotiations were concluded on 26 February 2009 with an agreement which was constructed on the basis of the calculations realized by the statisticians.[16] The negotiations were based on calculations simulating the financial impact of a €200 bonus on a variety of economic branches, and the collection of available information for this. Such work was of course technical. But the statisticians' work also contributed to the political mediation of the conflict. INSEE's regional department head was considered by the Paris negotiators to be part of the state administration and, off the record, he was an attendee of internal meetings held at the Prefecture. Such an integration was quite unusual, since INSEE's mandate of independence would normally require that it remained distant to the work of the Prefecture.

In parallel, a close link could also be established informally between the LKP and the INSEE. The CTU's union representative within INSEE, a statistician himself, knew that he could count on his department chief's cooperation during the negotiation. Sometimes, the LKP and the INSEE would even discuss urgent matters by telephone. Thus, the social dialogue relied on links that each party managed to establish with the INSEE, which on the one hand assisted in elaborating measures of economic policy based on quantitative data, and, on the other hand, sought to act in a mediating capacity in the conduct of the social dialogue.

Despite the existence of such a political mediation through numbers, the relations between the parties remained conflicted and unequal. The events following the agreement show that the process carried numerous uncertainties. The agreement turned out a posteriori to be much less favourable to LKP than it had seemed to be at the outset, because the collective had contented itself with too vague terms. In fact, the State had merely redeployed funds that had already been budgeted for a similar scheme, simply re-shaping old policy measures.[17] The 1.4 SMIC limit also created confusion as to whether the threshold amount was net or gross, which made the agreement actually more restrictive than it had appeared. As a result of the bonus, certain households passed into a higher tax bracket, thus reducing the sum they were expected to receive overall (Le RSTA moins avantageux que le RSA? Newspaper article, France-Antilles Guadeloupe, 15 May 2009; Verdol, 2010, p. 63). Lastly, from the end of the first year onwards, the system was no longer fully financed. It thus turned out that the social negotiation had been less favourable to the collective than it had seemed, in particular because the other negotiators had been better armed than the LKP to deal with the files and the evidence contained in them.

These different sequences show the multiple roles technicity and calculation played in the social dialogue and the struggle against the high cost of living in Guadeloupe. In the negotiations examined above, the LKP had indeed succeeded in imposing the establishment of new public policies, even if the State services, initially taken by surprise, were able to regain the upper hand. The collective showed that it was possible to defy the State's power on its own ground, and even to get the existing policies shifting. But its activism was also subject to an unequal relation. This raises the question whether or not, observing the ways in which economic policies were formulated before and after the conflict, LKP's activism succeeded in creating a lasting change in the relations of power in Guadeloupe.

THE LEGITIMACY OF PRICE AND MARGIN MEASUREMENTS

Did the conflict modify the calculation of prices and quantification and assessment of (high) living costs? Did the LKP succeed in making its voice heard, by changing the socially accepted methods of measurement, in the short and in the longer term? To answer these questions, this section will examine the means by which the unions' action and the social movement legitimized, or de-legitimized, new price measures, as well as ways of thinking about the question of high cost of living.

The Absence of Prices and Margin Measurements Before 2009

Prior to the 2009 crisis, the measurement of prices was in a paradoxical situation. In the French Overseas departments, the debate on price formation was at the centre of attention and socio-political relations, and the high level of prices was recurrently denounced. And yet the question of price levels gave rise to very scarce economic analyses and statistical follow-up efforts. One of the consequences of the 2009 movement is the questioning of this *status quo* in which the price question remains outside the scope of what can be discussed by public institutions.

When purchasing power became the centre of attention of Guadeloupe's boiling political scene in 2009, INSEE's most recent studies on the cost of living differential between mainland France and its Overseas departments (DOM) were surprisingly old. The last one dated back to 1992, and the one before that to 1985. Such deficiency surprises, because conducting such studies is theoretically required and constitutes a core part of INSEE's working programme. Once every ten years at least, INSEE is supposed to establish a "geographical price comparison" for the State to adapt a series of public policies towards the Overseas departments (DOM): in particular, the level of the bonuses granted to civil servants to make up for the cost of living gap (the famous "sur-remunerations") and other social transfers, such as the "territorial continuity" which subsidizes transportation to mainland France. In other words, price differentials between the different DOM and mainland France constitute key data for the State, but when the contestation broke out in Guadeloupe, INSEE had not measured them for a long time.

It is difficult to assess precisely why such a situation prevailed. The debates following the adoption of the Euro in 2002 were for sure a part of the explanation. The adoption of the Euro generated in particular very strong discontent in the Overseas departments (DOM). In the DOM, and to a lesser extent in mainland France, the changeover to the Euro was deemed to have entailed particularly heavy price increases, especially in the large-scale retail sector. Such increases could not be formally proven by existing surveys in Guadeloupe, but interviews with INSEE officials confirmed that closer scrutiny of the price factors could have revealed and confirmed these. INSEE's official speech, however, buried these increases inside an assessment of the general level of prices, which allegedly had remained stable (INSEE, 2002). The lack of official statistical data became the object of a public debate.

In La Réunion, where the price gap with mainland France reached 70% to 80% on many supermarket products at mid-decade (UCF/Que choisir, 2004, 2005), the pressure exerted by elected representatives and consumer groups sought to reinforce the establishment of better price analysis structures. Thanks to the relentless fight of a communist elected representative of La Réunion,[18] the decision to create regional price observatories in Overseas departments (DOM) had been adopted in 2000. These observatories would bring together consumer associations, administrations, social partners, chambers of commerce, etc., under the presidency of the Prefect. But until the controversy over purchasing power grew on the island in 2004–2005, the resistance of state services remained very intense (Sénat, 2005, p. 1725). According to witnesses of the creation of the observatory, this reluctance could be linked to the pressures exerted by the large-scale retail sector lobby (Le collectif pour l'observatoire des prix: Dix mille signatures pour Baroin, newspaper article, Le Quotidien de La Réunion et de L'Océan Indien, 5 September 2006). The implementation decrees were only adopted in 2007 (Doligé, 2009, p. 147).

In Guadeloupe, the price observatory held its first meetings in 2008, and its activities were expanding when the conflict erupted. The observatory intended to conduct a whole new series of analytical works, but at least during the first years of its existence, it had not been able to produce any important results (Favorinus, 2009).

The public authorities' reluctance to (re)calculate and analyse prices should also be understood in the light of the Overseas departments' social history. The matter is explosive and had caused the periodic resurgence of historical disputes. One matter was particularly explosive in this context: the 40% bonus (*sur-rémunération*) granted to civil servants in Guadeloupe, which they managed to obtain in 1953 following a tough fight and a 65-day strike. At that time, only mainland civil servants were granted such bonuses (similar to the compensations paid to the civil servants accepting to work in the colonies), and even when Guadeloupe had become a department in 1946, the native civil servants were not entitled to this premium (Dumont, 2010, p. 170).[19] Since then, and although the level of the *sur-rémunération* should theoretically be indexed to the observed level of prices (measured by the gap with the mainland), discussions about the adjustment of the bonus were often very risky, because they carried with them the nagging and conflictual question of equal treatment within the Republic, which became effective only recently (Burbank & Cooper, 2010; Forgeot & Celma, 2009; Mam Lam Fouck, 2006).

In theory, the price gap observed in 2009 could no longer justify a 40% bonus: INSEE's studies of 1992 placed the synthetic indicator at around 12%. Since the beginning of the 1990s, there had been regularly calls for a reform to reduce the *sur-rémunérations* (Doligé, 2009, p. 147; Fragonard et al., 1999; Laffineur, 2003; Ripert, 1990), but these recommendations, issued in Paris, never became effective, with the spectre of revolt apparently still on everybody's mind. In addition, the *sur-rémunérations* also played an implicit redistributionary role in a situation of great poverty prevailing in the Overseas departments (DOM: le Medef remet en cause la sur-rémunération des fonctionnaires, newspaper article, Journal de l'île de la Réunion, 11 August 2010; Sur-rémunérations, des avis plus contrastés à la Réunion, newspaper article, Journal de l'île de la Réunion, 12 August 2010), all of which made it a very sensitive subject.

In La Réunion, the inopportune release of a price study resulted in a very serious social unrest in 1997, as well as a "civil servants' revolt" (Conan, 1997). My interviews suggest that INSEE, by omitting to carry out this task in Guadeloupe for almost twenty years, avoided taking up a position on a question its officials felt was politically too sensitive. But in doing so, it also took the risk of having to react under pressure by starting a study at the very moment the debate would truly flare up.[20] And this is precisely what happened in 2009. To be exact, this position is in no way indicative of the statistics policy in the French Antilles, which rather tends to be maximalist, because there is a recognized need for more precise data on the *départements'* economies (Morel & Redor, 2008; Rivière, 2009; Sénat, 2009). Nation-wide surveys are often over-sampled to ensure their representativeness at the level of the *département*, and sophisticated macro-economic aggregates (regional accounts) are compiled. All in all, it can be held that on the eve of the conflict, the absence of price data was both the symptom of the conflictual nature of the matter in the public sphere and the impact of the status quo on the politics of distribution (symbolized by the pursuit of the *sur-rémunération* policy) which was also an obstacle to frank reflections on price levels on the island.[21]

This observation also relates to the surveillance of margins and competition practices, in particular in the mass-market distribution sector. The pricing practices of economic empires, such as those of Bernard Hayot or Alain Huyghues-Despointes, both "békés" (Antillean Creole term to describe a descendant of the early European, usually French, settlers in the French Antilles) from Martinique, were at the very heart of the 2009 protests. Mass market and import fortunes were built in the West Indies during the 1970s, with the help of state-sponsored policies. Coming from the plantation economy, merchants and other entrepreneurs found in the "catching-up" policies in place from the 1960s through to the 1980s numerous opportunities to save their assets from the historic collapse of the sugar-cane sector in the 1970s, in particular by taking advantage of the public subsidies supposed to stimulate investment in the new markets and sectors, such as tourism, or large-scale retail. These programmes generated windfall effects, as well as misuses and excesses. The negative consequences of these policies were discussed at length by the parliament (Jalabert, 2007, p. 75; Ripert, 1990).[22] And yet, though recurrent since the mid-1990s, the numerous calls for a serious re-evaluation of the *défiscalisation* (tax exemption policies), which were a continuation of the systems initiated in the 1960s, had never been successful. Recently the French Court of Auditors (*Cour des comptes*) and the Economic and Social Council (*Conseil économique et social*) took once again an interest in the question, but their recommendations were not implemented either.

In the same vein, since the DGCCRF lacked human and logistic resources, its agents undertook no serious study of competition struggles on the Islands, although such studies would have revealed the dominant position the main actors had managed to build in the large-scale retail sector (Doligé, 2009, pp. 132–133).[23]

The tense *status quo* around economic policies extends to economic analyses, measurements and evaluations of prices and margins, which remained understudied until the end of the 2000s. The 2009 struggle, with the massive general strike and the roadblocks which paralyzed the Island's economic activities for 44 days, got things moving. The social conflict and the existence of LKP's platform of protest forced the opening of a participatory debate, in particular through the organization of a large multi-stakeholder consultation under the auspices of the State's Overseas department (les Etats généraux de l'Outre-mer, from March to July 2009) which was followed by the establishment of an *Interministerial Committee for Overseas Departements* (Comité Interministériel pour l'Outre-Mer). In this context, the group working on price formation, purchase power issues and large-scale retail issued a series of recommendations largely inspired by LKP's claims. The group was headed by a former chief of the Guadeloupean regional office of the INSEE, Delile Diman Antenor, who was also very respected by LKP members for being a former leftist activist. The report produced under her guidance proposed the conduct of new economic and statistical analyses, enabling the setting-up of a fully fledged "transparency policy" in response to the social movement.

The particular measures provided in the agreement of the 4th of March 2009, which put an end to the general strike, included studies that were to be undertaken by the INSEE, the DDCCRF, the Price and Income Observatory, or workers' associations, such as the "Bureau of Labour Studies" (BEO) (see *Protocole d'accord du 4 mars 2009*). The "typical shopping trolley" and the "household shopping basket"[24] were meant to permit price tracking (and adjustment) in the large-scale retail sector. The spatial comparison of prices between mainland France and Guadeloupe conducted by INSEE was intended to evaluate the price gaps. The programme of competition audits to be undertaken by DDCCRF was destined to shed light on the practices of certain strategic sectors. In addition, a study of consumption patterns was to be launched, with the aim of boosting local production. Furthermore, the creation of a regional commission for economic and statistical information (CRIES) was considered (Diman-Antenor, 2009). These various studies were furthermore slated to be submitted to the Price and Income Observatory (*Observatoire des prix et revenus*). At least on paper, the response to the problem of purchasing power appeared to be ambitious and coherent.

However, studying the transformation of the ways in which prices were managed by numbers requires to go beyond examining the presentation of these plans on paper. The problem lies not so much in the questioning of the effectiveness of the implementation of this "transparency policy"— a large portion of the measures has been more or less implemented since 2009. It is about examining whether the price measurements gave way to new social practices and formed a base for new measurement conventions, which could be either seen as legitimate, or, on the contrary, as sparking debates and controversies. To examine these questions, the remainder of this chapter will investigate how three different modes of price quantification were used and put up for discussion. Firstly, the chapter will analyse negotiations of "voluntary reductions" of the prices of the products considered as "necessities" (*produits de première nécessité*), where LKP used very intuitive, but not very robust commercial margin estimates. These calculations became quickly delegitimized for their lack of precision, but they also helped improve the balance of power as they generated a debate on what constitutes legitimate levels of prices and margins. Secondly, the case of the spatial price comparison study carried out by INSEE is considered which highlights how a sophisticated study, evaluating the price gaps between Guadeloupe and mainland France, can come to be very negatively received and held to be socially and politically illegitimate, although such a study had been among the social movement's core demands. Lastly, by examining a diverse series of studies of prices and margins, I will argue that new legitimate price-setting practices eventually emerged. However, neither the price index nor the average price level measurements stood out as proper ways to address issues related to high living costs; rather, extreme values, such as examples of individual high prices, seemed to be better able to reflect the population's feelings of inequality and injustice in the face of abusive pricing practices. Here, new quantification methods took root as legitimate ways to describe and denounce such pricing practices.

The Quantification of Pwofitasyon: *Innovation and Tests of Reality*

The negotiations held to determine price reductions for the 100 products knowns as "necessities" reveal LKP's working methods, its approach to obtain margin reductions and its use of figures.

The negotiations were held from March to May 2009 under the supervision of the Prefecture (*Préfecture*) and the DGCCRF, and they were particularly tedious. One hundred product families were designated; a set of products had to be selected for each of these families. The various actors of the large-scale retail industry had successive discussions with the LKP, from local minimarket chains, such as *Huit à Huit*, to the very large *Bernard Hayot Group*. Every day discussions were held, from late afternoon to four o'clock in the morning, over a period of approximately three months.[25]

The collective of the LKP intended to get the upper hand by showing quantitative evidence (or at least what it held to be such evidence) of the existence of *pwofitasyon* and of the necessity to lower prices and margins. This was possible thanks to the series of calculations on prices it had carried out prior to the negotiations.[26] At that time, there was no other available information on commercial margins in the large-scale retail sector. No public institution could be blamed in this respect, since only a detailed audit of the sector, or a competition audit could have produced solid facts about profit margins, and such actions, though within the DGCCRF's competence, were not deemed necessary to be undertaken on a regular basis at that point.[27]

The technique the LKP collective had adopted under the guidance of its economist, Alain Plaisir, was rudimentary, but it made it possible to put a figure on the table. LKP teams prepared listings of prices observed in mainland France, using the *Carrefour* chain's website; for each product, a theoretical DOM price was then calculated by adding a fixed percentage to the mainland price to account for the costs of transportation, taxes and other logistics. The members of the collective considered that these costs could be estimated by adding a lump sum of 10% to the initial price. The resulting theoretical prices were then compared with the prices that were actually observed in the island's supermarkets; and the resulting differences equipped the LKP with estimates of "illegitimate margins" picked up as *pwofitasyon* by the companies.

LKP's ambition was at least twofold. First, it sought to expose the illegal profits, and, second, it used the figures in its price setting negotiations. Each brand brought its own price records and negotiated item by item. Armed with spreadsheets, the LKP thus cornered the large-scale actors and obliged them to justify the level of their commercial margins (see also Fig. 11.1, which provides an excerpt from the records that were used by LKP in the negotiations). Obviously, the method the LKP applied was very clumsy, and the obtained values were impossible to verify. Nevertheless, the figures reflected LKP's mental representation of the price formation, and they rested on the collective's "expertise". They were thus considered significant enough to uncover misuses and force the concerned players to admit their abusive pricing practices and to lower their prices.

The method worked. The negotiations led to a series of agreements providing for price reductions on the 100 "necessities". These agreements were binding for the companies concerned and gave rise to new control procedures. Announcements of price reductions had to be

Libellé	Prix Moyen France	+10% (transp.)	1 Destrellan			2 Cora			3 Milenis		
			Prix	Diff	Ecart	Prix	Diff	Ecart	Prix	Diff	Ecart
LAIT CONCENT SUCRE NESTLE 397G	1,65	1,82	1,85	0,04	2%	1,89	0,08	4%	1,96	0,15	8%
LAIT NIDO 28% 400G		0	5,31			5,4			5,3		
LAIT POUDRE LAICRAN 900G	6,9	7,59	8,05	0,46	6%	8,79	1,2	16%	7,99	0,4	5%
LAIT CROISSANCE CANDIA BRK 1L	1,85	2,04	1,59	-0,45	-22%	2,2	0,17	8%	1,85	0,19	-9%
LAIT 1/2 ECREME VIVA CANDIA 1L	0,9	0,99	1,1	0,11	11%	1,35	0,36	36%	1,1	0,11	11%
GLORIA LCNS 3X410G	3,6	3,96	4,73	0,77	19%	4,59	0,63	16%	4,73	0,77	19%
LAIT PPX 1 L	0,59	0,65	0,79	0,14	22%	0,75	0,1	16%	0,8	0,15	23%
LAIT ECREME REGILAIT BTE 300G	1,88	2,07									
ST HUBERT 41 250 G	1,9	2,09				2,19	0,1	5%			
BEURRE PLAQUETTE BOCAGE 250GR											
BEURRIER BRIDEL D/S 250G	2,05	2,26	2,36	0,11	5%	2,59	0,34	15%	2,35	0,1	4%
BEURRE DX 250G PRESIDENT	1,49	1,64	2,29	0,65	40%	2,65	1,01	62%	2,4	0,76	46%
BEURRE MOULE DEMI SEL U	1,55	1,71									
BEURRE DOUX 250G MAITRE LAITIER	1,49	1,64									
MARGARINE ASTRA 500G	1,4	1,54	2,37	0,83	54%	2,39	0,85	55%	2,35	0,81	53%
FRUIT D'OR VITALITE ALLEGE 250GRS	1,27	1,4	1,77	0,37	27%	1,79	0,39	28%	1,78	0,38	27%
MARG TOURNESOL ALLEG N°1 500G	0,76	0,84	1,15	0,31	38%				1,65	0,81	97%
MARGARINE à frire winny 1kg	1,95	2,15				3,33	1,19	55%	3,33	1,19	55%
S.EPA 4% UHT BRIDELIGHT 2	1,18	1,3									
CREME EPAIS.LEG.E&V 33CL+15%MG	1,47	1,62	2,04	0,42	26%	2,05	0,43	27%	1,99	0,37	23%
YOPI CHOCO 4X100G	0,98	1,08	1,7	0,62	58%	1,7	0,62	58%	1,69	0,61	57%
BRIE ROITELET POINTE 200GR	1,75	1,93	2,48	0,56	29%	2,45	0,53	27%	2,59	0,67	35%
VACHE QUI RIT 8 PORTIONS 128G	1,35	1,49	1,58	0,095	6%	1,59	0,11	7%	1,59	0,11	7%
EMMENTAL RAPE 100G ENTREMONT	1,24	1,36									
CAMEMBERT 45%MG 250GR BRIDEL	1,7	1,87	2,78	0,91	49%	3,19	1,32	71%	2,79	0,92	49%
EMMENTAL PLAQ PDT 220G	2,35	2,59				2,89	0,31	12%			
EMMENTAL RAPE 3X70G +1 GT PRESIDENT	2,7	2,97	3,21	0,24	8%	3,25	0,28	9%	3,15	0,18	6%

Fig. 11.1 Excerpt of the chart used by the LKP during the negotiations of the prices of the 100 "necessities" (Adapted from the original obtained by the author with permission from Alain Plaisir/LKP)

displayed visibly in all stores. The DGCCRF was in charge of verifying the enforcement of the agreement. It was also responsible for the monthly publication of a survey on large-scale retail prices. The negotiations thus led to actual results. However, at the same time, the LKP was taking a big risk by employing such a simple method. This could easily engender powerful resistance, as the figures could easily be invalidated. Both actually occurred very quickly.

The "quantification" of *pwofitasyon* proposed by the LKP was rapidly contested. Their use of numbers entailed certain weaknesses which came to be exposed in the course of the negotiations. Rather than proving the high level of prices, the numbers were also used to constrain companies to lower the prices.[28] Although some of the negotiators had true expertise in price formation, as for instance Alain Plaisir of the CTU union or Justina Favorinus of the *Consommation, Logement et Cadre de Vie* association, the negotiations revealed that LKP had missed some major components of price formation in their calculations. LKP's theoretical prices were thus grossly underestimated, so that some firms came even to be obliged to sell at a loss. According to members of the LKP, the unions became only in the course of the negotiations aware of this and the fact that a major part of the commercial margins escaped mass-distribution operators, and were instead distributed to other actors, such as importers and wholesalers. The important role of these actors had not been identified by the LKP experts before. More generally, the sharing of "gross margins" among a myriad of participants had never before been perceived by the analysts as a major cause of high prices (Favorinus, 2009). For a while LKP considered inviting these other actors—distributors, wholesalers, logistics and warehouse operators—to the negotiation table as well. But this turned out to be unfeasible, since this would have entailed more than 300 companies.[29] Therefore, the negotiations on necessities had to be stopped in the face of this obstacle.

In an apparent paradox, by undertaking efforts to quantify price differences and margins, the trade unions had made it possible for themselves to better understand the formation of prices, but such better understanding made it in turn impossible for them to demand a significant lowering of prices. LKP's initial analysis of prices had proven inaccurate. The pertinence of the notion of *pwofitasyon*, understood as the grabbing of commercial profit by a very limited number of actors, was also seriously challenged. Nevertheless, the negotiations brought the quantification of price formation to the centre stage of the public debate.[30] From this standpoint, it cannot be considered that LKP did not succeed in undermining the long lasting status quo around pricing practices.

An Expected but Socially and Politically Unacceptable Intervention by Public Statistics

The case of the work known as "spatial price comparison" produced by INSEE (Berthier et al., 2010) was the exact opposite of the situation described above. This study was supposed to be the highpoint of the "transparency policy" initiated in response to the social crisis. But although the study managed to finally quantify the price gaps with mainland France, thereby officially acknowledging the existence of such gaps, it failed to become the "instrument of proof" it should have come to be (Desrosières, 2008, p. 59). On the contrary, the study became an occasion for heated exchanges and controversy among the actors of the conflict.

First of all, the INSEE study was published relatively late, in July 2010, i.e. roughly sixteen months after the open conflict had ended. Its release had been postponed after a debate that had ensued among INSEE's specialists, and for the reason of being able to use what was considered to be the best-suited, but also cumbersome method: the purchasing power parity (PPP) method of comparison. A reference method used in international organizations (for instance, the International Comparisons Programme carried out under the auspices of the United Nations has used it since the 1970s),[31] PPP calculations mobilize expertise composed both of national accounting and price statistics. The head of the regional statistical office of Guadeloupe in particular had argued in favour of this methodology. Yet, in the end, the investment was deemed too costly and unwarranted in Paris, so preference was finally given to a much lighter method based on the data that had already been collected for the price indexes.

There were two good reasons behind opting for such a technique: first, the existing databases used for the calculation of price indexes immediately allowed for this type of analysis, no additional data collection was required; second, the method appeared completely natural and clear to the price statisticians in charge of the study, who, however, were not really at ease with the complex analyses of purchasing power parities. The choice was thus apparently technically driven, marking the reluctance of price statisticians to engage with a methodology perceived to be too complex, and involving national accounting approaches they could not master. This "technically" driven choice had however substantial consequences on the public reception of the study. The applied method was less suited to address the purchasing power question, and thus was also far from adequate to address the societal demands for more accurate data on price differentials.

The study compared the value of a basket of representative consumer goods observed in the DOM to what exactly would be the price of this same basket in mainland France; conversely, it compared a typical mainland basket to what its price would be in the DOM; and then, in a last step, it determined an average.

The applied method can be considered problematic for a balanced understanding of the purchasing power issue for several reasons. For example, nobody drinks whisky in the West Indies, and conversely, few people in France eat yam. The comparisons outlined above fail to adequately consider such (cultural) differences between consumption patterns, and are in the end not very accurate with regard to reflecting people's behaviours and preferences. In addition, the results are not very legible to the unversed: the final, synthetic gap indicator is computed with the help of a complex methodology (resulting in a geometric average, based on Fisher's law), which is not easily understandable (see also Table 11.1).

A further factor complicated the public reception of the "spatial price comparison". The synthetic indicator put forward by the statisticians in

Table 11.1   Excerpt from INSEE's 2010 report based on a spatial price comparison

%	Price differences DOM/mainland France (market basket of representative consumer goods in mainland France)	Price differences mainland France/DOM (market basket of representative consumer goods in the DOM)	Fisher's differences (écarts de Fisher) DOM/mainland France
Martinique	16.9	−2.9	9.7
Guadeloupe	14.8	−2.2	8.3
French Guiana	19.6	−6.4	13.0
La Réunion	12.4	−0.4	6.2

Translated by the author from Berthier, J. P., Lhéritier, J. L., & Petit, G. (2010). *Comparaison des prix entre les DOM et la métropole en 2010*. INSEE Première, n°1304. Paris: INSEE
Price differences between overseas departments (DOM) and mainland France in March 2010
Explanation: Taking the mainland consumer basket as a reference, prices in Martinique were on average 16.9% higher than in mainland France. Taking Martinique's representative consumer basket as a reference, prices in mainland France were 2.9% lower than in Martinique on average. Fisher's difference indicator, a geometric average of the two differences, shows that prices were overall 9.7% higher in Martinique than in mainland France. Coverage: Household consumption except fuel oil, town gas and rail transport
Source Spatial price comparison survey, INSEE

the study amounted to about fifteen percent, to be precise 14.8%. Estimating the price gap with mainland France at such a level could obviously create misunderstanding in Guadeloupe given that differences exceeding 50% had been so far mentioned in all studies that were based on the observation of supermarket shelf prices, in particular concerning many of the most common imported goods, such as food and household products. Representations of high living costs and *pwofitasyon* were thus based on estimated differences of 50% or more. The INSEE figure did not contradict these estimates, as it was based on a sample of different prices, including prices for which the difference was much smaller or even negative, such as rental prices, insurance premiums, etc., which explains the lower value of their indicator.

But the difference calculated by the INSEE was not socially acceptable, because it was not in line with the commonly shared representation of the "high level" of abuse existing on the island. Besides, claiming that there is a 15% price differential could also pave the way for a possible questioning of the 40% bonuses (the *sur-rémunérations*) granted to civil servants. We have seen before how politically sensitive these are. When INSEE's study was published in the summer of 2010, the spatial price comparison failed to settle the debate. Instead, it led to the creation of more controversy.

Among other things, the publication of the study was undermined by an untimely intervention of the Prefecture. INSEE's regional office had planned a press conference to accompany the release of the study in order to be able to publicly explain the results and guide interpretations to be attributed to the figures, to highlight what conclusions could (and could not) be drawn from the study. For example, they wanted to stress the considerably large increase of food prices and the disturbing growth of the synthetic gaps over the last ten years in this area—results which were in line with the movement's expectations. Yet, before the press conference could take place, the regional INSEE office was put under pressure by the Prefecture and by Paris.[32] The French Ministry in charge of overseas territories and the Prefecture did not want the results of the study to get any media coverage as the public release of the results coincided with the "unfreezing" of gasoline prices in July 2010. Thus the officials were worried that, in this context, INSEE's study might inflame matters further.

This statistical work produced by the INSEE was thus one of the salient points in the end-of-conflict agreement, but its publication went completely unnoticed. It even spread dissent. In mid-August 2010, a controversy began to grow, opposing in particular the MEDEF, the Regional council and the INSEE. The interventions drew a connection between the price gap indicator of roughly 15% which INSEE had identified and the level of the civil servants' over-remunerations, as if INSEE's study was linked to an alleged plan to question the over-remunerations, which of course was not what had been intended (Bellance & Coste, 2010; DOM: le Medef remet en cause la sur-rémunération des fonctionnaires, newspaper article, Journal de l'ile de la Réunion, 11 August 2010; Drella, 2010; Sur-rémunérations, des avis plus contrastés à la Réunion, newspaper article, Journal de l'ile de la Réunion, 12 August 2010; Sur-rémunération des fonctionnaires: les clés du débat, newspaper article, France-Antilles Guadeloupe, 16 August 2010).

Victorin Lurel, President of the region at the time, even accused INSEE of publishing studies "on the sly" in order to call into question the social gains of past struggles. A multiplicity of reactions followed, some in favour of, some against a questioning of over-remunerations, by associations, parties, newspapers, etc. (Erichot, 2010). Thus, the synthetic 15% difference, although produced by expert statisticians, failed to be regarded as a legitimate numerical representation of the department's high cost of living problems. Neither was it able to offer mediation in the social conflict.

INSEE's work had not been vain, though. Victorin Lurel himself, who meanwhile had become a minister, used the study a few years later as one of the pillars in his communications. He nonetheless noticeably changed the interpretation of the results. This is how he presented his draft economic regulation law for the overseas territories to the Senate in 2013:

> Inside these territories, the prices of most goods and services remain much higher than those of mainland France (a gap of 22% to 38.5% was measured by INSEE in 2010 on food products alone). Yet, at the same time, wages are notoriously lower there, with the median income below 38% [of that of mainland France], again in 2010 according to INSEE. (Lurel, 2012)

He thus uses the study as the cornerstone of a new strategy of communication on price levels. I will show in the following that this strategy succeeded in asserting itself by representing prices not through their average values or an average index,[33] but particular, extreme values, considered to be a fairer representation of the inequitable situation lived by many Guadeloupians.

Towards a New Articulation of Prices and Margins

This last section will show how new legitimate quantifications of prices and margins emerged in the wake of the social conflict. These quantifications did not consist of synthetic price indicators. To the contrary, these new quantifications focused on the reporting and denunciation of individual price abuses. Their emergence becomes particularly clear in a series of studies and opinions which investigated the question of price abuses in the months after the struggle had been resolved.

The Competition Authority's Report published in September 2009 (Autorité de la concurrence, 2009b) identified a long list of likely violations of competition law in the large-scale retail and import sectors. The report highlighted in particular suspicions of vertical anti-competitive integration. Importers representing certain brands appeared to own some of the main retail chains, opening the way to illegal exclusive arrangements, thereby blocking price competition. Likewise, agreements between local importers and mainland suppliers appeared to hinder new importers from entering the market, obliging the retail chains to deal with the brands' local representatives.

The Competition Authority Report of 2009 further considered that the difficult access to real estate on the island could act as a barrier to entry, preventing new distributors from finding land to establish their business. Conversely, local actors and descendants of old land-owning families had an advantage. The Authority thus asserted in its communication from 8 September 2009 (Autorité de la concurrence, 2009b, 2009c):

> In the DOM, the markets' small size and their distance from the main supply sources are natural obstacles to obtaining prices comparable to those noted in mainland France. [...] However, these particularities do not suffice to explain the price gaps on large consumption items between mainland France and the DOM. Price data from a sample of around 75 imported goods collected in the four DOM show that differences exceed 55% for over 50% of the sampled goods, a percentage that is too high to be explained solely by freight costs and dock dues ("octroi de mer"). Above all, the Authority identifies several features of the supply chain

in the DOM markets which enable the operators to partially escape the competitive game. (Autorité de la concurrence, 2009c)

These conclusions delighted LKP's members, in particular the most leftist trade unionists. The report meant indeed that the official Authority in charge of the most liberal economic regulations acknowledged the relevance of their analyses of the Guadeloupian markets. This is how LKP's main economist commented on the report:

> This report is truly devastating for the large-scale retail sector and for the importers. It explains that *pwofitasyon* is very strong in this sector. It explains that prices exceed those of mainland France by an average of 20 to 60%; some of them even by up to 100%. We had already said so, but this time it is [officially] written, in contradiction with the statistics of INSEE—another public body—according to whom the price differential is a mere 10%. This time, thanks to our work, they are obliged to tell the truth about the prices and about the margins, which are sometimes up to 100%. [...] These are centuries-old colonial ties. [...] A manufacturer can decide to grant exclusive rights to a company in Guadeloupe. [....] Thanks to these ties both parties make profits. [...] Such practices are illegal, and the report acknowledges it when it considers punishing anticompetitive practices. (UGTG, 2009)

LKP members thus saw the report as a legitimation vehicle of their own quantification methods aimed at attesting the existence of *pwofitasyon*.

Before the Competition Authority's report had been published, a fact-finding mission dispatched by the Senate during the movement had led to the publication of another report in July 2009 (Doligé, 2009). In a long passage titled "The crucial question of prices: A two-way solution, competition and, above all transparency" (Doligé, 2009, pp. 118–149), the report calls for a clear analysis of price formation, to reveal the specific cost items entailing high prices. It regrets that state services had not managed to ensure price surveillance, neither INSEE nor DGCCRF. In addition, the report supports the idea of the existence of predatory pricing (Doligé, 2009, pp. 121–123). To prove the existence of such pricing practices, it presents a large amount of information on individual prices and economic operators.

Regarding freight, for example, it denounces the monopoly held by the French sea-freight giant CMA-CGM, as well as the excessive prices it imposes on the market. To prove its assertions, it compares and contrasts cost figures presented by the Organization of French ship-owners with other expert estimates. Its conclusions are definitive: the data provided by the operators are shown to be false, and the report accuses the shipping companies of being responsible for 5–15% of the final retail price of large consumption goods (Doligé, 2009, pp. 141–143). In the same vein, the report accuses Air France of charging too high tariffs, by comparing them to other airlines' tariffs. It also presents the oligopolistic structure of large-scale retail by showing in a table which supermarket chain belongs to which old family, thereby highlighting the inherited dominant positions of the economic elites (Doligé, 2009, pp. 127, 129). In other words, the Senate's fact-finding mission took on the role of an informer, adopting a "naming and shaming" logic in its presentation of price data. It adopted its own methods to deal with the problem of price levels. In doing so, it legitimized the movement's position, and it endorsed LKP's inferences.

The price surveys carried out by the mission highlighted that price differences seemed totally random, thus excluding the possibility of explaining them by systematic factors, such as increased supply costs. In addition, the mission took account of individual cases to expose price differentials, including minute details as the following (see also Table 11.2):

> The price of 'Nesquik', an imported product, is considerably higher in the DOM: 42% in La Réunion, 75% in Martinique (although the product was on special offer there), 128% in Guadeloupe and 142% in Guyana. (Doligé, 2009, p. 126)

Thus, the report turned each price, as experienced by the consumers in their everyday life, into an indication of the existence of abuse and injustice, deserving to be discussed and publicly denounced in an official document. Shortly before, price aggregates had been deemed liable for the triggering of strong protests, and at least some considered it better not to discuss those in public. But in 2009, it appeared that individual prices could be considered meaningful events (Boltanski & Esquerre, 2017) proving the existence of unacceptable practices, and deserved to be known by the public.

Many other instances of this way of quantifying and debating prices and price differentials can be found after the 2009 conflict: for instance, in press articles (Vachert, 2010), in studies by consumer associations,[34] in surveys carried out by the LKP feeding the press (Témoignage, 2010),

Table 11.2 Extract from the reported price differentials by the Senate's fact-finding Mission (*Mission d'information parlementaire*) from 2009

Product	La Reunion	Guadeloupe	Martinique	French Guiana	Mainland France
Nido milk powder (2.5 kg, origin: mainland France)	31.80	24.05	26.95	24.85	X
Orange juice (2 litres, imported)	4.18	2.02	2.99	2.60	1.12
Cristalline spring water (6 × 1.5 L)	3.85	2.64	X	3.60	1.08
Nesquik chocolate powder (1 kg)	4.40	7.08	5.43	7.50	3.10
Strawberry jam No. 1 (1 kg pot)	1.66	2.81	2.39	4.73	1.27
Fresh chicken (local production, price per kg)	3.70	X	6.10	7.15	3.08
Rouelle of pork (fresh) (price per kg, local product, wrapped in cellophane)	8.34	10.20	8.20	9.90	4.70
Sweet potatoes (price per kg, local product)	1.50	2.60	2.50	3.50	1.50 (origin: Israel)
Tomatoes (price per kg, local product)	3.45	2.50	3.80	4.30	2.80
Bleach No. 1 (5 litres can, origin: mainland France)	3.99	2.30	2.99	3.15	1.05
Mir dishwashing liquid (750 ml, origin: mainland France)	4.20	3.22	3.15	3.20	1.46

(continued)

Table 11.2   (continued)

Product	La Reunion	Guadeloupe	Martinique	French Guiana	Mainland France
Detergent (Xtra, 27 washes, 2.5 kg)	10.66	12.07	12.37	11.50	6.07

Translated by the author from Doligé, E. (2009). Rapport d'information au nom de la mission commune d'information sur la situation des départements d'outre-mer (p. 125). Paris: Editions du Sénat

Extract from the price quotes reported by the Senate's fact-finding Mission (prices in Euros)

etc., which suggests a real change in how prices and price differentials were considered and quantified. This change is also confirmed by an episode that stayed in everyone's memory, because it caused hilarity on the island: Yves Jégo, shortly after his arrival in Guadeloupe, was shocked and started protesting against the "4 Euro toothbrushes". Jégo, then a minister of Overseas territories, suggested several concrete measures to fight such abuses: in addition to a surveillance unit, it was planned that a toll-free number would be installed for the receipt of instant complaints from consumers noting abusive prices in supermarkets. In this, his plans echoed LKP's very Trotskyist proposal to create "price brigades" charged with the enforcement of the agreements, which the collective had submitted to the Prefect in the aftermath of the conflict.

Against this background, it becomes clearer why INSEE's synthetic price index, which was based on the calculation of averages for the entire economy, was considered indistinct at the time, as it was not limited to certain symbolically significant basic items (such as toothbrushes) in an attempt to avoid any overstatement (Témoignage, 2010). Since then, we can observe a shift in the representations of price gaps which the mainland considered legitimate. Moral criteria were increasingly used to talk about price levels, and the denunciation of individually high prices seen as "abusive" became widely acceptable. In this context, magnitudes had to be sufficiently high to be deemed acceptable and fit understandings of unjust price differentials; for instance, "several tens" appeared to be in line with representations of levels of abuse; and measurement was supposed to get closer to control such abuses (i.e. prices came to be seen as something that must be controlled, audited and possibly denounced and acted upon).

Victorin Lurel, for example, as already shown above, used the INSEE study several years after its publication, when he presented his draft law for overseas economic regulation to the Senate. Here's what the Minister declared, following the passage cited above:

> [...] we are not talking about relatively bearable differences of 10, 15 or even 20%. No, we are talking about the chocolate powder all families in mainland France and overseas put on their breakfast table, which can be found at € 3.10 here in Paris, while it may be priced € 4.40 in La Réunion, € 5.43 in Martinique, € 7.08 in Guadeloupe, and even € 7.50 in Guyana!
>
> We are talking about four pots of plain yogurt, priced at € 1.15 in mainland France, and never less than € 2.30 overseas. Here again, a 100% difference for two identical everyday goods.
>
> I could continue the list of examples, which may seem harmless and trivial to you. But believe me, Ladies and Gentlemen of the Senate, they are the testimony of the striking injustice our overseas fellow-citizens feel and which can become a ferment for a feeling of abandonment. (Lurel, 2012)

Victorin Lurel clearly and explicitly expressed this new legitimate way of articulating the price question. The draft law he then presented focused on avoiding "inadmissible" practices: it affirmed the right to regulate basic product prices, trade margins and to sanction abusive practices by using a new tool, the "power of structural injunction" (Evrard, 2013). According to this legislation, a firm appearing to have built a dominant position on a given market could be forced to cede a part of its productive capital (land, shops, machines) in order to make the market more competitive (Venayre, 2015).

Conclusion

This chapter has studied the role of the measurement of prices and commercial margins in the 2009 Guadeloupian social struggle against high living costs. It first showed that calculation was central in the framing of the mobilization (Cefaï & Trom, 2001). The State and some leading economic actors (in particular large-scale retail and oil industry operators) used quantitative tools to manage or regulate prices. These practices triggered revolt on the island, because they were considered opaque and illegitimate by several political parties, unions and other associations. The actors who led the strike, grouped in the LKP collective, used their own quantification and economic analysis techniques to identify abusive pricing and margin-setting mechanisms, and to prove that pricing practices on the Island had enabled wealth extraction from Guadeloupian consumers—a situation they referred to as *pwofitasyon*.

Calculation was also central throughout the struggle and in the ensuing negotiations. This chapter described thus a "statactivist" (Bruno et al., 2014) movement in action, showing that the use of quantification was one of the best political weapons employed by the LKP collective. The LKP succeeded in challenging the State and powerful economic actors on their own ground by using quantified arguments. It showed that it was possible to use economic numbers and arguments to get existing policies shifting. INSEE attempted to play a mediating role in these negotiations.

However, at the same time, the chapter also showed that quantification and calculation can end up being one's Achilles heel—in this case LKP's. Although some members of the LKP were highly informed economic experts, in the end, it was not possible for them to compete with the State's calculative skills and expertise on an equal basis, and their pertinent use of numbers could only establish temporarily a favourable balance of power in the negotiations. This observation is important at a time when prices are measured through ever more complex statistical techniques. The possibilities to use quantification as an emancipatory device could be shrinking with the greater complexity of statistical tools (Jany-Catrice, 2019; Touchelay, 2014). By making such a point, and by documenting the use of numbers by trade unions, this chapter fills a gap in the literature on quantification and on "statactivism".

Finally, the chapter stressed the existence of multiple price level measurement methods, and the shifting legitimacies associated with each of them in the post-2009 Guadeloupian society. It showed that, although scientifically legitimate, INSEE's price indexes were subject to radical political criticism by a range of actors. By using an average, these indexes could indeed not account for the existence of the abusive prices that were targeted by LKP's mobilization. The high level of prices on some widely used consumption items was indeed considered as a form of political oppression, and INSEE's publications nurtured controversies by not singling them out: LKP actors, the press and Guadeloupian officials accused the statistical office of making the existence of such price abuses invisible.

Furthermore, the aggregate indexes displayed a price difference between Guadeloupe and mainland France of 10–15%, a magnitude that was widely perceived as contradicting the everyday experience of consumers, who, at least in some instances, experienced price gaps of 100% and more. The perceived illegitimacy of INSEE's numbers shows that there existed a different, generally accepted and naturalized, understanding of value in the Guadeloupian society at the time, based on consumers' experiences and on their imagination of what the price gap had been (in this case around 40% at least, also corresponding to the historical *sur-rémunerations* entitled to the civil servants in Guadeloupe).

After the 2009 struggle, the most legitimate quantifications of prices and price differentials in Guadeloupe were thus of another sort: abandoning averages, these singled out abusive prices (or pricing practices), either based on individual products or on groups of products. Since individual experiences of commercial abuses appeared politically significant (Boltanski & Esquerre, 2017), extreme prices were mentioned in many press articles, political speeches and administrative documents in the aftermath of the conflict. Such a use of numbers made it possible to quantify price differentials while meeting the social demand for a moral and political denunciation of abusive commercial practices. Its large adoption by administrations, journalists, political actors, activists and elected representatives strongly contrasts with the rejection that INSEE's indexes had generated. This new way of presenting prices in an official report (see Lurel, 2012 above) generated satisfaction among LKP actors, and was used in the years following the conflict by officials, such as the Overseas Territories Minister, to display their political engagement for overseas citizens. The evidence presented in this chapter suggests that this new way of problematizing and representing prices can be considered a socially validated way to account for the existence of high prices, and a new legitimate quantification of prices and price differentials after 2009.

NOTES

1. In the remainder of the text, the acronym DDCCRF refers to its outpost in Guadeloupe.
2. The term of "formula" is not used in official texts, which refer instead to "price structures". It was, however, a term generally employed by my interlocutors in Guadeloupe.
3. Didier Payen is the chief executive of an import company, PHP Trading, which holds the exclusive rights to import major brands (such as Danone, British American Tobaccos, Johnson).

4. The SARA has a monopoly for refining and supplying the market with imported goods. The policy allowing for this monopoly dates back to the choices made by General de Gaulle. Its principal aim was to ensure the autonomy of supplies.
5. Data obtained by the author from DDCCRF.
6. The majority worker's union in Guadeloupe, stemming from separatist movements. In total, the LKP included 49 organizations (Gircour & Rey, 2010, p. 101; Verdol, 2010, pp. 23–26).
7. Interviews with administration officials; see also Bolliet et al. (2009, pp. 12–13, 26).
8. To access the full platform of claims see http://ugtg.org/article_700.html (last accessed 15 July 2019).
9. *Pwofitasyon* has been promoted by the UGTG since 1997.
10. My inquiries did not afford me the possibility to interview SARA executives.
11. Such issues were also documented about France (Jany-Catrice, 2019; Touchelay, 2015).
12. See the case of the GONG, *Groupe d'organisation nationale de la Guadeloupe*.
13. According to certain counts, there were close to 80 dead, but the official count states that only five people were killed. The event also resulted in a political trial against 18 union leaders.
14. The expression "XXIst century movement" was taken from Julien Mérion, political scientist. I interviewed him in Pointe-à-Pitre in November 2010.
15. Interviews conducted with trade union and administration officials in Basse-Terre in 2009, and then again between August and November 2010.
16. The "Bino Agreement" was named after the union member killed on the roadblocks during the night of 18 to 19 February 2009. It provided for the payment of a €200 bonus to workers earning less than 1.4 times of the SMIC (guaranteed minimum wage).
17. Yves Jégo also declared that it was a strategy elaborated by Raymond Soubie, advisor to Nicolas Sarkozy at the time (Jégo, 2009, p. 121), as well as several other persons close to the negotiations. Fred Reno described this situation as a "triumph of the State" (Réno, 2012).
18. Elie Hoarau, in the context of the *Loi d'orientation pour l'Outre-mer* (LOOM) adopted by the French government in 2000.
19. A feature of the newspaper *Antilla*, from La Réunion, makes the connection between the 1953 and the 2009 sequels (Pied, 2010).
20. Interviews, conducted in March 2012.
21. Calculation possibilities are sometimes limited (Espeland, 1998; Maurer, 2007), but ignorance can also be deliberate (Henry, 2017; Hirschman, 2016).

22. According to Jalabert (2007), debates on these matters occurred within the National Assembly during the discussion of the 1973 Budget Law, and also within the *Commissariat général au plan* in 1980.

23. Etienne Pfister, Vice General Rapporteur of the French Competition Authority and Florent Venayre, University of French Polynesia, confirmed this during a roundtable meeting at the French Agency for Overseas Development's *Third Overseas Conference* on 25 November 2011 in Paris.

24. The "typical shopping trolley" is designed to monitor the prices of 50 among the most consumed goods. It distinguishes in particular between brand products, the lowest priced items, and distributor-brand products. The "household shopping basket" is composed of primary necessity goods. Both are mentioned at the end of the conflict agreement.

25. Interviews with various parties present at the negotiations which I conducted in Guadeloupe in August, October and November 2010.

26. My thanks go to Alain Plaisir, General Secretary of CTU, for the information supplied on this matter.

27. The DGCCRF initiated a series of audits in the wake of the 4th of March agreements, dealing with various sectors, such as distribution, fuel, telecommunications or banks.

28. Interview with administration officials.

29. Interview, Basse-Terre, November 2010. Alain Plaisir mentions LKP's attempt to negotiate with the wholesalers at the end of May 2009 also in Verdol (2010, p. 7).

30. See the report of the workshop on prices of the *Etats Généraux*, which mobilized experts from the INSEE, the DDCCRF and customs to explain that large-scale sector commercial margins and *pwofitasyon* were often over-estimated.

31. PPP calculations use the expense aggregates and not only the price index weights in order to examine consumption habits.

32. Interviews conducted in Guadeloupe in August 2010.

33. Various conceptions of price measurement led to fights (Neiburg, 2011; Stapleford, 2009). For a theory on the measurement through indexes and aggregates in economics see in particular Morgan (2012, p. 204) and Boumans (2005).

34. For example, the "*Consommation, Logement, Cadre de Vie*" (CLCV) association, which takes an active part in the Price observatory, together with several other associations, succeeded in setting up price monitoring via the internet (Lerondeau, 2013).

REFERENCES

Autorité de la concurrence. (2009a). *Avis n° 09-A-21, 24 June, 2009, relatif à la situation de la concurrence sur les marchés des carburants dans les départements d'Outre-mer*. Autorité de la concurrence (French competition regulator).

Autorité de la concurrence (2009b). *Avis n° 09-A-45 du 8 septembre 2009 relatif aux mécanismes d'importation et de distribution des produits de grande consommation dans les départements d'Outre-mer*. Autorité de la concurrence (French competition regulator).

Autorité de la concurrence. (2009c). L'Autorité de la concurrence recommande d'améliorer le fonctionnement des mécanismes concurrentiels des marchés afin de redynamiser le secteur de la grande distribution, seule manière de faire baisser les prix en faveur du consommateur. *Press Release*. Paris.

Bellance, A., & Coste, P. H. (2010, September 7). Faut-il encore croire en l'avenir des 40%? *France-Antilles Martinique*.

Berthier, J. P., Lhéritier, J. L., & Petit, G. (2010). Comparaison des prix entre les DOM et la métropole en 2010. *INSEE Première, n°1304*. INSEE.

Bolliet, A., Bellec, G., de Chalvron, J.-G., Cazenave, T., Sartre, T., & Clouet, N. (2009). *Rapport sur la fixation des prix des carburants dans les départements d'Outre-mer*. Inspection Générale des finances.

Boltanski, L., & Esquerre, A. (2017). *Enrichissement. Une critique de la marchandise*. Gallimard.

Bonilla, Y. (2010). Guadeloupe is ours: The prefigurative politics of the mass strike in the French Antilles. *Interventions, 12*(1), 125–137.

Bonniol, J.-L. (2011). Janvier – mars 2009, trois mois de lutte en Guadeloupe. *Les Temps Moderne, 662–663*(1), 82–113.

Boumans, M. (2005). *How economists model the world into numbers*. Routledge.

Braflan-Trobo, P. (2007). *Conflits sociaux en Guadeloupe. Histoire identité et culture dans les grèves en Gouadeloupe*. l'Harmattan.

Bruno, I., Didier, E., & Prévieux, J. (Eds.). (2014). *Statactivisme: Comment lutter avec les nombres*. Zones.

Burbank, J., & Cooper, F. (2010). *Empires in world history: Power and the politics of difference*. Princetone University Press.

Calimia-Dinane, N. (2009). *Chronologie. In La Guadeloupe en bouleverse*. Editions Jasor.

Cefaï, D., & Trom, D. (Eds.). (2001). *Les formes de l'action collective. Mobilisations dans des arènes publiques* (Raisons pratiques, n°12). Éditions de l'EHESS.

Chivallon, C. (2009). Guadeloupe et Martinique en lutte contre la 'profitation': du caractère nouveau d'une histoire ancienne. *Justice Spatiale/spatial Justice, 1*, 1–14.

Conan, E. (1997, August 7). Réunion: les dessous du volcan. *L'Express*.

Daniel, J. (1997). L'Espace politique martiniquais à l'épreuve de la départementalisation. In F. Constant, & J. Daniel (Eds.), *1946–1996: Cinquante ans de départementalisation outre-mer* (pp. 223–259). l'Harmattan.

Desrosières, A. (2008). *Pour une sociologie historique de la quantification*. Presses de l'Ecole des Mines de Paris.

Diman-Antenor, D. (2009). *Rapport de l'Atelier Formation des prix, circuits de distribution et pouvoir d'achat*. Rapport des Etats généraux de l'Outre-mer.

Doligé, E. (2009). *Rapport d'information au nom de la mission commune d'information sur la situation des départements d'outre-mer*. Editions du Sénat.

DOM: le Medef remet en cause la sur-rémunération des fonctionnaires. (2010, August 11). *Journal de l'île de la Réunion*.

Drella, P. (2010, August 14–26). Les 40% de vie chère, une question récurrente à régler d'urgence. *Antilla*.

Dumont, J. (2010). *L'amère patrie. Histoire des Antilles françaises au XXe siècle*. Fayard.

Erichot, G. (2010). Communiqué du 17 août 2010. *Parti communiste martiniquais*.

Espeland, W. N. (1998). *The struggle for water: Politics, rationality, and identity in the American Southwest*. Chicago University Press.

Evrard, C. (2013, June 25). La Loi Lurel contre la vie chère: une efficacité à prouver. *France-Antilles Martinique*.

Favorinus, J. (2009). *Compte-rendu de la Réunion du 7 octobre 2008*. Observatoire des prix et des revenus de la Guadeloupe, Groupe de travail «Evolution des prix à la consommation».

Forgeot, G., & Celma, C. (2009). *Les inégalités aux Antilles Guyane: Dix ans d'évolution*. INSEE in collaboration with CAF.

Fragonard, B., Raymond, M., & Soubeyran, D. (1999). *Les Départements d'Outre-mer: un pacte pour l'emploi*. Secrétariat d'Etat à l'Outre-mer.

Gircour, F., & Rey, N. (2010). *LKP, Guadeloupe: le mouvements des 44 jours*. Editions Syllepse.

Guadeloupe: An agreement to reduce oil prices triggers the removal of the roadblocks. (2008, December 11). *20 minutes* (www.20minutes.fr).

Henry, E. (2017). *Ignorance scientifique et inaction publique. Les politiques de santé au travail*. Presses de Sciences Po.

Herzfeld, M. (1992). *The social production of indifference: Exploring the symbolic roots of Western bureaucracy*. University of Chicago Press.

Hibou, B. (2012). *La bureaucratisation du monde à l'ère néolibérale*. La Découverte (English edition, Palgrave Macmillan, 2015).

Hirschman, D. (2016). *Inventing the economy. Or: How we learned to stop worrying and love the GDP* (PhD thesis). University of Michigan.

INSEE. (2002, October). L'impact du passage à l'euro: Pas d'emballée des prix en 2002. *AntianEco, 54*.

Jalabert, L. (2007). *La colonisation sans nom. La Martinique de 1960 à nos jours*. Rivages des Xantons.

Jany-Catrice, F. (2019). *L'indice des prix à la consommation*. La Découverte, Collection Repères.

Jégo, Y. (2009). *15 mois et 5 jours entre faux gentils et vrais méchants*. Grasset.

Laffineur, M. (2003). *Rapport d'information n° 1094 sur la fonction publique d'Etat et la fonction publique locale Outre-mer*. Commission des finances, Assemblée nationale.

Larcher, S. (2009). Les Antilles françaises ou les vestiges de l'Empire ? Les aléas d'une citoyenneté Outre-Mer. *La Vie des idées*, https://laviedesidees.fr/Les-Antilles-francaises-ou-les.html.

Le collectif pour l'observatoire des prix: Dix mille signatures pour Baroin. (2006, September 5). *Le Quotidien de La Réunion et de L'Océan Indien*.

Lemercier, C. (2008). Statistiques et 'avis divers': l'Etat, les chambres de commerce et l'information des commerçants (vers 1800-vers 1845). In D. Margairaz & P. Minard (Eds.), *L'information économique XVI-XIXe siècles, Comité pour l'Histoire Economique et financière de la France* (pp. 335–369). Minard.

Lerondeau, E. (2013, May 11). An internet site for consumers. *France-Antilles Guadeloupe*.

Le RSTA moins avantageux que le RSA? (2009, May 15). *France-Antilles Guadeloupe*.

Les pêcheurs exigent de la transparence. (2006, August 18). *France-Antilles Guadeloupe*.

Lurel, V. (2012, September 26). *Discours prononcé lors de la discussion générale du projet de loi sur la régulation économique Outre-mer au Sénat*. Paris.

Mam Lam Fouck, S. (2006). *Histoire de l'assimilation, des « vieilles colonies françaises » aux départements d'Outre-mer. La culture politique de l'assimilation en Guyane et aux Antilles françaises (XIXe et XXe siècles)*. Ibis rouge.

Maurer, B. (2007). Incalculable payments: Money, scale, and the South African offshore grey money amnesty. *African Studies Review, 50*(2), 125–138.

Morel, B., & Redor, P. (2008, July). *Rapport du Groupe de travail « Statistiques DOM-COM »*. Conseil national de l'information statistique, n° 109.

Morgan, M. S. (2012). *The world in the model: How economists work and think*. Cambridge University Press.

Neiburg, F. (2011). La guerre des indices. L'inflation au Brésil (1964–1994). *Genèses, 84*(3), 25–46.

Payen, D. (2009). *Rapport sur les prix des produits pétroliers en Guadeloupe. Version définitive du 10/01/2009*. Observatoire des prix de la région Guadeloupe/Conseil économique et social régional.

Pied, H. (2010, September 2–9). Martinique méconnue... la grève de 1953. *Antilla*.

Réno, F. (2001). Qui veut rompre avec la dépendance? In M. Abraham, & D. Maragnes (Eds.), *Autrement: Guadeloupe, Temps incertains* (pp. 236–249). Autrement, n° 123.

Réno, F. (2012). l'Etatisation du movement social. In J. C. William, F. Réno, & F. Alvarez (Eds.), *Mobilisations sociales aux Antilles. Les événements de 2009 dans tous leurs sens* (pp. 341–358). Karthala.

Ripert, J. (1990). *L'égalité sociale et le développement économique dans les DOM: Rapport*. Ministère des départements et territoires d'Outre-mer. La documentation française.

Rivière, F. (2009). Développement ultra-marin et dépendance à la métropole. In *L'Outre-mer français: où en sommes-nous? Regards sur l'actualité, 355*. La Documentation française.

Ruffin, F. (2009, November). Une flammèche obstinée a embrasé la Guadeloupe. *Le Monde Dimplomatique, No. 668*(11), 4–5. https://www.monde-diplomatique.fr/2009/2011/RUFFIN/18428.

Sénat. (2005, July 3). Creation d'un observatoire des prix à La Réunion – Question orale sans débat N°11725S de Mme. Anne-Marie Payet (La Réunion - UC). La Réunion: JO Sénat.

Sénat. (2009, April 7). *Audition de M. François Lequiller, chef de l'Inspection générale, et M. Philippe Doumergue, inspecteur général de l'INSEE*. Sénat.

Stapleford, T. A. (2009). *The cost of living in America. A political history of economic statistics 1880–2000*. Cambridge University Press.

Sur-rémunération des fonctionnaires: les clés du débat. (2010, August 16). *France-Antilles Guadeloupe*.

Sur-rémunérations, des avis plus contrastés à la Réunion. (2010, August 12). *Journal de l'île de la Réunion*.

Tarrade, J. (1972). *Le commerce colonial de la France à la fin de l'Ancien Régime, Tome 1: l'évolution du régime de « l'Exclusif » de 1763 à 1789*. Presses universitaires de France.

Témoignage. (2010, August 14). Le LKP conteste les résultats de l'étude de l'INSEE. *La Réunion*.

Touchelay, B. (2014). Les ordres de la mesure des prix. Luttes politiques, bureaucratiques et sociales autour de l'indice des prix à la consommation (1911–2012). *Politix, 105*(1), 117–138.

Touchelay, B. (2015). La fabuleuse histoire de l'indice des prix de détail en France. *Entreprise et histoire, 79*(2), 135–146.

UCF/Que choisir. (2004). Enquête: Prix 2004. La Réunion: UCF/Que choisir.

UCF/Que choisir. (2005). Enquête: Prix 2005. La Réunion: UCF/Que choisir.

UGTG. (2009). *Rapport de l 'Autorité de la concurrence sur la grande distribution. Le point de vue d'Alain Plaisir*. UGTG.

Vachert, F. (2010, April). Pourquoi les DOM restent Hors de Prix? *Linéaire, 257*.

Venayre, F. (2015). L'efficacité du pouvoir ultramarin d'injonction structurelle en question. *GREDEG Working Paper, No. 50*. Nice: GREDEG.

Verdol, P. (2010). *Le LKP, ce que nous sommes*. Editions Ménaibuc.

Volle, M. (1982). *Histoire de la statistique industrielle*. Economica.

CHAPTER 12

"La donnée n'est pas un donné": Statistics, Quantification and Democratic Choice

Robert Salais

This contribution focuses on the use of quantification in the new governance techniques that emerged for the most part in the 1980s, first in the United States and Britain, before spreading to Europe and the rest of the world under the auspices of international organizations such as the World Bank, OECD or the European Union (OECD, 1994). In this use, "quantification" refers to maximizing quantitative objectives to be achieved through definition, implementation and supervision of policies, either by management rules in organizations and companies, or by measures adopted in the context of public policy. Such techniques have profoundly altered the practice and final purpose of quantification. Far from underpinning statistical observation of reality, quantification is now expected to serve political measures that are proposed, or already decided. Quantification is expected, not only to test them, but is also enjoined to demonstrate their efficiency over time or in comparison with other policies. Its master concept is performance, which witnesses an inversion of priority in the use of data in politics: not only aimed at measuring,

R. Salais (✉)
Institutions et dynamiques historiques de l'économie et de la Société (IDHES), École Normale Supérieure Paris-Saclay, Gif-sur-Yvette, France

© The Author(s) 2022
A. Mennicken and R. Salais (eds.), *The New Politics of Numbers*,
Executive Politics and Governance,
https://doi.org/10.1007/978-3-030-78201-6_12

379

but, above all, performing. The historical connection of statistics with implementing public policies (as described by Desrosières, 1998 [1993]; Hacking, 1990) is taken over by performance indicators that both orient and evaluate outcomes, objective by objective.

Governance seeks to place effectiveness at the core of collective action, whether in organizations or in government administrations. This effectiveness is measured by performance indicators. These indicators pertain, directly or indirectly, at resources and means granted, to results expected from better management of the company or the administration. These reforms are justified by the stated (and debatable) claim that "more" is the equivalent of "better". It is the role of quantification to show that this is indeed the case, by internalizing this definition as a key step in reform.

The mainstream economy wholeheartedly applauds this targeted focus on efficiency. It took time for practitioners of political science to become aware of the issue, at least partially, especially for those not interested or not familiar with data construction's subtleties.[1] The sociology of quantification (in particular the branch derived from the sociology of science) sees in this phenomenon a field of research related to its habitual domain, with description of these new quantification instruments and practices, particularly within states, and according to the nature of these states. In this book, and elsewhere, one finds significant contributions to the sociology of quantification in these areas.

Introduction: Towards Governance-Driven Quantification

There is nonetheless something different, something more than the focus on effectiveness; there is a sort of "revolution" of quantification, when it is governance-driven. Three aspects emerge, which, we will see, challenge democracy as a government procedure and collective practice.

First, the objectivity of figures is used as a political argument. Figures do not lie. In itself a figure tells the truth of the moment on the question at hand. This truth is of course approximate, because truth is beyond our reach in this world. But this approximation is taken to vouch for the seriousness of the figure, and of the arguments based on the figure. Debate is developing mostly on the existence and amplitude of the margin of error, and not on the political relevance of the figure produced (that resides in the details of their modes of definition and calculation).

Second, the qualitative complexity and diversity of social and historical processes are reduced to a few quantitative scales of appraisal. The notions of equivalence and comparability, and a static perception, are introduced as tools of analysis, whereas many phenomena are singular, not commensurable to others, and are part of a dynamic group process.

Last but not least, the rational construction of data by the implanted quantification processes tends to be formatted for proving that the policy implemented is appropriate and successful. This capacity of self-producing the politically expected data is one of the most significant innovations of governance techniques, one that is surprising and hard to understand for the non-specialists.

To apprehend this development we will take a *long view of the social history of quantification* by looking at a specific example, that of its role in the emergence and decline of unemployment as a social category. For the effects of governance are not limited to quantification instruments. By capillarity this governance—as an emerging phenomenon, rather than as a rational project—gives birth to another political, social, financial and economic world that alters the way governance sees itself, how it frames and analyses problems, how it acts and evaluates its action.

By comparing the role of quantification (when it was known as *statistics*) during the "invention" of the unemployment category, and then during its decline, we can see what is different and what is new in today's *governance-driven quantification*. The core of the changing between the two is the status and the role given to democracy and participation of people, both in political and quantification processes.

In brief, the purpose of statistics is *to build general knowledge* "extracted" from the plurality and variety of *social conventions* people use in daily life to understand their world, to coordinate with others, to pursue their aims and try to achieve their ends; and on its knowledge basis to define policies apt to meet these conventions (see also Desrosières, 2011). By fabricating cognitive proofs that things are going the right normative way, governance-driven quantification becomes part of the political process: producing knowledge becomes the oriented by-product of politics. The purpose of governance-driven quantification is to find ways *to rationally transform social conventions* towards some pre-given political objective, judged by the Centre as optimal.[2] So in this case, as we will see below, such quantification could be best defined as "inverted statistics". In my view, the what-works approach that led to evidence-based policies (see in particular Davies et al., 2000) has been the premise of such an orientation which, afterwards, has been developed into technologies of management by performance.

The history of unemployment as a social category comprises two periods, its rise, and its fall. Unemployment emerged in Europe, roughly from 1880 to the 1950s, as a social category to be elaborated, measured and targeted by public policies. At the same time, what was then called *statistical science*, collective reflection and thinking about the instruments and uses of statistics, emerged. After a period of stability, through the 1980s, this social category began to decline. For Europe, the progressive relegation and probable future disappearance of this category as a public social concern are the paradoxical fruit of the European Employment Strategy (EES). It is not that this disappearance is deliberately intended, quite the contrary, the EES aims to increase employment levels, by raising the employment rate in the population of working age. Along with other factors (and the evolution of the labour market) it is the result of a choice made by the European Union in the 1990s, to *monitor* employment policy using performance-based governance techniques. As employment policy remains the prerogative of states, the European Commission invented a system of voluntary coordination of national policies, called the Open Method of Coordination (OMC). It has been extended at the European level to other social domains: social inclusion, pensions, health/long-term care, and to the Broad Economic Policy Guidelines (BEPG). The launching period 1997–2006 was crucial in that the European institutions boosted collective learning by national senior civil servants of the method and, more generally combined with other influences, of New Public Management methods.

In the next section (Part I), we review how far producing and interpreting data rely upon institutional machineries and, often neglected, upon the participation of inquired people. In the subsequent section (Part II), lessons that can be drawn from the socio-historical invention and deconstruction of the category "unemployment" for three European countries, France, Germany and the UK, will be discussed. The turn from statistics to governance-driven quantification is illustrated by the way European institutions deconstructed the category. The following section (Part III) draws the implications for democracy from the turn towards governance-driven quantification. It emphasizes the political move towards "a-democracy". The final section (Part IV) explores ways by which social criticism can oppose this turn by taking on board justice expectations into quantification processes and, in so doing, make way for reintroducing democracy. To be just quantification must be *correct and fair* is the message implicitly sent by Amartya Sen when he puts forward his concept of informational basis of judgement in justice.

Producing and Interpreting Data is a Collective Undertaking

Most often, if not always, quantitative data are taken at prima facie by users. Data present themselves as evidence. For users data are "real", or tend towards a pure reflection of this "real". Thanks to them collective decisions are evaluated, undertaken and followed. Such beliefs neglect the fact that data are produced and interpreted along a chain of several steps, in specific configurations of actors in which statisticians or quantifiers are involved with other actors, in deliberative arenas. Any data process should be viewed from two sides: the institutional machinery organizing the process on one side; and, on the other side, (very often neglected or even forgotten) the people who, through their answers, are the object and support of the searched data.

The institutional machinery could be directly that of the state, or that of a firm or any collective organization requiring data. In this second case, the state is indirectly present through public regulation and law. The main components of the machinery are: the conception along which the state is built; the questioning and its tools (organization of the questionnaire; the type of inquiry and its methodology or administrative requirements in case of data as by-products of administrations or management services); the instructions for coding the answers; the production of statistical tables (which requires nomenclatures, categories to classify answers and rules to aggregate individual answers in order to put every person into one case and only one). All these components play their role along a chain of production with many steps; each of them open to several possible technical options and to different interpretations; all managed by sets of organizational rules.

As outcomes of this chain, users have at their disposal a wide statistical material: variables, tables, correlations between variables, dispersion figures, indicators, and so on. They can quietly assert, for instance, that "the number of unemployed people is this", "the rate of unemployment is that". In so doing, they neglect that the data they use have been produced along a chain of production in which many not neutral technical choices are to be made. They also neglect the second side of data, namely that their primary resource, like coal or iron, has been worked out, is constituted of persons who have to answer or to be classified.

The beliefs about questioned people oscillate between two extremes. In one extreme, they have no margin, except to provide the expected right answer. They are viewed as passive resources or sites automatically responding to some external stimulus, like in behavioural models of experience. At the other extreme, they are viewed as pure rational cheaters who have to be severely controlled. These are both dire mistakes against which quite nobody (be it politicians, technocrats, civil servants, economists and statisticians) can be taken, at diverse degrees, as protected.

Firstly, such beliefs impede us to see what one could call "the democratic paradox" in our (until now) democratic societies. To understand such a paradox, it is necessary to have, in data production, a wider view of "democratization" than usual in politics. The first step is to be aware that asking people to respond to questionnaires basically means that, "somewhere", their answer has some intrinsic value and should be collected as such. Not as pure and transparent carriers of some pre-existing underlying reality, the standard view, but as active interpreters bringing some practical experience and knowledge of enough value to be used in collective choices. When collecting their world views and experience of the domain at stake, persons become *active mediators* and go-between between the supposed real and the data. Their experience has to be considered as having a knowledge value. The second step is to take into account what they have to say on them when defining categories and methodologies.

Secondly, data are built upon the "official understanding" of the investigated domain.[3] However, depending on their situations of life and work, their biography and life course, people have varied experiences with regards to this domain. There are many personal or collective understandings of the same reality, each being a priori as effective and relevant as the others. Sometimes the major part of these understandings could differ from the "official understanding" which forms the basis of the questioning. Basically, the intrinsic value of individual answers does not depend on their good will to answer, or on their correctly answering in the sense of adhering to the official meaning. This value is elsewhere, in its potentiality to reveal *distances* between different understandings for the same "object", which leads us to the third point.

Thirdly, and not the least, such intrinsic value is in essence democratic. For it has the capacity to put the spotlight on the distance (and to open a window on its meaning) between the understanding a person has of her situation and the questioning incorporated in the questionnaire.[4] These distances or gaps between individuals and official understandings on the same domain signal the existence of the plurality of possible relevant questionings (Boltanski & Thévenot, 1983; Thévenot, 1983). The official one is one among others. No data, especially aggregated data, can be said to be the truth, not only because they are deeply linked to the series of both technical and political choices made along their chain of production, but basically because among a range of possible choices, one path only has been chosen.

One will see the huge impact of all these factors on the nature of data produced in the three countries we will review below: France, Germany and the UK.

Inventing and Deconstructing Unemployment as a Category: The Role of Quantification

Almost at the same historical period (the turn of the twentieth century), unemployment as a social category and as a procedure to count those to be classified as unemployed was invented in the major European countries: France, Germany and the UK. Such inventions lasted half a century or more. The national processes and their outcomes were very deeply anchored into national specificities. They brought to people and their political communities new resources to understand "their" real, to act within it, to form expectations and projects, to legitimate decisions, disagreements and conflicts. Invention has followed the road of statistics as we suggested in the introduction. Statisticians were involved in diverse deliberative arenas, and were at the initiative to create both the category and the methodology. Deconstruction is following the road of governance-driven quantification (for a detailed demonstration see Salais, 2007). It disqualifies ancient and familiar resources that offered stable anchors for people, without, until now, providing alternative types of resources.

The Invention of Unemployment: Comparing France, Germany and the UK

The "invention" of the category unemployment at the turn of the twentieth century demonstrates how far data (categories, procedures, numbers) are worked and re- worked all along the process by the actors, in a sort of joint production.[5] There was a kind of double plurality at work, the plurality of institutional machineries among national states and within them, on one side, and the plurality of indigenous categories among people on the other side. In each country, the data to be produced had to meet a demand for information which was linked not only to public policies but, above all, to their specific conceptions of collective objectives and to the way the state should intervene for their achievement. Formats, specifications, levels of collecting and using data, even the need to collect or not, all will depend—for the same domain of observation— on these state specificities, which were, and still are, very diverse among countries and over time. States were more or less inclined to systematic and general quantification, more or less open to democratization of data production. They required different types of data and of their "production system". From the people's side, there was another type of plurality, one of the principles of justice considered as legitimate on which to build the category.[6] In the final two sections (Parts III and IV) we will connect democratization and justice.

In France, whose state has been historically built along top-down, systematic and central intervention through general categories, the search for defining unemployment has been undertaken directly by elites surrounding the central state administration. Lawyers and economists (at that time trained into the same faculties), statisticians, economic and social actors, members of the parliament, public officers, tried to have their word, using their own knowledge and experience. They met in different assemblies, circles and savant societies (Didry, 2002). In the 1890s, the state created a special institution, named "Office du travail" which launched inquiries, monographs, collected professional advices to have a clear understanding of the various work conventions especially with regards to periods of no work (Luciani, 1992). All together were able to define a general and practicable category of unemployment which was incorporated into the census and administered to the whole population, for the first time, in 1896. All French administrative levels were progressively required to use the same category and to produce the same types of statistics and tables at all administrative levels. However, the disparities in the rates of unemployment among regions, professions or labour statuses reveal durable traces of other conceptions, especially homeworkers and independent workers, employees of local small firms, craft workers who have their own conception of the primacy of individual, local or craft responsibility in social compensation or in job search organization (see Salais et al., 1986, chapter 3).

In Germany, a federal state, such a national unification of the category failed. There were already a series of local definitions and conventions, depending on the professions, the unions and the towns. These definitions were founded on specific principles of justice that led to various principles to identify unemployed people: for instance, belonging to local crafts; being citizen of the local town; being registered on local social help bureaus (Zimmermann, 2001, pp. 126–138). Land statisticians, convoked to Berlin, were unable to agree to a common definition. It is only in 1927 that some unification was achieved, thanks to the national social insurance system which was eager to generalize insurance to unemployed situations. Yet, being centred on previous craft insurance systems, it tended to exclude workers that did not belong to craft unions. A categorization of the unemployed appeared for the first time only in the 1931 population census. This was not renewed by the Nazi regime in the 1936 census. Beyond the failure to generalize, it shows that, except for the Nazi period, Germany as a national entity is built along with a different conception of the state, mostly that of one we call a "situated" state,[7] a concept we develop in Salais and Storper (1993, fourth part). Such a state gives precedence to collective autonomy over national top-down intervention. It tolerates diversity; the responsibility to define the common good at stake and to take care of it can be left to various levels, especially the Land, the city, the profession or the economic sector. Statistics can have different frameworks and tables for the same domain, which leaves some collective freedom to choose the relevant principle of justice for building the data.

In the UK, the historical picture was also another one, a long and uncertain battle between at least three conceptions of unemployment, implying the state only indirectly, and of the assignation for responsibility: poor laws, trade unions or the market. All these systems had their own statistical categories and data which were not consistent with each other. Poor law, the oldest system, was placed under the sovereignty of the King, but managed at the very local level of the parishes; unemployed people were not differentiated from the poor and treated as such. Trade unions had their own system for their members. Unemployed members were supported by friendly societies which did not differentiate between the lack of work due to unemployment or strike; both were financially helped (Phillips & Whiteside, 1985). At the turn of the twentieth century, social reformers (the most famous, among many, being William Beveridge) were hostile to these systems. They pleaded—with more or less success—in favour of the creation of labour offices which could rationally construct a true national labour market. Such a market not only should work as a perfect market, but have the tasks to clear the market from the unemployable (sent to other social policies) and to teach workers to be individually responsible for their situation and their future (Mansfield, 1992). Regarding unemployment, the UK has thus implemented contradictory conceptions of the state, valid at certain levels and for some organizations, but not at others: interventionist for constructing from the top the perfect market, but in competition with local autonomy and professional diversity which would have been best taken in charge by a "situated" state. Several principles of justice are in competition to define and observe unemployment, presumably in some unstable compromises even today, based on, respectively: the deserving poor, the acknowledgement by peers, the morally regular worker (Whiteside, 2014).

Governance-Driven Quantification as Inverted Statistics: Europe and the Reversal of the Pyramid

However, a new actor appears on the field of employment in the 1990s: Europe, its institutions and political frameworks (for a historical perspective on building Europe see Salais, 2013). It added complexity, more uncertainty in the definition and observation of unemployment, and in the meaning of data. Basically, it contributed to blurring the boundaries within established categorizations and to deconstructing them. Especially, short unworked periods are less and less considered as "unemployment", but as transitions—that have to be the shortest possible—between two jobs or tasks. In practice for part of the population it corresponds to precariousness, but precariousness is not recognized as a valuable category of social policy and not counted as such (see Standing, 2014).

European institutions introduce new public management reforms through a specific method, called the open method of coordination (OMC). This method constitutes a fascinating illustration of the social and political impact of quantification when internalized into governance schemes. It reveals its basic specificities. We will pass in review five of them: the reversal of the statistical pyramid; a new target for employment policies; statistical tables as driving forces; the set of indicators as embedded norms and guidelines as justificatory covers; a cooperative game between rational actors.

The Reversal of the Statistical Pyramid

Governance-driven quantification operates a *reversal of the pyramid* which, in classic statistics, links its large basis (the multiplicity of individual experiences and the mobilization of their social knowledge of situations and problems) to its top (the producing of aggregated data, via the progressive reduction to numbers by aggregating individuals' answers). Governance-driven quantification puts the pyramid not on its basis, but on its top. It starts from the top data (the quantitative global performance) to be maximized at all costs whatever the means used to achieve this objective. It tries, through a descending movement, to produce the required basis of the pyramid able to generate the expected global outcome. Quantification rules of measurement, organizational rules of political schemes are adjusted in order to fabricate, if not individual behaviours themselves, at least answers, or statistical treatments that fit with the quantification objectives. The underlying utopia of quantification and, as a consequence, of governance by numbers (see Miller & O'Leary, 1987; Miller, 1992; Supiot, 2015), is that social subjects are expected to create by themselves a reality that complies with the objectives. They would, eventually, spontaneously produce the required data. In general these are only answers that, through several organizational means, at the end begin to fit with maximizing the scores. Such utopia to make people spontaneously creating an "optimal" social reality must not be confused with the ordinary faking of statistical data, frequent on sensible domains like unemployment statistics.

A New Target for Employment Policies

In the European employment policies promoted since the end of 1990s, European Union authorities took the global rate of employment as one of its major macroeconomic indicators.[9] They substituted the search of Keynesian full employment for the maximizing of the rate of employment as their main target. In so doing, a "job" is no longer what it promised to be in the model of full employment. In that model, any employment guarantees minimum standards of remuneration, of security in the face of unforeseeable events based upon social and economic rights. What Europe now guarantees to its citizens was only to have a task, whatever it could be and under the condition they accept it.

In practice, to measure the national rate of employment, the European authorities recommended applying the definition that is used by the ILO to build international statistics on employment: "Employed persons consist of those persons who during the reference week did any work for pay or profit for at least one hour, or were not working but had jobs from which they were temporarily absent" (see, for instance, European Commission, 2006). Statistically speaking, applying this definition is simply following the ILO definition.

But it takes on a very different meaning when it is translated into political action. It means that, whatever the task is in terms of quality (wage, working conditions, duration, type of labour contract), it can be considered as employment if it lasts at least one hour a week. All other characteristics were deemed irrelevant when creating employment data. One should call this "the convention of employment without quality". This convention is far from trivial. Employment without quality is a task stripped of all legislative guarantees (in terms of recruitment, protection against unfair dismissal, minimum starting wage) and social provisions (social and economic rights). By removing quality features when comparing and putting in competition their social systems by means of such single quantitative scale, the Member States are encouraged to water down the quality of their employment conventions in order to improve their quantitative performance.

Statistical Tables as Driving Forces
One should pay attention to what is ordinarily taken for neutral, hence unproblematic, that is the collection of statistical tables that, for each yearly report, national administrations are required to fulfil in the areas using the OMC. One must suspect that, to a large extent, these tables are the driving forces "behind" the formalism, not only for data, but, beyond, for political discourse (vocabulary and syntax). Tables also act as rhetorical justifications of the normative background imbedded and for most people dissimulated in data, especially in the selected indicators.

Contrary to the standard view, a table is not only a collection of figures (one in each box, for instance, as in a double-entry table), some being higher and others lower, from which one can *directly* draw conclusions like "the female rate of employment in 2005 is higher in the UK than in France". A table is, above all, a procedure for aggregating individual situations, for instance, relating to employment and the person's position in the labour market as built by nomenclatures. All situations compiled in the table which are considered as identical with regards to these two nomenclatures are placed in one box. They are considered as equivalent according to the corresponding properties. In other terms, filling a table by combining individual data requires conventions of equivalence,[10] which decide about what should be considered as similar. These conventions ensure the passage from the particular to the general (what Luc Boltanski & Laurent Thévenot (2006 [1991]) call the rising into generality).

Generally speaking, conventions of equivalence are ignored or misunderstood by the ordinary users. From the above statement on female rates, users will spontaneously conclude that "women work less in France than in the United Kingdom". But this conclusion is valid only if the legal, statistical and social definitions of what should be considered as a "job" are identical in the two countries. In practice, the UK is using a "softer" definition of part-time work than France, which results in women who work very few hours a week being considered as having a job and driving them into such jobs. The situation is even reinforced with the invention in the UK of the zero hour contract. Applicants are asked to stay available at home for whatever task and at whatever moment their employer decides. They are considered as employed even with zero worked hours. This helps maximize the rate of employment in which the UK is champion (which, as a counterpart, corresponds to one of the highest poverty rates in Europe).

The Set of Indicators as Embedded Norms—Guidelines as Justificatory Covers

Conventions of equivalence govern what we select, what we exclude and what we construct. Thus, the requested description becomes not far removed from a normative evaluation of the situation under review.

The basic issue with the Open Method of Coordination—and more generally governance-driven quantification—is not immediate strategic action, or neoliberal ideology[11]; it is about the cognitive conventions that are selected to drive the political process. The selected set of indicators frames the normative background of the political decision-making process. It is neither malignity nor political cunning. It is the mere consequence of the fact that any indicator selects what is worth to be known or not and, in so doing, basically builds the reality that is relevant both for the deliberative process preceding the decision and for the action to be undertaken.

The set of monitoring indicators selected by the European Employment Strategy (EES) focuses on the supply side of the labour market, which is the work offer by the manpower. It expresses the norm that work offer should be the highest, the most flexible and adaptable to economic hazards as possible. Employability is the main concept. The higher it is for an individual, the more he would have access to job opportunities. At first glance, there is no problem here. But complete labour market models emphasize a second concept at the same level of relevance, the one of vulnerability to job losses. The more you are vulnerable to job loss, the less you could access a stable job. So employability should go hand in hand with job security (or at least stability) as objectives for employment policies. There is nothing like this in the EES. Furthermore, the monitoring indicator for evaluating employability is the rate of return to employment. The fastest it is, the best it is for the EES. But improving employability is wider than increasing performance, for it has qualitative aspects that, normally speaking, should be taken on board by public policies, but are not.

Here appears the mismatch between the political rhetorical justification one can see in the wording of guidelines, and the effective policies that are driven by their monitoring of performance indicators. The search for consistency between data and discourse is, in effect, part of the global drift from politics to management. It tries to be achieved through the connection between quantitative monitoring indicators and guidelines that are expressing the objectives corresponding to the different indicators. There is a rather subtle, but essential shift of normative requirements from guidelines to indicators. The *formal* normativity is provided by the guideline, the *effective* normativity by the indicators to maximize. One will take the example of the EES guideline "Ensure inclusive labour markets", introduced in 2006 (European Commission, 2006). It asks the Member States to develop "active and preventive measures including early identification of needs, job search assistance, guidance and training as part of personalized action plans, provision of necessary social services to support the inclusion of the furthest away from the labour market and contribute to the eradication of poverty". Such wording sounds perfect in ethical terms.

But what does it mean in practice? The answer is provided by the tool and its real use: the corresponding monitoring indicator, called "New start" is calculated as being the "share of young/adults becoming unemployed in month X, still unemployed in month X+6/12, and not having been offered a new start in the form of training, retraining, work experience, a job or other employability measure" (see EES 2006 Guideline "Ensure inclusive labour markets") (European Commission, 2006). National implementation, aimed at increasing performance, puts incentives and pressures on the unemployed to take any available task, whatever it is and as soon as possible. The European definition of what to count as a "job" is rather vague and extensive. It leaves room for free interpretation at national level allowing for the inclusion of new schemes. Maximizing such indicators cannot really improve inclusion in labour markets: it mostly increases precariousness.

A Cooperative Game Between Rational Actors (the Member States and the Commission)

It follows that the EES operates as if it was a cooperative game between rational actors. Such a game sounds like this. Its mechanism is familiar to economic theory. Take the Commission and the Member States as the players. The aim of the game is to maximize the key indicators, those intended to evaluate the policies being followed. Actors know in advance the formatting of future evaluation of their actions. Insofar as any learning outcome takes place, it is of a rational order and likely to affect the procedure. Cooperation consists, for each Member State, in manipulating the rules of its own measures and their implementation to meet the requirements of European indicators. In the cooperation, there are invisible but known conventions between actors not to go beyond what each actor was ready to accept. It is not a collective action aimed at genuinely improving employment in Europe. Due to the limited competences given to the European level, Member States are not held responsible for a substantial improvement in European employment, nor do they feel themselves accountable to such improvement when they define their employment policy actions and coordinate with the others in the EES framework. The only constraint is that they have agreed—and this commitment derives from the management by objectives of the OMC—to be accountable vis-à-vis the Commission with regard to their national scores over the whole set of indicators.[12]

This whole process fabricated positive quantitative outcomes. The global rate of employment (in the European definition) has risen between 1997 and 2005 for the three countries: +2.9 for France; +3.8 for Germany and +2.1 for the UK. Table 12.1 tries to compare these results with the evolution of a full-time equivalent rate of employment. It

Table 12.1 Trends in the overall rate of employment (age 15–64), 1997–2005, in France, Germany and the UK

	1997	2004	2005
EUROSTAT employment rate (from Community Labour Force Surveys)			
France	59.6	63.1	63.1
Germany	63.7	65.0	65.1
United Kingdom	69.9	71.6	71.7
OECD employment rate[1] (from national accountings)			
France	60.2	63.3	63.1
Germany	67.3	71.0	71.1
United Kingdom	70.2	72.3	72.3
Annual number of hours effectively worked by person[2] (from both Community Labour Force Surveys and OECD)			
France	1559	1531	1542
Germany	1537	1468	1464
United Kingdom	1697	1631	1635
OECD adjusted rate of employment (corrected from the evolution of hours worked by person from 1997)			
France	60.2	62.2	62.4
Germany	67.3	67.8	67.7
United Kingdom	70.2	69.5	69.6

Source Data collected and compiled by Odile Chagny (Centre d'Analyse Stratégique, Paris). This information was kindly provided to the author

Notes

[1] Employment data is provided by OECD and is calculated per person and not per job. The source of the population data is also the OECD. For Germany, OECD data is provided by the Institut für Arbeitsmarkt- und Berufsforschung and includes mini-jobs; the EUROSTAT data do not include these jobs

[2] For 2004, the annual number of hours effectively worked comes from the table produced by Bruyère et al. (2006). The trend has been interpolated from previous OECD series of the annual number of hours worked

corrects the global rate of employment with the decrease of the annual number of hours effectively worked by person between 1997 and 2005. The difference between the two roughly estimates the impact of the increase of short-term and precarious jobs, among them the subsidized schemes of return to jobs (for instance the mini jobs in Germany): +0.7 for France; +3.4 for Germany and +2.7 for the UK. Beyond approximations, the impact is notable, more important in the countries already engaged in the move like Germany and the UK, than in France which at that time appeared reluctant. The computation made by a team of

researchers (Bruyère et al., 2006, pp. 363–370) was overwhelmingly difficult (in particular for hours effectively worked by a person). To my
knowledge, it seems that such an undertaking has not been renewed,
though it would be extremely relevant.

Just a (significant) anecdote to conclude this section: at its own
expense, the Belgian employment administration was worried to discover
the very low rank of Belgium among European countries in the national
benchmarking along the "New Start" indicator. The reason was not the
bad functioning of Belgian labour markets, but the Belgian definition of
inclusion. To be considered as included in the labour market, the job
found must have lasted at least two months. When this was not the
case, people remained classified as "unemployed", which led to a higher
registered unemployment duration. The Belgian administration quickly
corrected this "mistake" by cancelling this constraint on employment
duration. Its quantitative performance improved at the satisfaction of all
European and national officers, except Belgian unemployed people who
were now compelled to accept any task as a job.[13]

Quantification: Contrasting Rational Governance with Democratic Choice

Comparing the two phases of the history of unemployment (emergence,
deconstruction) offers some incidental views on the differences of democratic choice versus rational governance. In both cases quantification plays
a central role, though in very different ways.

Democracy and the Emergence of the Category "Unemployment"

Democratic choice does not consist simply in putting into place optimal
procedures for making a choice, or in asking an assembly, even a democratically elected one, to vote. What must be achieved is a free and
pluralistic process of public debate, taking the time to weigh all aspects
of the choice to be made, without rushing to come to a conclusion. In
such a debate, the establishment on the subject in hand of a knowledge
basis that should be collectively considered as just and fair is a key dimension, often underestimated. Just in the sense of not forgetting any relevant
information, fair in the sense of obeying some shared principle of justice.
It is thus, above all, a multifarious social and historical process, driven by

many social forces, and not simply the construction of a rational choice operated by the Centre.

Regardless of the country and specific forms that ensued, the "unemployment" category, especially, emerged and developed itself roughly between the 1880s and the 1950s. It was the occasion of a vast and long public debate encompassing contrasting and opposing views, with peak moments at certain points in time. In each country, in its own way, this debate took place in different arenas (political, economic, social, intellectual, statistical) propelled by organizations and their modes of expression (reviews, scholarly societies, public events and demonstrations, etc.). The debate was pursued at different levels and on different scales, in parallel, or in coordinated fashion, within local and regional entities, sectoral, professional and trade groups, and internationally. This process preceded or accompanied the creation of legislation, regulations and institutions. Most of the collective structures where at the time these debates took place were hardly democratic, properly speaking, if "democratic" is taken to mean that the bodies are duly elected and entrusted with a specific mandate to debate issues and propose measures. They were rather the result of a need for collective expression that arose at the time, whether under an authoritarian regime like the German empire or regimes with democratic leanings as in France and the United Kingdom, whether the right of freedom of speech existed or not.

A democratic process of choice cannot be decreed from above, or from outside. This process is often messy, not controlled, nor foreseeable. However, as we said before, democracy is intimately linked with inquiry. The answers people give to an inquiry have an intrinsic democratic value, for they have the capacity to reveal gaps between citizens' understandings and official intentions for the domain under scrutiny. These gaps underscore disagreements and the plurality of social experience of the same reality, hence the possibility of several relevant questionings, other than the official one. Both between countries and within them, these disagreements emerged as to how to understand unemployment and to count the unemployed.

Due to the plurality of relevant judgements on a given situation, the most important moment in democratic choice processes is not the final step, the decision, but the preceding phase, the reaching of an agreement between actors on the "reality" of the situation, on what is at stake, and on the relevant features to take into account when framing the decision to be made. So, a major democratic concern is to enable people and stakeholders to reach, at least partially, an agreement on the pertinent reality that matters for their choice; that is what Amartya Sen calls the informational basis for judgement in justice (IBJJ), as we will see in the final section further below (Part IV).

Governance-Driven *Quantification* and *"A-Democracy"*

The deconstruction process undertaken by European authorities is in contrast with what we might envision as elements of democratic choice in the earlier process of invention of the "unemployment" category. The big change is that, instead of being the fruit of long-term collective debates implying a variety of actors at different levels, the informational bases that pilot the choices are now predetermined from the top by the Centre without any serious deliberation; they incorporate norms into quantification processes before discussion and choice. Such bases orient the decisional processes towards some prefixed types of political outcomes, the ones that the most "naturally" comply with the embedded normativity of the data. These norms are mostly incorporated into technicalities (definition of operational categories; rules of management implementing political schemes; exploitation of the data produced, and so on). Remember the political recourse by the European Commission to ILO statistical categories; and the set of indicators that offer biased models of labour market functioning, or the subtle ambiguity between guidelines and indicators.

Political parties and collective organizations become involved in discussions whose questions, informational bases, and agenda have been prefixed before, on which they have no grip (and often no true understanding of the stakes). Classical representative political democracy and social democracy, too, are circumvented and their role weakened.

I will call "a-democracy" a political regime that maintains the formal procedures of democracy, but impedes, not formal participation of citizens and actors, but any palpable outcomes positive for them (meaning by positive outcomes those that truly improve their situation). Several trends progressively reinforce the efficacy of such a political regime, viewed from the point of view of the political elite and professional politicians. We will point out three aspects of such self-enforcing trends: creating cognitive ambiguity; fabricating quantitative proofs and justifications; generating difficulties to articulate alternative legitimate claims.

Creating Cognitive Ambiguity

The "veil of ignorance"[14] surrounding the statistical conventions used to produce the figures creates a situation of cognitive ambiguity. This ambiguity acts like a smokescreen, allowing the conventions adopted as benchmarks for public policy to be changed without any awareness or protest on the part of the public. For example, if the employment rate goes up, ordinary citizens conclude that their chances of finding a job (corresponding to *their* criteria for a good job) are going to improve. But the European authorities may well—and in fact do—ascribe a different meaning to the notion of employment, one that resonates with the labour market deregulation policy they are pursuing, which obviously works against the expectations of the ordinary citizen. Since it is difficult for citizens, who have nothing but their individual and local experience to test general categories, this situation may last. In a situation of cognitive ambiguity, the task of the authorities consists in maintaining discursive consistency between the established meaning and the new meaning they assign to each category. Public administrations and politicians both are incited to follow this opportunity to maintain such discursive continuity, as it provides them with better justifications. Referring to Austin (1962) (as mobilized by Bohman, 1996, p. 204), one could say that, while employing the same discourse, the European Commission is acting to modify all the possible worlds in which the language convention ("to have a job") is valid. Believing they have remained in the same world, citizens looking for a job according to the established categories in their world, are confronted by a world in which the same terms are interpreted differently and refer to other actions.

Fabricating Proofs of Effectiveness and Efficiency

What is more, through its self-referential logic, this political method produces justifications of its efficacy that are not only theoretical or discursive but also quantitative. The change in the rules of public policies (employment policies here) does not aim to improve actual social situations but to directly boost scores on performance indicators. The ratings go up without any real improvement in social situations. In fact, those situations may even deteriorate under the impact of standard, short-term measures that cost little per beneficiary because they are designed to affect as many people as possible. The management of public agencies—from the national to the local level—is reorganized according to the logic of performance criteria (Salais, 2010). As a result, the data based on management and on assessing operating rules show progress is being made. They may even be used to demonstrate the veracity of the policy position. In other words, even if it was not their initial goal, reforms tend to establish a direct connection at every level between management and the production of evidence—in other words, self-fulfilling justifications.

Generating Difficulties to Articulate Alternative Legitimate Claims
Creating an environment of procedures of information and of evaluation adequate to predefined political goals (ultimately, a system self-producing proofs) leads to growing difficulties to articulate legitimate alternative claims. As figures and procedures are seen by most of the people as guaranteeing truth by their mere existence, they allow for the endorsing of political credibility. Even if the public debate begins to be fed with such fabricated data (without any professional or democratic control of their process of production), which raise scepticism, it nevertheless means for people that the "facts" are already there. As already existing evidence, these "facts" format the public debate. So it becomes harder to set claims which have not been the object, not only of cognitive elaboration but, more deeply, of common knowledge. For to be heard, claims need to be backed by other socially produced facts; facts that could constitute the basis for shared understanding within the political community and can successfully contest the "official" facts. Following Dewey (1927), such understanding should not be purely intellectual, but also embedded into the engagement of people into "publics".

If not, the path for democratic expression is cut, even if, formally, democracy remains. The social foundations for active political participation and of citizenship would be undermined, the value of them disappearing for a growing part of the population. By the same process, quantifiers and their demanders are trapped in self-referential loops in which data is taken as the right mirror of reality and, finally, as the reality itself. So the "real" disappears below its quantitative representation which, being taken as the true real, becomes the basis for defining and implementing management reforms and, more generally, public policies. But losing a grip on political and social reality is dangerous for the political credibility and effective performance of policy makers and politicians. A-democracy is the ultimate step of the diffusion of such political methods.

Such political trends call for alternative solutions that correct their negative outcomes. What needs to be put in place to ensure pluralism? What about those who are vulnerable (e.g. citizens with disabilities who cannot easily articulate their opinion)? What should be the relationship between lays and experts in such debates? How far should participation go? How should deliberation be organized? And what are the pitfalls? These questions are beyond the scope of this chapter, mostly because they are waiting for a relevant effective political agenda that does not yet exist. To be possible, it requires, above all, collective learning on the subtleties of social processes of quantification. The first step, in our view, is to be able to develop an approach to quantification that is open to the social critique of its use in governance issues. This is the object of the next section (Part IV).

Social Criticism, Justice and Plurality of Quantification Regimes

The avenue taken by most of the social critics today is the Foucauldian one, especially in English language literature. To quote only one, the work of Wendy Brown (2015) is exemplar. Her book develops a radical and implacable criticism of all aspects of the turn towards a new political governance. At first glance one cannot be but in close agreement with her title "undoing the demos" and her arguments. I discover at work in the governance-driven processes of quantification what I call a-democracy, that is the progressive remoteness of the demos from any effective participation in collective choices. But is it the same as "undoing the demos"? Brown's subtitle, "Neoliberalism's Stealth Revolution", and her demonstration of the omnipresence and omnipotence of neoliberalism leave no room for any collective reaction, or for any counteracting possibilities. Why to exclude any possibility for the demos to survive and find issues? I would like to suggest that our approach to analysis of the relationship between quantification and democracy helps to clarify the point.

In our view, social criticism today must cope with a new element: the emergence of political strategies whose effectiveness lies in acting through the choice of "optimal" informational bases of judgement. Such strategies are perverse, because they distort collective choices in favour of the interests of the central power (and its supporters) at the detriment of citizens and communities. The main worries are that citizens and communities' aspirations and needs are not correctly represented by the categories, methods of inquiry and data produced that construct the informational bases used to pose and solve collective choice. Above (see Part II), we became aware of such distortions in the case of employment in Europe. European authorities modified the meaning of what should be counted as employment, chose an informational basis centred on the rate of employment and its maximization, all of this pushing the deregulation of labour markets and job precariousness, without any public debate. Evidence is that there is a denial of democracy, biased participation in collective choice and social injustice (with regards to peoples' aspirations).

Introducing Justice and Democracy

The only way to cut the Gordian knot is to introduce preoccupations with social justice into quantification matters. In every collective choice implying human activities, two objectives should be involved: economic efficiency and social justice, and not only one, efficiency, as in rational governance.[15] These objectives should be considered to be at the same level of importance. It follows that, in one way or another, people submitted to quantification in some domain should be asked, or inquired, or adequately represented by movements, associations, political parties at the collective decision levels, on what they consider as social justice for them. This is not a simple thing. We all know—because we experience such moments of feelings of justice or injustice—whether in given circumstances or activities we are well treated or not by others (or by the institution we are facing). But to jump from such personal evaluation to a general principle of justice that would be agreed or accepted by all is another matter. It, no more, no less, requires democracy, an effective one in the making of collective choices. So justice and democracy cannot but go hand in hand in quantification processes. As we have seen above (in Part II), the historical emergence of unemployment statistics in France, Germany and Great Britain reveals some presence of such requirements of justice and democracy that have had unequal collective expressions due to national specificities. Furthermore, in each country several principles of justice competed with each other and had to search for compromise or, at least, for some unstable coexistence at the national level as, in Britain for instance, the deserving poor, the acknowledgement by peers, the morally regular worker.

Remembering such past circumstances today does not mean that the past was better in itself; all the more as social, economic, political realities as well as the people themselves have changed. It nevertheless under-lines—not a small thing—that true participation of people, taking into account (to a varied extent) their say and experience of the domain object of public policies, fortunately, is possible. These cannot be excluded. It follows that it is no longer enough today to denounce the governance distortions that are both unjust and non-democratic. One must produce alternative data founded upon just and correct representations of situations and aspirations of people. A different quantification on the same issue should be achieved, based upon another collective "understanding" of the problem to be dealt with. Such quantification has to become legit-imate in terms of both *fairness and correctness* of the data produced; and these data are to be offered to public debate in all their dimensions. Becoming objectively and politically legitimate is the necessary condition to be accepted in the public debate and to be opposed to the "official" basis promoted by the Centre. There is, at the same time, a need to develop a collective social movement able to take charge of the process and to oppose the Centre.

The "Informational Basis of Judgment in Justice" (IBJJ)

The only economist (and social philosopher) that I know for his deep concern about social justice in quantification matters is Amartya Sen. There are others, however, in my view Sen's works are the most appealing and enlightening ones for us to go further.

The crucial point in Amartya Sen's approach lies in his emphasis on the informational basis of judgement in justice (IBJJ), which determines the content and methods of collective choice in a democracy. Sen main-tains the need for an objective assessment of the state of persons (against the dominant trend of purely ordinal rankings in theories of justice). Sen's accent on objective assessments connects his approach to quantifi-cation issues. Sen introduces in these issues, as soon as human beings are involved, the need to provide as grounds for agreement between people (and for disagreement, as we shall see), tables and indicators that must be just, in the twofold sense of objectively right and socially fair. If so, tables and indicators will cover what, in a genial intuition, Sen calls "the factual territory" over which considerations of justice would directly apply:

> The informational basis of judgment identifies the information on which the judgment is directly dependent – and no less important – asserts

that the truth or falsehood of any other type of information cannot *directly* influence the correctness of the judgment. The informational basis of judgment of justice thus determines the factual territory over which considerations of justice would *directly* apply. (Sen, 1990, p. 111)

This definition of an IBJJ has been introduced by Sen in the context of a dispute with Rawls within the theoretical field of theories of justice. I will just say a brief word on this debate. Sen argues that:

Interpersonal comparisons that must form a crucial part of the informational basis on justice cannot be provided by comparisons of holdings of *means* to freedom (such as "primary goods", "resources" or "incomes"). In particular, interpersonal variations in conversion of primary goods into freedom to achieve their life objectives introduces elements of arbitrariness into the Rawlsian accounting of the respective advantage enjoyed by different persons; this can be a source of unjustified inequality and unfairness. (Sen, 1990, p. 112; italics in original)

It is worth noting that for Sen the freedom to achieve should be an actual freedom, not simply a formal one. People should have access to means calibrated to offer them true possibilities, though it is up to them to realize these possibilities, or not. It implies that quantification objectives and methods cannot be but defined in coherence with the objectives and implementation rules of the corresponding policies (see also Salais, 2008).

One will not follow Sen in his debate with Rawls further. Their principles of justice are ones among others. But we will insist on the tight connections with our discussion on quantification. While it was not the direct purpose of Sen, in practice he severely questions the concepts of "fact" and of "objectivity" as usually understood and implied in governance-driven quantification. Most often, the fact is reduced to the status of evidence, something that is not contestable. For a given problem in a given situation, there is only one valuable set of facts, those that pass the test of evidence. No need for justice considerations. By contrast, Sen demonstrates that to be truly objective, an informational basis—in other terms a quantification—should satisfy criteria of fairness (like, in his case, "the *actual* freedoms enjoyed by different persons–persons with possibly different objectives—to lead different lives that they can have reasons to live" (Sen, 1990, p. 112)). Thereby, in introducing the notion of factual territory, Sen implies that for a given problem in a given situation, there can be several different factual territories, depending on the principles of justice that are applied. It follows, first, that all these factual territories are a priori valuable for posing the terms and purposes of a collective choice on the issue at stake; second, that to evaluate how far the data produced are right requires two things, that they have been produced along rigorous methodologies (correctness) and, too, that judgement and agreement (or at least satisfying compromises) have been achieved between the involved persons and actors on the chosen principles of justice (fairness).

Deliberative Inquiry as Data Processing

The fecund intuition of Sen regarding deliberation from the point of view of social criticism is what is at stake is not prior deliberation over which norm is the right one (a conception based on a hypothetical ontological *plurality* of norms), but deliberation suited to an adequate grasp of the social reality (a conception based on the observation of a *variety* of situations from the point of factual *territories of justice*). Due to the impossibility to objectively decide between ontological norms, an approach in terms of plurality of norms falls into an endless "reconciliation through the establishment of justificatory equivalences" in line with Boltanski and Thévenot (2006 [1991]). In so doing, as Pellizzoni rightly points out, social criticism becomes unable to pose any foundational opposition. Especially, to return to our object of analysis, it would fail to address the "regimes of truth" established by governance-driven quantification. As we have seen above (in Part III), such regimes of truth are precisely fabricated so that "even contesting parties are compelled to accept [them] and to channel their dissent within specific boundaries and on a specific plane" (Pellizzoni, 2012, p. 10; see also Pellizzoni & Ylönen, 2016; and the conception of deliverative inquiry in Bohman, 2004).

It follows that social criticism should give priority to building social facts that, fairly and correctly, represent the *territory of justice* that the community judges relevant to the collective objective under consideration. Considering the variety of these territories for the same collective objective, the search of the relevant levels to build these facts, the cognitive categories to be used at these levels and the methodologies of inquiry are open questions to be posed and solved. As the members of the community possess the ultimate practical knowledge of the concrete reality of situations, they themselves only can provide access to what remains inaccessible even to the smartest researcher or observer, the data coming from their experience of the situation. Without their participation, it would be impossible to bring out—or to closely approach—the complete internal and external relevant features of their "factual territory". These data are not evidence reflecting reality; they are elaborated by people through the prism of their own feelings on what is or is not justice and injustice.

It means that access to such data is not only a question of inquiry in the classical social sciences conception; it has to do with an "extraction" from the people of intimate practical knowledge *that they know without knowing that they know it*; which means that they should deliberate with researchers all along in the process of inquiry. Such inquiry should be defined as a deliberative inquiry. Its specificities are that its levels, cognitive categories and methodologies, as well as its participants should be "produced" along the processing of the data itself. There is no a priori standard recipe, but something multifaceted (mobilizing people, reflexive awareness, political and scientific) to invent collectively.[16]

Claiming for Another State

While those developing counter-quantification processes may be not fully aware of their expectations, at the horizon of their action is the perspective of another type of state. Let us return to the two sides of quantification processes discussed in the first section (Part I), the "quantifiers" and the "quantified". Two correlated questions, political and methodological, have to be addressed: the conception and legitimacy of the authorities who lead the process of quantification (the "quantifiers"); and the nature of the deliberative process that surrounds the quest for answers of quantified people. In the context of a plurality of possible data buildings, what one could call the cognitive moment appears more complex than the simple technical administration of some questionnaire or pure imposition from above. To what extent and how do the quantified have some voice in the choices? How far should the cognitive moment be understood as belonging to a deliberative process? These questions largely remain terra incognita, and they require the possibility of a plurality of types of states. We have already seen in above (see Part II) that the respective role of expectations about state intervention versus collective autonomy differed between France, Germany and the UK for the quantification of unemployment and associated policies. France has the most interventionist top-down state, imposing the same rules to all levels. Germany is historically more open to collective autonomy and diversity at the lower levels (Lander and cities for instance) and the UK is navigating in between. European authorities adopted a French-type interventionist style when they imposed the same panel of indicators to all countries for liberalizing the labour market.

In Salais and Storper (1993, 1997),[17] we tried to formalize several types of state supported by different conventions between persons and actors. Such conventions allowed us to understand historical examples. Applying a conventions approach (see also Diaz-Bone, 2018; [2015]; Eymard-Duvernay, 1989; Lewis, 1969) means that these types of states are realized, renewed and made stable through common expectations between people and the authorities. They hold by the virtue of shared beliefs that become deeply rooted in institutions. Such an approach helps to define, at least,[18] two types of quantification processes, depending on the state that is object of mutual beliefs. It is worth noting that, if one "partner" (quantified or quantifiers) moves towards another convention of the state, political tensions and conflicts arise. A road is potentially open to social criticism for claiming other public policies, provided it organizes its counter-quantification around another convention of the state than the one already implanted.

In the first convention of the state,[19] evaluated people devolve to the central authority the whole task of building the quantification process (modalities, what and how to measure). One can imagine several ways to legitimate such devolution: such tasks are accepted as technical, so no need for voices to be expressed (the European conception again); or, through their representatives, evaluated people are asked to indicate if they agree with the choices made by the central authority. The applied procedure is similar to the one which is used in standard representative democracy. But, for Europe at least, are we still in a democracy or in a move towards what we call a-democracy? Such a convention seems today being replaced by governance by numbers and a-democacy. In such a regime of truth, objectivity is reduced to standardization (Porter, 1992).[20] As we have asserted for Europe above (see Part II), in practice evaluated people have no say on choices on the informational basis (the set of indicators); they cannot be truly committed to take the evaluation procedure as their practical benchmark. In a-democracy, such a question becomes irrelevant, because the problem is no more to achieve an effective substantial evaluation, but only to betray current beliefs and representations by producing data apparently supporting them.[21] In contrast to the following second convention, there is no need for true deliberation in a-democracy.

In the second convention,[22] the authority and evaluated people choose to build a part or the whole of the procedure together, including questions of what and how to measure issues. In practice, it requires that both sides commit themselves to deliberative procedures, which are aimed at achieving deliberate decisions. Such a conception of the state is for us the most fitting for social criticism developing counter-quantification. In contrast to strategic decisions obeying instrumental rationality, deliberate decisions are decisions that both sides have the effective intention to afterwards apply. One will not go further, except to note the proximities with the concepts of subsidiarity[23] and of deliberative democracy.[24] People should have their say and be mobilized for imposing their views. One cannot expect from central authorities that they spontaneously enter into such a demanding coordination. In his works, John Dewey (1927) has explored the political conditions making such frames of coordination possible more in-depth. Dewey understands democracy as a collective practice led by collective movements that struggle for creating what Dewey calls publics. Publics are to be built along a process that progressively gathers people together to defend a cause (a common good for instance). But such a process is not political in its standard understanding. Political movements mostly conceive such a process as based on ideological or strategic arguments. For Dewey, it consists of a collective learning process anchored in the collective search for the knowledge relevant for implementing the cause at stake. It is, more or less, for people the search for their "true" common world in our pragmatic meaning of the concept. The ultimate stake for them remains not only to publicly oppose their understandings and proposals to those of the authority they are confronted with (which is necessary), but also basically to generate in their community (also necessary) whatever it is, an openness towards conceptions, pragmatic compromises or agreements taking on board their true common world.

CONCLUSION: IMPLICATIONS FOR RESEARCH ON QUANTIFICATION PROCESSES

The development of governance-driven quantification processes creates opportunities to have a fresh look at factors which previously were taken for granted and not considered problematic. For they introduce to the fields of research and social practice of quantification new concerns about democracy, participation in collective choice, and social justice. The possibility of a plurality of "data makings" for *the same situation* becomes now visible, thanks to the different relationships of social cognitive practices to politics. Where are their respective scientific and political legitimacies? Should we consider the potentiality and even existence of a plurality of quantification regimes? In line with Sen's conception of informational bases of judgement, introducing considerations of justice into quantification processes should become relevant and, even more, necessary for better efficiency. One knows how far the right coordinationbetween people depends on their expectation to be fairly treated by others and by institutional or regulatory frameworks that surround their activities. There are several principles of being fairly treated, in other terms of justice. If such an assumption of plurality is relevant, it would extend to the objectivity of data. It also means that a regime of quantification can be validly contested by another one; such contestation should be conceived as a necessary component of any democracy. It opens the road to social criticism based on the creation of alternative informational bases, all being politically and scientifically relevant and legitimate.

Acknowledgements My many thanks for helpful comments by Andrea Mennicken and Ota De Leonardis.

NOTES

1. For instance, in his remarkable, internal and procedural analysis of the OMC and its impact on national social policies, Zeitlin (2009) never mentions the impact on quantification and evaluation.
2. In that respect, the USSR and the People's Republic of China appeared as pioneers in developing such utopia. See the contributions by Tong Lam and Martine Mespoulet in this volume.
3. See below the subsequent section which discusses the domain called "unemployment". What does it mean to be "unemployed"? The official

understanding today and everywhere make reference to the ILO definition: actively searching for a job; having no job; to be immediately available to take a job. One will recall that, historically, and depending on the country, to be unemployed was not clear and took time to be so for people.

4. An example is, in France, the fact that, until the 1950s, female homeworkers, though knowing periods of no work each year, did not produce in the population censuses answers allowing to classify them as "unemployed". Similarly urban craft workers did not register them as unemployed at manpower bureaus, considering this as an insult to their dignity.

5. Here we draw lessons from a series of researches, starting independently from each other in the 1980s. See here in particular Phillips and Whiteside (1985), Salais et al. (1986) (reprinted in 1999); Keyssar (1986), Piore (1987), Luciani (1992), Mansfield (1992), Topalov (1994), Mansfield et al. (1994), Whiteside (2007, 2014), Zimmermann (2001), Salais (2011) and Latsis (2006).

6. As demonstrated by the example of craft workers who do not register in unemployment bureaus, but have their own systems.

7. See also Storper and Salais (1997) and Salais (2015). One takes this opportunity to rectify a misunderstanding in Thévenot's contribution to this volume (see note 15) who speaks of "some familiarity with orders of worth". The foundations for our worlds of production have not much to do with those of orders of worth. They are centred on the product, at the crossing of production and market, precisely two basic economic principles (economies of scope vs economies of scale for the productive organization; risk vs uncertainty for the market; and not on disputes). Furthermore, the state is present as a specific convention with regards to the common good. The only resemblance, is the use of pluralism, which is a brand mark of the economics of convention since its beginning. We already used it in Salais et al. (1986).

8. For one of his inventors see Telo (2002); see Kröger (2009) for to which we intend to answer here.

9. As a statistician, my first surprise, even incredibility, was about what the European Commission was doing with the European Employment Strategy and the "abnormal" way it uses data and indicators (see Salais, 2004, 2006).

10. Alain Desrosières has posed and used this concept in his seminal book (see Desrosières, 1998 [1993]; but see also Desrosières, 2008). Espeland and Stevens (1998) speak of commensuration as the process that makes objects and persons commensurate, i.e. reduced to the same quantitative scale.

11. Here I disagree with a radical Foucauldian interpretation, which is inclined to see the paw of the monster Neoliberalism everywhere. See also the last section of this chapter (Part IV).

12. This analysis can be found in Salais (2004) and in Salais (2006). It took time for political scientists specialized in the European domain to understand the complexities of the game. They took the EES as if it were a purely political procedure, with virtual disregard for other factors (especially for the status and formats of numbers). Most of the studies have focused on the wide range of actors for whose involvement the European texts contain provision and on the procedures laid down to organize their complex interactions; this is the famous "multi-level governance". In the English literature, studies of such gaming and ranking can be found in Bevan and Hood (2006), Hood et al. (2008) and Hood and Dixon (2010).

13. Raveaud and Salais (2002) analysed all the problems connected to the calculation of European Employment indicators. A more detailed draft is available on request from the author.

14. To draw on Rawls' famous concept, which is well suited for the issues described here.

15. NPM defenders would also say that they are not only concerned with efficiency, but also with effectiveness and outcomes, i.e. to what extent performance meets the stated objectives of a policy, which can include objectives of enhancing equality, fairness, etc. The problem, however, is how such objectives are then made "governable"/measurable through indicators that are quite removed from the original goals (as we have shown before for the example of unemployment in the EU).

16. A wonderful illustration of this can be found in the contribution of Boris Samuel to this volume. See also the experiment led by Stavo-Debauge and Trom (2004) and the literature on statactivism (Bruno et al., 2014).

17. See Salais and Storper (1993, pp. 326–346) and Storper and Salais (1997, pp. 207–223). For further developments see Salais (2015).

18. In practice, we define four conventions of the state (see Salais & Storper, 1993; Storper & Salais, 1997).

19. This conception corresponds to the convention of the external state.

20. For a powerful critique of the current conception of objectivity see Sen (1993).

21. Michael Power (1997) developed the same conjecture for audits, namely that they mostly support current beliefs.

22. Which corresponds to the conventions of the situated state.

23. The best presentation I know for the concept of subsidiarity is Millon-Delsol's (1992), unfortunately in French. She established that the European authorities confuse subsidiarity with decentralization. For more detail see Salais (2015).

24. See Bohman (1996, 1999) and for a rather convincing heterodox development, Besson (2003).

REFERENCES

Austin, J. L. (1962). *How to do things with words.* Harvard University Press.

Besson, S. (2003). Disagreement and democracy: From vote to deliberation and back again? In J. Ferrer, & M. Iglesias (Eds.), *Law, politics and morality: European perspectives* (Vol. 1, pp. 101–135). Duncker & Humblot.

Bevan, G., & Hood, C. (2006). What's measured is what matters: Targets and gaming in the English public health care system. *Public Administration, 84*(3), 517–538.

Bohman, J. (1996). *Public deliberation.* MIT Press.

Bohman, J. (1999). Deliberative democracy and effective social freedom: Capabilities, resources, and opportunities. In J. Bohman & W. Rehg (Eds.), *Deliberative democracy* (pp. 321–348). MIT Press.

Bohman, J. (2004). Realizing deliberative democracy as a mode of inquiry: Pragmatism, social facts, and normative theory. *The Journal of Speculative Philosophy, New Series, 18*(1), 23–43.

Boltanski, L., & Thévenot, L. (1983). Finding one's way in social space; a study based on games. *Social Science Information, 22*(4–5), 631–679.

Boltanski, L., & Thévenot, L. (2006 [1991]). *On justification: Economies of worth.* Princeton University Press (French edition, 1991).

Brown, W. (2015). *Undoing the Demos.* Zone Books.

Bruno, I., Didier, E., & Prévieux, J. (Eds.). (2014). *Statactivisme: Comment lutter avec les nombres.* Zones.

Bruyère, M., Chagny, O., Ulrich, V., & Zilberman, S. (2006). Comparaisons internationales de la durée du travail pour sept pays en 2004: la place de la France. *Données sociales (La société française).*

Davies, H., Nutley, S., & Smith, P. (Eds.). (2000). *What works? Evidence-based policy and practice in public services.* The Policy Press.

Desrosières, A. (1998 [1993]). *The politics of large numbers: A history of statistical reasoning.* Harvard University Press.

Desrosières, A. (2008). *Pour une sociologie historique de la quantification.* Presses de l'Ecole des Mines de Paris.

Desrosières, A. (2011). The economics of convention and statistics: The paradox of origins. *Historical Social Research, 36*(4), 64–81.

Dewey, J. (1927). *The public and its problems.* Holt.

Diaz-Bone, R. (2018 [2015]). *Die "Economie des conventions". Grundagen und Entwicklungen der neuen französischen Wirtschaftssoziologie.* Springer VS.

Didry, C. (2002). *Naissance d'une convention collective. Débats juridiques et luttes sociales en France au début du XXè siècle*. Editions de l'EHESS.

Espeland, W. N., & Stevens, M. L. (1998). Commensuration as a social process. *Annual Review of Sociology, 24*, 313–343.

European Commission. (2006). Joint Employment Report 2005/2006. Time to move up to a gear. Annex to the Communication to the European Council. COM(2006) 30 final/annex 25 January 2006. Brussels: European Commission.

Eymard-Duvernay, F. (1989, March). Conventions de qualité et pluralité des formes de coordination. *Revue Economique, 2*, 329–359.

Hacking, I. (1990). *The taming of chance*. Cambridge University Press.

Hood, C., & Dixon, R. (2010). The political pay-off from performance target-systems. *Journal of Public Administration Research and Theory, 20*, i281–i298.

Hood, C., Dixon, R., & Beeston, C. (2008). Rating the rankings: Assessing international rankings of public sector performance. *International Public Management Journal, 11*(3), 298–328.

Keyssar, A. (1986). *Out of work: The first century of unemployment in Massachusetts*. Cambridge University Press.

Kröger, S. (2009). The open method of coordination: Underconceptualisation, overdetermination and de-politicization and beyond. *European Integration online Papers (EIoP), 13*(5).

Latsis, J. (2006). Convention and intersubjectivity: New developments in French economics. *Journal for the Theory of Social Behaviour, 36*(3), 255–277.

Lewis, D. (1969). *Convention*. Harvard University Press.

Luciani, J. (Ed.). (1992). *Histoire de l'Office du travail 1890–1914*. Syros.

Mansfield, M. (1992). Labour exchanges and the labour reserve in turn of the century social reform. *Journal of Social Policy, 21*(4), 435–468.

Mansfield, M., Salais, R., & Whiteside, N. (Eds.). (1994). *Aux sources du chômage 1880–1914. Une comparaison interdisciplinaire entre la France et la GrandeBretagne*. Belin.

Miller, P. (1992). Accounting and objectivity: The invention of calculating selves and calculable spaces. *Annals of Scholarship, 9*(1–2), 61–86.

Miller, P., & O'Leary, T. (1987). Accounting and the construction of the governable person. *Accounting, Organizations and Society, 12*(3), 235–265.

Millon-Delsol, C. (1992). *L'Etat subsidiaire*. Presses Universitaires de France.

OECD (1994). *The management of performances into administration: Measure of performances and outcomes-oriented management*. OECD Special Studies, 3.

Pellizoni, L. (2012, February 3). *Ways of searching for the common good*. Paper presented at the EHESS Seminar held by Francis Chateauraynaud "De l'alerte au conflit – Logiques argumentatives et trajectoires des mobilisations", EHESS Paris.

Pellizoni, L., & Ylönen, M. (2016). Hegemonic contingencies: Neoliberalized technoscience and neorationality. In L. Pellizoni & M. Ylönen (Eds.), *Neoliberalism and technoscience: Critical assessments* (pp. 47–74). Ashgate.

Phillips, G., & Whiteside, N. (1985). *Casual labour: The unemployment question in the port transport industry, 1880–1970.* Oxford University Press.

Piore, M. (1987). Historical perspective and the interpretation of unemployment. *Journal of Economic Literature, 25*(4), 1834–1850.

Porter, T. M. (1992). Objectivity as standardization: The rhetoric of impersonality in measurement, statistics, and cost-benefit analyses. *Annals of Scholarship, 9*(1–2), 19–59.

Power, M. (1997). *The audit society: Rituals of verification.* Oxford University Press.

Raveaud, G., & Salais, R. (2002). A study on indicators for employment policies: Objectives, methods, proposals. A preliminary report for the French Department of Employment and Social Affairs. *Note Research Centre Institutions and Dynamiques Historiques de l'Economie (IDHE, n° 145/02, 2 November).*

Salais, R. (2004). La politique des indicateurs. Du taux de chômage au taux d'emploi dans la stratégie européenne pour l'emploi (SEE). In B. Zimmermann (Ed.), *Les sciences sociales à l'épreuve de l'action: Le savant, le politique et l'Europe* (pp. 287–331). Éditions de la Maison des Sciences de l'Homme.

Salais, R. (2006). Reforming the European Social Model and the politics of indicators. From the unemployment rate to the employment rate in the European Employment Strategy. In M. Jepsen, & A. Serrano (Eds.), *Unwrapping the European Social Model* (pp. 189–212). The Policy Press.

Salais, R. (2007). Europe and the deconstruction of the category of unemployment. *Archiv Für Sozialgeschichte, 47*, 371–401.

Salais, R. (2008). Capacités, base informationnelle et démocratie délibérative. Le (contre-)exemple de l'action publique européenne. In J. De Munck, & B. Zimmermann (Eds.), *La liberté au prisme des capacités. Amartya Sen au-delà du libéralisme* (pp. 297–326). Editions de l'Ecole des Hautes Etudes en Sciences Sociales, Raisons pratiques 18.

Salais, R. (2010). La donnée n'est pas un donné. Pour une analyse critique de l'évaluation chiffrée de la performance. *Revue Française D'administration Publique, 135*, 497–515.

Salais, R. (2011). Labour-related conventions and configurations of meaning: France, Germany and Great Britain prior to the Second World War. *Historical Social Research, 36*(4), 218–247.

Salais, R. (2013). *Le viol d'Europe. Enquête sur la disparition d'une idée.* Presses Universitaires de France.

Salais, R. (2015). Etats extérieurs, absents, situés, une revisite à la lumière de la crise de l'Europe. *Revue Française de Socio-Économie, 2nd Semester*, 245–262.

Salais, R., Baverez, N., & Reynaud, B. (1986). *L'invention du chômage.* PUF.

Salais, R., & Storper, M. (1993). *Les mondes de production*. Ed. de l'EHESS.

Sen, A. (1990). Justice: Means versus freedoms. *Philosophy and Public Affairs, 19*(2), 111–121.

Sen, A. (1993). Positional objectivity. *Philosophy and Public Affairs, 22*(2), 126–145.

Standing, G. (2014). *The precariat: The new dangerous class*. Bloomsbury.

Stavo-Debauge, J., & Trom, D. (2004). Le pragmatisme et son public à l'épreuve du terrain. In B. Karsenti, & L. Quéré (Eds.), *La croyance et l'enquête. Aux sources du pragmatisme* (pp. 195–226). Editions de l'EHESS (Raisons pratiques n° 15).

Storper, M., & Salais, R. (1997). *Worlds of production: The action frameworks of the economy*. Harvard University Press.

Supiot, A. (2015). *La gouvernance par les nombres*. Fayard (English edition, Verso 2016).

Telo, M. (2002). Governance and government in the European Union: The open method of coordination. In M.-J. Rodrigues (Ed.), *The new knowledge in Europe* (pp. 242–271). Edward Elgar.

Thévenot, L. (1983). L'économie du codage social. *Critiques De L'économie Politique, 23–24*, 188–222.

Topalov, C. (1994). *Naissance du chômeur 1880–1910*. Albin Michel.

Whiteside, N. (2007). Unemployment revisited in comparative perspective. *International Review of Social History, 52*(1), 35–56.

Whiteside, N. (2014). Constructing unemployment: Britain and France in historical perspective. *Social Policy Administration, 48*(1), 67–85.

Zeitlin, J. (2009). The open method of coordination and reform of national social and employment policies: Influences, mechanisms, effects. In M. Heidenreich & J. Zeitlin (Eds.), *Changing European employment and welfare regimes: The influence of the open method of coordination on national reforms* (pp. 214–245). Routledge.

Zimmermann, B. (2001). *La constitution du chômage en Allemagne. Entre professions et territoires*. Editions de la Maison des Sciences de l'Homme [in German. 2006. Arbeitslosigkeit in Deutschland. Zur Entstehung einer sozialen Kategorie. Frankfurt a. M.: Campus].

CHAPTER 13

Free from Numbers? The Politics of Qualitative Sociology in the U.S. Since 1945

Emmanuel Didier

It has been well established now that quantification is not only a means to produce knowledge, but also a means of power. This insight has given rise to the famous and important body of works on *social studies of quantification* (Daston, 1988; Desrosières, 1998 [1993]; Espeland & Sauder, 2007; Gigerenzer et al., 1989; Krüger et al., 1987; Porter, 1995), which studied in many diverse fashions the historical conditions of the production of numbers and their social effects, denaturalizing quantities while at the same time re-specifying their authority. Most of these works suppose that, first, there was a state of affairs without numbers; second, that measures have been applied on it; and third, that the situation has finally become quantified. Desrosières (2008), who can rightfully be taken as the primary representative of this tradition, states this idea in a very clear equation: "quantification = convention + measurement" (Desrosières, 2008, p. 10). The crucial insight of this proposition is that this process is

E. Didier (✉)
Centre Maurice Halbwachs - SNRS/ENS-PASL/EHESS, Paris, France
e-mail: emmanuel.didier@ens.fr

© The Author(s) 2022
A. Mennicken and R. Salais (eds.), *The New Politics of Numbers*,
Executive Politics and Governance,
https://doi.org/10.1007/978-3-030-78201-6_13

417

a social one–and not a natural or straightforward one–that deserves to be problematized and understood with the tools of the social sciences.

Yet, the contemporary excitement around "big data" makes one wonder if the problem should not be reversed. We hear today that the planet is increasingly populated by digital data (for example: "90% of the data harvested since the beginning of humanity have been generated in the last two years" [Dupont, 2015]). But we know that it is only an exaggeration in the long history of people being mesmerized with the mechanized production of the quantitative–to which the "big data" phenomenon belongs since it comprises many numbers, if not anything else. One must not forget that the decades 1820–1840 already witnessed an "avalanche of printed numbers" (Hacking, 1982). The invention of the Hollerith machine at the end of the nineteenth century and its adoption by bureaus of public statistics all over the world produced a "revolution in data processing" (Austrian, 1982). With the development of polls and sample surveys, The New Deal was a period during which the U.S. was entirely "statisticized" (Didier, 2009). Every period has had its own quantitative revolution related to technologies of data production and to creativity in the use of data. The fuss around big data proves only that our current era makes no exception: it is, as it was, filled with quantities.

Thus, since society is quantitative through and through, the real mystery might not be the amount of data that circulates and governs but on the contrary, the existence of social spheres pretending to remain free from numbers. If the world has already been quantified since at least the first half of the nineteenth century, are there some spheres that could remain exceptions, and how is this possible? What does the activity of purifying a social sphere from numbers consist of? What are the political endeavours associated with such a goal? Or, to put it differently, how can we account for the political production of the border of qualitative enclaves which exclude quantities?

To tackle these questions, I will go back to the history of what is now called *qualitative sociology*. Indeed, sociology is a discipline in which the great founders never chose between quantification and non-quantification. In France, Emile Durkheim and Gabriel Tarde, who were opposed in every respect, had two main points in common: first, each was the leader of a powerful current of sociology and strove to institutionalize it according to his own definition (against the other's), and second, both relied on quantitative reasoning among other arguments, as *Suicide* on the one hand and *The Laws of Imitation* on the other attest

(Durkheim, 1986 [1897]; Tarde & Parsons, 1903). In Germany, Max Weber, along with his definition of the longstanding "verstehen", also performed quantitative surveys (Brain, 2001; Pollak, 1986). Finally, in the U.S., the Chicago School of sociology never *chose* between the two (Abbott, 1999; Chapoulie, 2001). Sociology was founded as a science commonly using quantification as one of its diverse cognitive tools and methods. It entertained a "relaxed" relationship to quantities and qualities (to use Glaser and Strauss's (1967) expression). Thus, the branch of U.S. sociology that came to be labelled "qualitative sociology" during the 1970s made an astonishing move, apparently of the ascetic sort, in defining a discipline that would be freed from quantities.[1]

Why would one distinguish a sub-discipline by its absence of numbers? How did the conceptual pair "qualitative vs. quantitative" come to settle within sociology? What were the conditions in which sociology was produced and the publics it addressed that might explain this link? Finally, is it even possible to eradicate quantification and stay with conceptions encompassing qualities only?

Using the methods of the sociology of quantification, I will pay attention to both the epistemic and political forces that participated in the production of the border between qualitative and quantitative in sociology.[2] I will inquire into the political worth of the qualitative. It was within a very specific power field, ranging from the constitution of the Welfare State after WWII to the radicalism of the 1970s and finally ending in the liberal 1980s, that those who would ultimately defend a "qualitative" sociology forged and used their epistemic arguments separated from the quantitative. I will pay special attention to how these two aspects of the story were intermingled.

These questions can best be understood when it is clear from the onset that here "quantitative" has two different meanings. We will see that "quantitative" analysis had been defined by mainstream sociologists as one single method, that of *survey sampling or polling*. This is a first definition of quantitative, the one of our "actors" or "members". But we can see furthermore that there have long been many other methods of quantification, many uses of numbers, and, as has been proven by the late Alain Desrosières, that these different methods of quantification are consistent with different political endeavours (Desrosières, 2003).

Finally, it is worth mentioning that this paper is a sequel to the question of the appearance and legitimization of quantitative surveys in the American Government during the New Deal (Didier, 2009, 2020). Here,

I follow the later fate of this method and trace how after WWII it came to be criticized. I aim to sketch the whole social life course of a statistical method, from its appearance to its decomposition. This paper is also an inquiry into the relationships between sociology and politics. The position born with surveys during the 1930s and the 1940s, of the sociologist as an expert advising political power, is here contrasted with that of the sociologist as a critic of any association with the power elites, the sociologist as a radical, a position that fully developed after WWII and came to be closely associated with "qualitative" methods. Finally, this paper is also a contribution to the "sociology of quantification". Rather than asking how qualitative things are quantified, I reverse this question and ask how it is possible, if ever, in a world already filled with quantities, to try and purify portions of it in the hope of establishing a "qualitative" enclave.

My first point will consist in emphasizing the seminal role played in the 1950s by Herbert Blumer and Aaron Cicourel in the fight against Lazarsfeld's definition of qualitative analysis. Both opposed a specific statistical method—surveys for the first and official statistics for the other, and they were not against quantification in general, which they in fact practiced. They opposed a specific political use to which the statistical method was associated. Then, we will see how their conclusions were refurbished by the young radicals in the 1960s and 1970s as a means to fight against the elite of the Welfare State. Finally, we will see that "qualitative sociology" as such appeared only during the 1970s as a weird association between the Lazarsdfeldian promoters of surveys and the neo-radicals opposed to it.

Excluding Quantities?

The two main sociologists embodying the tradition of "qualitative sociology", as far as they explicitly addressed their relationship to quantification, were Herbert Blumer and Aaron Cicourel. I will analyse their conception of the border between quantitative and qualitative research. I will thus clarify their critique of numbers and the social context in which they were expressed. Especially, I will clarify their relationship to the work of Lazarsfeld.

C. Wright Mills, in his *Sociological Imagination* (Mills, 1959), had a very influential critique of "abstracted empiricism" as a kind of sociology which, while transforming itself into a gigantic bureaucracy, turned the American public into a series of *masses*. Unfortunately, Mills died too young (1962) to take part personally in what later came to be called "Qualitative sociology" and actually, in his writings, never used the dichotomy qualitative/quantitative at all. So in our story, we shall treat his work as a resource for our actors, but not as an actor by himself.

Interpretation and Determinism

Herbert Blumer is credited with the invention of *Symbolic Interactionism*. This approach to human group life is deeply influenced by the philosophy of George Herbert Mead and the American pragmatist tradition. It locates the social primarily in situations of interaction between humans and between humans and objects. It focuses on the fact that members' action is guided and formed by a *process of interpretation* of the situation in which they are involved. This process of interpretation is an active one, and not a passive submission to outside forces. In Blumer's own words, members' "behavior with regard to what it notes is not a response called forth by the presentation of what it notes but instead is an action that arises through the interpretation made through the process of self-indication" (Blumer, 1969, p. 14). Placing the concept of interpretation at the heart of his concepts, Blumer has today among sociologists an "image as purely qualitative" (Abbott, 1999, p. 51). Indeed, *symbolic interactionism* became one of the core components of qualitative sociology.

The history of the growth of Blumer's opposition to quantification is quite complex. One has to keep in mind that until WWII, Blumer was in a very powerful situation in the American sociological field. He was a Professor of Sociology at the University of Chicago's Department of Sociology, one of the most distinguished and powerful departments in the country. From this position, he witnessed the fairly quick establishment of the partisans of statistical surveys, especially at Columbia.

The American Soldier

Blumer's powerful position was questioned in particular by the publication of *The American Soldier* edited by Samuel Stouffer and colleagues (Social Science Research Council (U.S.), 1949), a five-volume sociological study of the Army during the war. As Schweber (2002) shows, this book not only encountered huge public success, but was also heralded as the example to a *new* approach of social science, making important use of statistics. It bore on trends that began in the 1930s with the growing importance of polls on the one hand, and of the quantification of surveys on the other, associated with the growing power of welfare institutions, which were the primary users of this kind of knowledge, both at the local and the national level. *The American Soldier* was seen as the symbol of the will to promote statistics as *the* authoritative method in sociology. And, also problematic from the point of view of Blumer, it was associated with Harvard, since Stouffer, who earned his PhD from Chicago, had been hired by the University located in Cambridge, Mass., in 1946.

A panel was organized in 1949 by the American Sociological Association to discuss the book. Blumer was invited, and apparently criticized the book vehemently. The authors of the 5th and last volume of *The American Soldier* wrote that he adopted a "rivalrous posture" stated in a "vigorous negativism, which leads to the extreme attitude we have designated as *diabolic*" (Merton & Lazarsfeld, 1950, p. 227). His talk has apparently not been published, but Howard Becker (1988) states that the arguments were very close to his 1948 paper on polling, later re-published as the last chapter of *Symbolic Interactionism* (Blumer, 1969).

In this article about polls, Blumer does not attack quantification as such. He even states that he uses numbers himself, but in a very peculiar way: "I shall indicate by number the [six] features to be noted" (Blumer, 1969, p. 198). It is not that common to read a text composed in six parts!

He expresses two main criticisms of polls. First, polling does not *define* *"public opinion"*, *its object*. It suffices itself by applying a technique, which indeed produces data, but it never takes time to define the concept on which data is produced. On the contrary, it relies on the "narrow operationalist position that public opinion consists of what public opinion polls poll" (Blumer, 1969, p. 197).

A second criticism is exposed in six points. The argument is that polling does not respect the actual "realistic" structure of public opinion formation. In particular, there are "key people" who play an important part in the production of public opinion. Yet, these processes through which public opinion is expressed are not consistent with the *sampling techniques* used by polls:

> In my judgment the inherent deficiency of public opinion polling certainly as currently done, is contained in its sampling procedure. Its current sampling procedure forces the treatment of society as if society where only an aggregation of disparate individuals. Public opinion, in turn, is regarded as being a quantitative distribution of individual opinions. This

way of treating society and this way of viewing public opinion must be regarded as markedly unrealistic. (Blumer, 1969, p. 202)

Blumer admits later in his text that polls did succeed in predicting the elections (of Roosevelt in 1936). But, "a ballot cast by one individual has exactly the same weight as a ballot cast by another individual. In this proper sense, and in the sense of real action, voters constitute a population of disparate individuals" (Blumer, 1969, p. 205). In the case of elections proper, the structure of the electorate is realistically comparable to that of a sample. But this is not the case outside of this very rare case.

Thus, Blumer argues first that opinion polling is "logically unpardonable", because it does not define its object of inquiry, and second that it does not respect the body of knowledge derived from empirical observation and from reasonable inference that one already has about the nature of public opinion. There is a third scandal in the eyes of Blumer, which is kept implicit in his text. It is that, given the success that these techniques encounter, the very key players in the formation of public opinion, to whom he gives such an important role, seem nonetheless to adopt and use polls in their endeavour.

He himself sees the social role of sociologists very differently. He served as an arbitrator for the steel industry during WWII. Arbitrators, in his view, are not "experts" advising the Government, but act as facilitators helping both parties finding a settlement in their dispute. As Cantril (1939) interestingly writes (since he was one of the founders of opinion polls), this role presupposes "objectivity" in a very different manner than that of the expert adviser.

These criticisms from Blumer can indeed be transposed to the surveys used in *The American Soldier*. An army, being strictly hierarchical, is anything but a population of disparate individuals. The "opinion" of an army is not defined in the book. Finally, for these very obvious reasons, it must have appeared very strange to Blumer that the commanders of the Army might appreciate the book. The opposition between the two kinds of sociology became even more violent when Stoufffer's book was used as a weapon for a direct and nominal attack against Blumer.

The Qualitative as Propaedeutic

Quantifiers replied to Blumer. In 1951, Henry Zentner, a young assistant professor at Stanford, published a paper (Zentner, 1951) in which he unearthed a contribution of Blumer about "Morale" published during the war (Blumer, 1943). He presented it as "the most careful and systematic conception" of morale at the time when it had been written, and proposed "to test, against the data reported in *The American Soldier*, the validity of Blumer's conception of the generic nature of group morale" (Zentner, 1951, p. 298). Zentner extracted information from the charts of the book and compared them to Blumer's analysis. He pinpointed what he saw as many weaknesses and went on to argue that Blumer's conception of morale was "grossly inadequate" (Zentner, 1951, p. 306). He concluded that morale was better defined by opinion surveys than by Blumer's methods.

Blumer felt compelled to comment. He wrote "why Mr. Zentner believes that he refutes my analysis is mystifying" (Blumer, 1951, p. 308). His own contribution was about the morale of the civilian population when Stouffer's book was about the army. Hence, Zentner's paper "does not even test my analysis much less refutes it", since "a theory or proposition is tested empirically by applying it to an instance of what the theory or proposition logically covers, not by applying it to something that falls outside of such a logical class" (Blumer, 1951, p. 308). There was clearly an attack but, argues Blumer, it did not hit. As he had stated earlier about polls, the object of inquiry is ill-defined and in this case it creates catastrophic confusion.

It is important for our purpose to note that the question of quantification as such is entirely absent from the debate.

The attack was bold coming from a young man such as Zentner, and maybe too bold since he seems to have completely disappeared from the field after the bout. But he expressed an idea that would have very important consequences: that Blumer's analysis was "essentially speculative and propaedeutic" and still needed to be empirically tested to gain actual authority (Zentner, 1951, p. 297).

This furrow is precisely the one that, since the 1940s, Paul Lazarsfeld was digging. Lazarsfeld repeated essentially the same message: "There is a direct line of logical continuity from qualitative classification to the most rigorous forms of measurement" (Lerner & Lasswell, 1951, p. 155). Or, stated slightly differently a few years later: "Not only is qualitative analysis large in volume, but it plays important roles in the research process, by itself and in connection with quantitative research" (Lazarsfeld & Kendall, 1982, pp. 239–240). His argument was first and foremost that qualitative and quantitative social science existed as two extremities on a continuum of methods. Lazarsfeld uses the pair of concepts with a frequency not encountered anywhere else—in particular, it must be insisted upon that Blumer never used it. Lazarsfeld is the one who decisively introduced the conceptual pair in sociology, and thus insisted also on the importance of the qualitative. It is most probable that his own sources, even though they are not explicit in the literature as far as I can tell, are in the Vienna Circle from where he came. He brought the dichotomy with him while emigrating to the U.S.

But it was only to subordinate qualitative research to quantitative research. He gives a biographical explanation to this hierarchy: as an assistant to Bühler in Vienna before immigrating to the U.S., he worked on the "qualitative attributes" of categories. And after arriving in the U.S. he discovered it would have helped him to use the "statistical methods" found in America (Zeisel, 1950, p. xvi). But he also gave many scientific justifications to the hierarchizing of the two kinds of research.

First of all, what he calls qualitative research is a necessary propaedeutic. One cannot directly begin any sociological work with statistics. Qualitative research is a first obligatory passage point (to use an awfully anachronistic concept):

> The operations of qualitative analysis which are raised essentially prior to quantitative research [are]: observations which raise problems, the formulations of descriptive categories, the uncovering of possible causal factors or chains of causation for a particular piece of behavior. (Lazarsfeld & Rosenberg, 1955, p. 267)

Thus, the qualitative steps in research are necessary for two reasons: they help establish the categories of further quantitative analysis—and categories must logically precede quantification. And they indicate or suggest possibilities of further relations between factors. The uses of "these operations [are to] stimulate and focus later quantitative research, and they set up the dimensions and categories along the stub of the tables, into which quantitative research may fill the actual frequencies and measurements" (Lazarsfeld & Rosenberg, 1955, p. 267).

But at the same time, the qualitative is essentially defined by the fact that it is "unsystematic", "impressionistic", not "objective" enough (Lazarsfeld & Rosenberg, 1955, p. 166) (1951, 166), it "remains an art" (Lazarsfeld & Rosenberg, 1955, p. 250):

> Research which has neither statistical weight nor experimental design, research based only on qualitative descriptions of a small number of cases, can nonetheless play the important role of suggesting possible relationships, causes, effects, and even dynamic processes. (Lazarsfeld & Rosenberg, 1955, p. 261)

The qualitative is defined by its essential incompleteness as regards the scientific endeavour, which only the quantitative can fulfil. The qualitative is systemically associated with the subjective, the personal, so that to become fully scientific it has to be made quantitative, that is, independent of any personal perspective, fully *objective*. As Daston (1992) put it, numbers help to produce "aperspectival objectivity"—a "view from nowhere"—where the places and persons are extracted from their use. Numbers also permit "mechanical objectivity" (Porter, 1995), a set of rules about how to make and deploy numbers that contain the discretion and biases of those using them.

Then, in the process of quantifying the qualitative, some variables remain what was called "qualitative" because they did not refer directly to a quantity. For example, the sex variables (male, female), race (Caucasian, Blacks, etc.), even modalities built from a quantitative variable (income brackets, etc.) are said to be qualitative. These types of variables were called qualitative but still, they allowed a statistical treatment.

Lazarsfeld became undoubtedly the star of sociology in the 1950s. He was a professor at Columbia, and earned very important research contracts thanks to the Bureau of Applied Social Research. He was also advising political figures. To give an example, it could not have escaped Blumer that Lazarsfeld had been invited to the Stanford symposium on "policy science" financed by the Carnegie Foundation on which the 1951 book is based, and Blumer was not. The new quantitative sociology, to use Lazarsfeld's vocabulary, was eclipsing the old Chicagoan. The pollster was stepping on the ground of the arbitrator. Blumer could not let it happen, especially since he was weakened even in his own university (Abbott, 1999).

Interpretation Cannot Be Overlooked

Blumer could not accept that his sociology was to be turned into a servant of an allegedly more objective one. He replied, and chose to name his enemy "variable analysis". This expression indicates the aim "to reduce human life to variables and their relations" (Blumer, 1969,

p. 127). An independent variable is identified and the analyst aims at measuring its effect on a dependent variable. Concretely, it implies the use of a questionnaire and of survey methods to gather field data that is to be transformed into variables. The variable is not *necessarily* quantitative though, even if it is indeed most of the time. This definition of the variable is a clear attack against the propositions of Paul F. Lazarsfeld.

Against the "application of the variable analysis to human group life", Blumer saw three "shortcomings" and one "crucial limit" (1969, p. 132). The first shortcoming is that there is apparently no "limit to what may be chosen or designated as a variable" (1969, p. 128). The sociologist can choose anything to be a variable that acts upon another variable, to the effect that often they do not address the real problem that is at hand in the situation studied. The second shortcoming is that often the variables are not generic and thus lack any abstract character. Most of the time, variables are in fact "bound temporally, spatially, and culturally" (1969, p. 130) and thus cannot provide any theoretical grasp of the situation. Finally, the variables rarely give the "fuller picture", the "context" in which members interact, even though for Blumer the latter is crucial to understand their action. These are shortcomings, because they are not necessary consequences of the variable analysis, they are simply often observed in practice.

Much worse, there is a limit within variable analysis that was not overcome until the publication of his paper. It does not account for the actual process that takes place in between the action of the independent variable at the beginning of any social process, and the dependent variable as the terminal part. "The intervening process is ignored or, what amounts to the same thing, taken for granted as something that need not be considered" (1969, p. 133). "One is content with the conclusion that the observed change in the dependent variable is the necessary result of the independent variable" (1969, p. 134).

But Blumer insists that any modification of the dependent variable has necessarily occurred through a process of interpretation. "The interpretation is not determined by the variable as if the variable emanated its own meaning. If there is anything we do know, it is that an object, event or situation in human experience does not carry its own meaning; the meaning is conferred on it" (1969, p. 134). The variable analysis simply discards the very core of any social action. Blumer concedes that, sometimes, it happens that interpretations are stabilized, it "occurs and recurs". But this must be verified each and every time since "anything that is defined may be redefined" (1969, p. 135). Finally, Blumer states that "the question of how the act of interpretation can be given the qualitative constancy that is logically required in a variable has so far not been answered" (1969, p. 136). More generally:

> In the area of interpretative life, variable analysis can be an effective means of unearthing stabilized patterns of interpretation, which are not likely to be detected through the direct study of the experience of people. Knowledge of such patterns, or rather of the relations between variables which reflects such patterns, is of great value for understanding group life in its "here and now" character and indeed may have significant practical value. All of these appropriate uses give variable analysis a worthy status in our field. In view, however, of the current tendency of variable analysis to become the norm and model for sociological analysis, I believe it is important to recognize its shortcomings and limitations. (Blumer, 1969, p. 137)

Thus, the variable analysis is content in studying the part of social life in which the interpretative process is either absent or stabilized. But for Blumer, this seems to be obviously a very small part of life, and the less interesting one, the part of life that is completely *deterministic*. He criticizes variable analysis for its incapacity to account for interpretative operations performed by humans, part of what has been called much later "the creativity of action" (Joas, 1996). Now, does this criticism of variable analysis mean that Blumer rejected any quantification and was purely "qualitative"? I would like to prove the contrary.

The Quantifier Blumer

Blumer was against surveys, but he was for quantification, conceived very differently. First, it is important to keep in mind that Blumer did not accept the dichotomy qualitative vs quantitative that appeared in the writings of Lazarsfeld. He rarely used the word "qualitative", and avoided elaborating on the dichotomy itself. He chose to criticize "the variable analysis" and not the quantitative techniques. Variables were not doomed to be limited in scope and impetus; they were so only in the hands of limited sociologists whose investigations are limited in scope and impetus.

These precautions were not only rhetorical. Some of Blumer's first publications were two books that both came out in 1933. One was *Movies and Conduct* (Blumer, 1933), and the other *Movies, Delinquency and*

Crime (Blumer & Hauser, 1970 [1933]). The second book was co-authored with Philipp M. Hauser, a master in quantitative techniques who would eventually become the Director of the Bureau of the Census (1949), and this book was full of figures and tables! Both books asked the question whether the movies, which in the 1930s had become one of the most popular entertainment industries, lead youth to crime because crime is depicted in motion-pictures, or on the contrary, whether motion-pictures protect them from becoming criminals, because they show its condemnation? Both books utilize data which was gathered from nearly two thousand students through interviews, observations and students' "motion-picture autobiographies" in which informants were asked to write in narrative form their motion-picture experiences. In addition, a survey questionnaire was distributed to two populations: a sample drawn from high-school children and a sample drawn from young inmates (male and female). The surveys are analysed only in the book with Hauser. Thus, even though it is clearly Hauser who performed the quantitative analysis, Blumer did publish some quantitative analysis under his name.

Apparently, Blumer did not remember it as an error of his youth, but quite on the contrary, as twenty years later, when in 1952 he would leave Chicago, where he had lost much of his personal influence, to join Berkeley, he tried for several years to recruit Hauser with a "formidable salary". Upon Hauser's refusal, he made comparable offers to quantitativists Leo Goodman and Otis D. Duncan, who also ended up refusing (Abbott, 1999, p. 51).

These events prove that Blumer really thought that it was possible to produce interesting quantitative analyses, even in the case of methods involving "variable analysis". He fought hard to colour quantitatively the team that he had been dreaming to build up in California. Abbott (1999) argues rightly that this team was the result of a community of a Midwestern habitus. Sure, but this community would have been discarded if their sociology had been incompatible.

Even after his teamwork with Hauser was over, Blumer continued to produce research using numbers, but of a completely different nature than the one he addressed in his criticism of "variable analysis". Apparently, Blumer did not participate in the conduct and analysis of surveys anymore, but in his empirical work he always listed the things that were indicated and interpreted by the members of the interaction he observed. And Blumer counted the elements of these lists.

An impressive example can be taken from a posthumous book on industrialization (Blumer, 1990). In the introduction to the book, the editors Maines and Morrione insist on the fact that it is not possible to determine the exact date when its content was written, because the book comprises a collection of essays which were published at different times. But Blumer first wrote on industrialization when he was in Brazil in 1958 and chapter six of the book was published as an article in 1971. Thus most of it has probably been written after *Symbolic Interactionism* (Blumer, 1969) was published.

In this book, Blumer asks how to conceive "industrialization in terms of how it operates on group life" (1990, p. 42). Strikingly, Blumer goes on listing nine lines, or dimensions—nothing more and nothing less— through which industrialization entered group life:

> In its gross aspect, industrialization is the introduction or expansion of a manufacturing system of production. As an agent of social life, it has to enter into group life. This sets the very important tasks of identifying the lines of entry, instead of merely juxtaposing the manufacturing system to group life. [...] My analysis leads me to identify nine lines of entry that are important, common to industrialization and, I believe, reasonably comprehensible of what occurs in industrialization. [...] The scheme brings us out of the vagaries and confusion that encumbers scholarly conceptions of industrialization. The scheme is definitive, it is tied to the manufacturing scheme of production, and it allows an empirical tracing out of what happens socially in industrialization. (Blumer, 1990, p. 49)

Later in the book, he questions whether there could be one more dimension, only to reject it. Thus Blumer holds on firmly to the number 9. This example is striking. Not only is it rare to insist on the number 9, but it is not an isolated case in his writings. Very often Blumer looks for the entities that are "taken into account" in an interaction, and actually counts them for the sake of clear and distinct conception of the process. It is impressive to note how often (should I count?) he uses the rhetorical figure of numbering the elements contained in lists. See, for example, the chapter on polls that we analysed earlier where he mentioned six critical features, or the chapter on Mead where he counts to five the consequences of his conception of objects (1969, p. 68), and to six those of his theory of joint action (1969, p. 71). Blumer appears as a *canvasser* of elements, all of which more or less abstract, must be "noted", "taken into account", "indicated" in an interaction for it to be interpreted by the members, and only secondarily by the sociologist. Blumer had *bricolaged* his own specific quantitative method that was compatible with interactionism: the canvass of concepts and their quantitative identification.

Finally, we find ourselves with two opposite views. On the one hand, there are the Lazarsfeldians coming from a positivistic model of action and science. They conceive social actors as affected by causes, of which they are not necessarily aware, determining their behaviour. These behaviours once aggregated, might create social problems, as proven by the Great Depression. The government, being on a higher level of action than the actors, can act on these causes, using work and social projects, as during the New Deal. The sociologist produces objective information about the causal mechanisms at hand in using statistical survey techniques, and advises the government thanks to this specific knowledge (Didier, 2009).

On the other hand, there are Blumer and the Symbolic Interactionists, influenced by the American Pragmatists. Here, the actors' main characteristic is their ability to confer meaning upon their environment. Certain entities to be found in the environment of the actors find "lines of entry" into these actors' lives, and the latter react to them according to how they interpret them. Sometimes, several actors are led into conflicts of interpretation, which might become actual social conflicts. In this case, an arbitrator helps finding a settlement—which is a mode of action opposite to that of the government in the preceding model, because the actors are the agency, not the passive objects, of causal forces. The sociologist might himself be an arbitrator, or might take part in the arbitration, because he knows how to identify the pertinent entities in the context. To this aim, he indeed might use numbers, but of a specific kind. Numbers count pertinent social entities or lines of entry, but not humans, and they are used as their identifiers. These two models of society both have a conception of actors, of the government, of the social role of sociology, and of quantification, but they organize these specific "actants" in an opposite manner (to use an expression from semiotics).[3]

Questioning the notion of the "variable" and "variable analysis", Blumer did not refer to the dichotomy between the quantitative and qualitative. He refrained from using the very vocabularies of enemies that he saw becoming powerful enough to weaken his own position, epistemologically as well as socially. He saw important shortcomings in the actual practice of survey analysis and experimental design, and argued that these

methods were limited to the restricted part of group life where interpretation is stabilized so that interaction *looked like* a determination—an
argument which was actually on par with C. Wright Mill's "massification"
and which remains very powerful today. Manifesting a "besieged mental-
ity" (Katz, in Emerson 2015), he fought against the pretention of the
pollsters to speak objectively about the world, and was scandalized by
the fact that so many opinion leaders would listen to pollsters, arguing
that they were in fact reducing everything to a false determinism. But his
enemy was not quantification in general, only its use by the Lazarsfeldians.

Ethnomethodology Between Accounts and Official Power

The fight against the Lazarsfeldians was not only in the hands of
the symbolic interactionists. Ethnomethodology, originally developed by
Harold Garfinkel in the mid-1950s, elaborated another criticism of statistics (and also of symbolic interaction), which bears on their political
consequences. Aaron Cicourel, a pillar of this strand of sociology, is
responsible for this.

The situation of the ethnomethodologists in the 1960s was completely
different, nearly contrary, to that of the symbolic interactionists. The
ethnomethodologists had no strong institutional base; they were only a
small group of young scholars not fully united, working mainly in California, and thus in universities much less powerful than Chicago or those
of the East Coast, and these scholars were striving to be recognized.
They had few allies, since symbolic interactionists varied in their opinion
towards ethnomethodology, from indifference for a strand of research
that they saw redundant to a respectful but fairly distant interest. Still,
their criticism of quantification had wide consequences and was very often
used by those identifying themselves as "qualitative sociologists" afterwards. As we will see, first through the study of the work of Garfinkel
and then that of Cicourel, their criticism did not oppose all and every
quantification.

As demonstrated by Heritage (1984), ethnomethodology was a reaction against Parsons' model of scientific action, and bore on Alfred
Schütz's phenomenological sociology. It is a comprehensive sociology
and, like symbolic interactionism, it converged strongly with the American
pragmatists. Ethnomethodology was interested in how actors theorize *by
themselves* their *Lebenswelt*, and in understanding how action is based on

mundane cognition. As Heritage (1984, p. 36) put it, ethnomethodology's "proposal to develop a 'generalized social system built solely from the analysis of experience structures' thus presented a direct attack on the very domain which Parsons had omitted from consideration: the realm of approximate judgments and reasonable grounds which constitutes the common sense world". One of the ways to know about society is obviously statistics, and thus ethnomethodologists did not take long to launch studies of this kind of object.

Statistical Accounts

In 1954, shortly after having completed his doctoral dissertation at Harvard, Harold Garfinkel had been hired by the sociology department at UCLA, and he began field work in UCLA's hospitals. Aiming to create a sociology of the way group members produce day-to-day knowledge and *account* for it, very early on he had the idea to study the production of hospital statistics *as a sociological object*. He coined the expression "rate producing process" as early as 1956, meaning the study of the process through which quantitative rates are produced. Cicourel acknowledged that "the conception of the 'rate-producing' processes as socially organized activities is taken from the work of Harold Garfinkel, and is primarily an application of what he terms the 'praxeological rule'" (Kitsuse & Cicourel, 1963, p. 132).

Expanding his questioning on the production of rates, Garfinkel focused mainly on three aspects of quantification (Heritage, 1984). First, following the work of cognitive psychologist Eleanor Rosch, he questioned the categorization performed by statistical coders (in the context of a psychiatric institution). He observed that coders, even when a set of rules is provided to them, tend to proceed independently of that rule, through "ad hoc" practices so that the code chosen fits best their understanding of the whole situation of the case at hand. Garfinkel coins this as "interpretative realism", by which he means that the coders treat the data as signifying the whole social order. This is a capital point for his demonstration that members do indeed have a theory of the macro level of society: they, too, are able to generalize. Second, Garfinkel became interested in the ways in which "aggregate responses to questionnaire items" were used, especially when they seemed contradictory. Once again, Garfinkel highlights the fact that "the questionnaire user has to bootstrap a way beyond the literal 'face value' of the response in order to see them as evidences of a whole social arrangement" (Heritage, 1984, p. 166).

Finally, Garfinkel addresses the problem of "official statistics". He points here to three levels of "anxiety" about their use. First, their insufficiency (the fact that they might lack enough information on the cases), second, the extent of the error, especially in sampling, that they may contain, and third, the limited adequacy of the definitions and procedures to the topic at hand (Garfinkel, 1967).

In these studies, Garfinkel is not "nihilistic", to use Heritage's (1984) phrasing. Garfinkel does not oppose quantification nor does he advocate "the abandonment of coding" but, on the contrary, he recognizes that "the unavoidable gap between data and its sense is unavoidably and irreversibly bridged, at least in part, by a coding process having unknown characteristics", which deserve to be inquired into by the sociologist (Heritage, 1984, p. 162). When aggregated responses are contradictory, Garfinkel is "insistent that he is not criticizing, ironizing, correcting" the data (Heritage, 1984, p. 167). Rather, he is looking for a way to understand what their properties and deeper meanings are. And the observation that official rates are "made out socially" leads him to think that "an immense array of accounting practices and their organizational exigencies, previously occluded from the view by the preoccupation with accuracy, are laid open as possible avenues of investigation" (Heritage, 1984, p. 175).

Garfinkel gave several examples in two chapters of his *Studies in Ethnomethodology* (Garfinkel, 1999 [1967]) of how he thought his analysis of statistics as a social object could be productive for the use of statistics as a cognitive tool; how his analysis could help in using quantitative tables. He also showed that studies would allow us to deepen our understanding of the social processes through which members produce knowledge about the society they live in:

> The actors' account – whether they take the form of questionnaire responses or of the statistical rates produced by bureaucratic agencies – cannot be unproblematically treated either as disembodied descriptions or as the 'relaxed' or 'loose' versions of objective states of affairs which can subsequently be tightened up by the judicious application of social scientific methodology. On the contrary, no matter how firmly such accounts are proposed [they] still await an analysis which situates them, with all their exigencies and considerations, within the socially organized worlds in which they participate as constituting and constituted elements. (Heritage, 1984, p. 178)

Thus, Garfinkel was interested in the *epistemic* consequences of his findings, but always remained suspicious about their political consequences. He argued for "ethnomethodological indifference", which meant for him that he did not want to make any judgement on whether "members" did say the truth or not. Yet, later, especially in the 1960s, this was interpreted by many readers as *political* indifference.

Ethnomethodology has had important consequences in American sociology, especially within conversation analysis. Douglas Maynard, in particular, at the University of Wisconsin, built on this research approach in analysing the conversations between interviewers and interviewees in surveys and polls in a very inspiring and consequential manner (Maynard et al., 2002). More generally, every scholar working in the field of sociology or history of quantification listed earlier in my introduction to this paper owes something to the seminal work of Garfinkel. And one unexpected consequence (to Garfinkel himself) of his work has been that it helped shape a very strong criticism against quantification itself.

Measurement by Fiat

In the beginning of the 1960s, Aaron Cicourel prolonged Garfinkel's argument about statistics into an actual criticism epitomized by the expression "measurement by fiat" (Cicourel, 1964, p. 12), even though, interestingly enough, he took this expression from a statistics handbook. The author of the latter explained that sometimes there was no scientific knowledge on a fact or characteristic to be measured. It was thus necessary to use an "arbitrary definition" of the fact, which led to a "measurement by fiat" (the name of a legally binding command or decision entered on the court record by the judge) (Torgerson, 1958). Cicourel turned this practical argument into a criticism. The quantities he had in mind were not survey data produced through questionnaires, but official statistics produced in the course of the bureaucratic treatment of public problems.

Among other things, he and his co-author John Kitsuse analysed official statistics on criminality and deviance. Together, they stressed that a difficulty arises "as a consequence of the failure to distinguish between the social conduct which produces a unit of behavior (the behavior-producing processes) and the organizational activity which produces a unit in the rate of deviant behavior (the rate producing process)" (Kitsuse & Cicourel, 1963, p. 132). Kitsuse and Cicourel highlight that actors, in daily life, account for some behaviours as being identical and others as being different. But there is no reason to believe that the categories used by the official administration engaged in the "rate producing process" respect necessarily those of the actors. On the contrary, "what such [official] statistics do reflect, however, are the specifically organizational contingencies which condition the application of specific statutes to actual conduct through the interpretations, decisions, and actions of law enforcement personnel" (Kitsuse & Cicourel, 1963, p. 137). Criminal categories are imposed, as if it were by fiat, by official institutions upon social life:

> In modern societies where bureaucratically organized agencies are increasingly invested with social control functions, the activities of such agencies are centrally important 'sources and contexts' which generate as well as maintain definitions of deviance and produce populations of deviants. Thus rates of deviance constructed by the use of statistics routinely issued by these agencies are social facts par excellence. (Kitsuse & Cicourel, 1963, p. 139)

The official rates are not a valid indication of everyday practice and the beliefs of members, but they are facts that have been isolated from the social setting they pretend to represent. Official statistics belong to the arsenal of control of bureaucracies. According to Kitsuse and Cicourel (1963), these pretend to aim for the *welfare* of the weakest elements of the population, but in fact they produce by fiat the population of deviant people, of the unemployed, of the poor, etc. Yet, despite this very powerful, critical conclusion, Cicourel still remained interested in the use of statistics.

The Quantitativist Cicourel
Cicourel did not reject quantification, but proposed a better use of statistics. He became interested in fertility in Argentina and, being well-trained in mathematics, launched research on the topic using a survey method— that is, an ad hoc questionnaire that he had written himself on the topic (Cicourel, 1974). The objective of this survey was to capture "the actor's theory and method of accounting for and producing his everyday social organization" related to fertility. Cicourel established a very cautious methodological procedure, in which respondents were interviewed several times successively, so that the interviewer could be either changed, if he or she did not fit to this precise family, or get acquainted with them, and fixed-choice questions were avoided as much as possible. "The type

of interviewing conducted was intended as an alternative strategy to the conventional survey" (Cicourel, 1974, p. 87). The aim was to take into account the interviewer–interviewee interaction and to capture the accounts of day-to-day action scenes as articulated by the interviewees. The survey would not impose its own categories onto the respondent, but adapt to the ones of the interviewee. The result is a book with lots of methodological statements, important analyses of direct observations and field notes taken during the interview, and a whole load of tables and charts analysed at length. Much later, in an interview, Cicourel made plain that he does not oppose quantification in general, but only certain methods of quantification: "I am not opposed to quantification or formalization or modeling, but I do not want to pursue quantitative methods that are not commensurate with the research phenomena addressed" (Witzel & Mey, 2004). Those who, like Cicourel, really grapple with quantification, do not reject it as a whole; they sort methods out.

Much later, Kitsuse wrote a presidential address to the Society for the Study of Social Problems (Kitsuse, 1980) that helps qualify the political consequences of Cicourel's epistemological position. Kitsuse had a very personal experience with the authoritarian tendencies that inhabit any state, and the American one in particular, since as a second-generation Japanese American, he was imprisoned in an American internment camp in 1942–1943. He shows that Cicourel remains in an epistemological scheme, first identified by Gouldner (1968), coherent to the Welfare State, in which sociologists attribute to deviants "a vulnerability and subordination to the moral authority of what is commonly characterized as white, middle-class, protestant culture and society" (Kitsuse, 1980, p. 6). This conception implies that the sociologist, like the state, sees the deviant as "the passive 'man-on-his-back' seemingly incapable of resisting or opposing the inexorable process of attribution of abnormality and inadequacy, stigmatized as morally defective, progressively excluded and subordinated as deviant" (Kitsuse, 1980, p. 7). The deviant remains essentially politically passive in his treatment by both the state and the sociologist. And I would add that this remains true even in the work of adapting categories proposed by Cicourel when he was working in Argentina.

The scandal inherent to the theory of "measurement by fiat" comes from the implicit presupposition that statistical categories do in fact succeed in formatting the deviant. The latter is supposed to have no effective means to fight back, bend the categories or destroy them. Due to the "official" nature of these statistical categories, they are supposed to have enough inherent power to indeed impose themselves. Opposing such a view, Kitsuse proposed that sociologists should notice that in the 1980s, it became clear that deviants were "coming out all over" to "publicly demand their rights to equal access to institutional resources" (Kitsuse, 1980, p. 3).

The actor's first feature, for Cicourel, was his ability to produce his own account of social reality, even of its macro-structure. Not only the sociologists have a conception of the whole social order, but anybody within society. The government, when it pretends to help or re-educate those that it calls "deviants", in fact produces the category, and subordinates those that are categorized, especially through epistemic tools such as official statistics. The role of the sociologist is to unearth the accounts of the deviants, to help make their worldview visible and respectable. Thus, on the one hand, he criticizes official statistics imposed on the existence of "deviants", and on the other, he produces, among many methods, his own quantitative methodology provided that it remains commensurate to the research phenomena.

Blumer's and the ethnomethodologists' criticisms represent the two main strands of critique addressed to quantification by those who would later on be associated with "qualitative research".[4] As we have seen, these critiques take place in the wider context of developing theories of society, accounting for the characteristics and agency of social actors, of the government, and of the sociologist. They also comprise a definition of the good and bad uses of quantification. Thus the criticisms are addressed in fact to specific methods of quantification and are complemented by alternative quantitative practices. Within sociology, these two sets of criticisms were emitted from two completely opposite positions in terms of audience and power. Symbolic interactionists were initially dominant, and tried to prevent being overwhelmed by the new quantitativist contender; ethnomethodologists were on the contrary minuscule and fought a battle as bravely as they could, surfing on the recognition they were enjoying.

The sociologists in question were aware that they were not entirely condemning quantification, but only certain methods, as can be inferred from the fact that, in the 1950s and well into the 1960s, they did not use the dichotomy "quantitative vs. qualitative" sociology. It belonged to the very heralds of surveys, led by Lazarsfeld, who crafted the label "qualitative research" as a propaedeutic to quantitative analysis. Therefore, the next question that we have to answer is why and how symbolic interactionism and ethnomethodology finally ended up being considered "qualitative". Why is it that this label took consistency, when it initially belonged to the enemy? The first step to answer this question is to take into account the appearance of a new actor on the sociological scene, the coming-of-age "young radical", during the 1960s.

Radical Sociology, Quantification and the Welfare State

The beginning of the emergence of the 1970's spirit of radicalism within American sociology can be backdated quite precisely, to the 1968 annual meeting of the ASA, the cornerstone of sociological orthodoxy. President Philipp Hauser, who had co-written with Blumer *Movies, Delinquency, and Crime* (Blumer & Hauser, 1970 [1933]), which involved quantitative materials, and who was later appointed Director of the Bureau of the Census—and thus one of the, if not *the*, most prominent figure in the use of statistics in sociology—had invited Wilbur J. Cohen, Secretary of Health, Education, and Welfare, to give the keynote presentation at this conference. This invitation demonstrates the strong association that existed at the time between the quantitativists, who held top positions within the sociological academic world, and the political elite of the American Welfare State. The invitation provoked a fierce opposition from young sociology students who called themselves "radicals". They were:

> as rejecting of those who purvey sociological research on underdogs to the overseers of the welfare state as they are of caterers to the warfare state. To the Sociology Liberation Front, Cohen's 'guest of honor' status was an unacceptable example of what Gouldner (1968) has called the 'blind or unexamined alliance between sociologists and the upper bureaucracy of the welfare state'. (Roach, 1970, p. 228)

The meeting ended up in a mess and gave rise to a schism within the professional organization of the sociological field (Roach, 1970). This emerging radicalism in sociology was under a paradoxical influence. On the one hand, the new "radicals" were deeply influenced by the simple desire to reject the templates of the past: family, state interventionism, sobriety, war. This rejection is well embodied by Abbie Hoffman's book *Revolution for the Hell of It* (Hoffman, 1968), which does not propose much, except the joy and amusement of destroying everything from previous generations, including, as far as this paper is concerned, the University system. It was also associated with a fierce opposition to any alliance with the institutions of the Welfare State. This is also exemplified by the fact that Howard Becker was wearing a T-shirt at the ASA annual meeting depicting an unkempt, hairy, cartoon hippy saying "Hey Kids, Let's Fuck the State"; an ironic proposition mixing destruction and fun (discussion with Jack Katz).

On the other hand, sociology was a discipline that could provide intellectual tools to understand the system, its injustice and boredom, and thus help either fix it or destroy it. Since institutions and "the system" were identified as the problem, sociology seemed to be a straightforward answer to it. Thus, sociology was at the time attracting a large number of new students eager to change society *through sociology* (Turner & Turner, 1990).

So, in this conflict, how was quantification seen? How was "qualitative sociology" transformed in this turmoil? Behind the widespread non-articulated contempt and suspicion towards quantification (called "oversimplification" by the heralds of quantification Reitman, 1978), there were in fact two quite different strands of argument. The first one built on the post-Marxist tradition of the Frankfurt School, and here sociologists tended to be influenced more by Cicourel's arguments. The other strand was more "Blumerian", and stood thus more in the tradition of the American pragmatists.

Are Quantities Fascist?

One immediate consequence—next to the creation of the highly influential journal *Social Problems*—of the radical sociologists' actions was the foundation of *The Insurgent Sociologist* in 1969 (which later would become the journal *Critical Sociology* from 1988 onwards). Influenced by C. Wright Mills, neo-Marxism, and radical feminism, the initial goals of the journal were to organize the actions of the different activist groups, among other things, to ease communication between the Western Union and the Eastern Union of Radical Sociologists, and the Sociology Liberation Front, and to help define what radical sociology should look like. The first issues of the journal looked like street pamphlets with very short, explosive papers, unsigned, and full of images and caricatures. In contrast to previous generations, what was exhibited here was a completely different style of sociology. One cartoon ironizing the use of figures has been reproduced below (see Fig. 13.1). This caricature was published in the second issue of the journal.

The "Mo-Jan" system depicted in Fig. 13.1 stands for *Mo*rris *Jan*owitz, one of the founders of military sociology. One can see that in the text published next to the image of the rocket a certain positivist and quantified tone in sociology ("82.5%") is mocked. The reference to the rocket and "Camelot special" ironizes the role of the army in financing research (as we will see below), and finally Janowitz's book proposing an "urban control of racialized riots" with the help of the disciplinary tools of sociology are at the heart of the students' exasperation.

Soon thereafter, the *Insurgent Sociologist* published a paper entitled "Accidents, Scandals and Routines: Resources for Insurgent Methodology" (Molotch & Lester, 1973) addressing the role of quantification. Examining the news from a Garfinkelian perspective, Harvey Molotch and Marilyn Lester argued that ethnomethodology provided methods to suspend the belief that an objective world exists. They showed that the news content of the mass media is the "result of practical, purposive, and creative activities on the part of news promoters, news assemblers and news consumers" (Molotch & Lester, 1974, p. 101). Noticeably, the proposition that statistics measured reality "by fiat" played a key role in their argumentation. As they wrote: "Cicourel (1964) makes an analogous argument with respect to the creation of a juvenile delinquent" (Molotch & Lester, 1974, p. 103). Ethnomethodology was used by the authors as a tool to criticize not simply a fabricated reality, but a *politically biased* fabricated reality. According to Molotch and Lester, ethnomethodology helped to avoid "be[ing] duped into accepting as reality the political work by which events are constituted. Only by accident and scandal is that political work transcended, allowing access to 'other' information" (Molotch & Lester, 1973, p. 10). The politicization of quantities highlighted by Cicourel was thus ushered into a general criticism of a reality fabricated by the ruling elite.

Interestingly enough, soon afterwards, the same journal published a paper entitled "The New Conservatives: Ethnomethodologists, Phenomenologists and Symbolic Interactionists" which was influenced by neo-Marxism. Here, among other things, it was argued that the approaches at stake—especially ethnomethodology—are inherently conservative, and therefore not radical, for two reasons. First, they "implicitly deny the generalizability of any theory of social change", thus are opposed to the notion of revolution. Second, they "picture men

Fig. 13.1 "Breaking out of the Hothouse" (*Source Insurgent Sociologist*, Vol. 1, No. 2, p. 8. Reprinted with permission from *Critical Sociology*. Scan gratefully provided by the University of Michigan Library [Special Collections Research Center])

as individual entrepreneurs, and use the language of the market-place in extending laissez-faire individualism to contemporary social theory" (McNall & Johnson, 1975, p. 49). Both these features are associated with the tendency to mainly use data about *individual cases* and, the authors regret, very rarely "samples and replicable measurement techniques" (McNall & Johnson, 1975, p. 62). Radicalism was definitely still the object of a conflict of definition from within, as much as the roles of statistics in it.

But neo-Marxism was not entirely opposed to ethnomethodology. David J. Sternberg (1977) proposed a radical rereading—as the title of his book attests—of the concept of measurement by fiat. He deals with the famous F-scale invented by Theodor Adorno and others (Adorno et al., 1950). Given the huge impact of this work, it is important to explain its role within the question of quantification. Adorno discovered the practice of statistics when he first reached the USA in 1938 and—through Horkheimer—worked under Lazarsfeld at Columbia. He hated the experience. As he wrote, "I collided with the positivistic habits of thought" (Adorno, 1998, p. 220). But later, in the 1940s, after having settled in California, he began to work on a project that would eventually lead to the book about *The Authoritarian Personality* (Adorno et al., 1950). The book achieved a successful conjoining of Marxism and Freudism in trying to identify the psychological roots of Nazism. Besides, it was a methodological rarity, since it made large use of statistical surveys, and of the conceptual pair qualitative vs. quantitative (the expression "qualitative analysis" is in the title of the 4th part of the book). This time, Adorno deeply enjoyed the experience. He loved the atmosphere in which he worked: "the kind of cooperation in a democratic spirit that does not get mired in formalities […] was for me probably the most fruitful thing I encountered in America" (Adorno, 1998, p. 232).

Likewise, he praised the scientific achievement of the research, particularly because the "teamwork spirit" made possible an intelligent use of statistics: "The aporia – that what was discovered purely by quantitative means seldom reaches the genetic deep mechanisms, while qualitative discoveries can just as easily lose their generalizability and therefore also their objective sociological validity – we tried to overcome" (Adorno, 1998, pp. 232–233). In particular, the F-scale, a tool measuring the individual propensity to authoritarianism, was invented in Berkeley in a "free and relaxed environment […] in a manner that by no means coincided with the usual image of the positivism of the social sciences" (Adorno,

1998, p. 233). Afterwards, Adorno became generally suspicious towards empiricism and never used statistics again. But he nonetheless did not publish any general argument against quantification, most probably to stay true to this happy experiment (Genel, 2013, p. 91). And, in the 1970s, he left the U.S. in an *ambiguous* overall stance towards statistics.

Sternberg, in his book (Sternberg, 1977), gave an example of how the positions of the Frankfurt School could be radicalized. Discussing Adorno's F-scale and Cicourel's measurement by fiat, he argued that the F-scale was used widely by many American official bodies of administration, and he concluded: "the F scale has to do with fascism all right, but not in the sense its designers intended it. *Its* findings, *not* the people that it finds, are Fascist" (Sternberg, 1977, p. 43). Sternberg pushes Cicourel's argument to the point of arguing that statistics as a whole, even Adorno's F-scale, are fascist, insofar as they impose categories of social control upon society.

Sternberg is a good example of how the criticism of quantification made by ethnomethodology was radicalized by many scholars of the New Left, associating surveys with state authority and concluding that they are therefore fascist—even when discussing Adorno's work, to whom such a qualifier must have seemed quite strange! But the reception of Steinberg's book was far from laudatory. Reviewers qualified Sternberg's book as involving a "simplistic approach" (for example Reitman, 1978), and the overall judgement of this book and those alike was that such an inference could not be taken seriously. It was stepping outside the range of the sociologically admissible. It was definitely hard to call Adorno a fascist! And, indeed, it must be said that Sternberg did not make a career in the discipline of sociology. Rejecting all quantities as fascist did not hold. Symbolic interactionists, for their part, constructed another argument about figures, to which we will now turn.

Light Travelling: Numbers as Gleanings

Howard Becker and Louis Horowitz can be taken as representatives of the interactionist trend in radical sociology. Becker, directly influenced by both Blumer and ethnomethodology, had crafted "labelling theory" which shifted the focus from the causes of peoples' deviant behaviour to the definition of people and behaviour as deviant. In 1972, the *American Journal of Sociology* organized a remarkable symposium entitled "Varieties of Political Expression in Sociology", which was published as a special issue in June that year. The collection does not comprise any explicitly critical or Marxist sociologists, but papers, such as the article by Merton (1972), or the paper by Lipset and Ladd (1972) which presents an analysis of data from a comprehensive survey of 60,000 academics to explore "the actual political views of sociologists" (Lipset & Ladd, 1972, p. 68), and many other fascinating contributions. Also, Becker and Horowitz were invited to this symposium and took side with radicalism: "Both because of our own political position and for the sake of congruence with current discussion, we will take the tack of sociologists who conceive themselves, or like to conceive themselves, as radical sociologists" (Becker & Horowitz, 1972, p. 59). Their argument makes perfectly clear how they see the link between statistical methods and politics.

In their contribution to the symposium (Becker & Horowitz, 1972), they began by claiming that radical sociology can be good sociology. They define the latter as being "true to the world", especially when it analyses the causes of events, even in the most limited sense of the term "cause". Especially, and most important for our purpose, they insisted that, in principle, all the known methods of the discipline can be useful: "With all their faults, interviews, participant observation, questionnaires, surveys, censuses, statistical analysis, and controlled experiments can be used to arrive at approximate truth" (Becker & Horowitz, 1972, p. 50). It has to be said that in his whole career, Becker never expressed rejection of quantification. In a collection of methodological papers of his, he noticed that during fieldwork observation, "the observer will also find it useful to collect documents and statistics (minutes of meetings, annual reports, budgets, newspaper clipping) generated by the community or organization" (Becker, 1970, p. 79). Thus, like the ethnomethodologists, he insists that the quantities found in the field are interesting objects of study. And later, he highlighted that between "qualitative and quantitative" methods "the similarities are at least as, and probably more, important and relevant than the differences. [...] The same epistemological arguments underlie and provide a warrant for both" (Becker, 1996, p. 53; but see also Becker, 1958).

Thus, the specificity of radical sociology does *not* lie in its methods. It lies in its "distinctive contribution to the struggle for change" (Becker, 1996, p. 53), as on the one hand it provides the knowledge to critique inequality and lack of freedom, and on the other hand it provides the basis for implementing radical utopias. As Becker (1996, p. 53) put it, "the constructive aspects are rooted in the positivist tradition, and the critical aspects in the Marxist tradition".

One of the core concerns in the struggle for change is the attribution of causes to the events. All events have an infinity of causes, beginning with the presence of air that allows the humans to breathe. Thus "the assignment of causes to events has a political aspect", because "when sociologists link a cause to an event or a state of affairs, they at the same time assign blame for it" (Becker, 1996, p. 58). It is the specific causes chosen by the sociologist that make him radical. As Becker writes:

> In general, radicals will judge a sociological analysis as radical when its assignment of causes, and thus of blame, coincides with the preferred demonology of the political group making the judgment. (Becker, 1996, p. 59)

For the radicals, a shocking example of conservative attribution of causes was what came to be known as the "Moynihan Report". In 1965, Daniel Moynihan, then Assistant Secretary of Labour in the U.S., issued a report entitled *The Negro Family: The Case for National Action*. It was an entirely statistical report dedicated to understanding the causes of poverty in black families. In fact, the report attributed poverty to the disorganization of the black families themselves (Rainwater & Yancey, 1967). This argument provoked a huge intellectual controversy, because implicitly it was *Blaming the Victim* (Ryan, 1971). Radicals (and others) were shocked that such an important representative of the Welfare State could produce arguments that neglected so obviously the oppression exerted by white people on black people, and that a self-described "liberal" could engage in such a conservative political assault.

Having such a counter example did not help the radical in identifying the pertinent causes of any social process, those causes that are at the same time true to the world and belong to radical demonology. Becker & Horowitz (1972) argue that there are three "obstacles" to a radical sociology, three specific elements that oppose the pursuit of its objectives. These are:

> (1) The conservative influence of conventional technical procedure, (2) Commonsense standards of credibility of explanations, and (3) The influence of agency sponsorship. (Becker & Horowitz, 1972, p. 62)

Let us review each of these shortcomings in turn. (1) Research means testing the deductions made from existing theories on data suitable for making such a test (cf. the controversy between Blumer and Zentner mentioned above). This is done through a method, statistical or not, that restricts the range of causes to be tested to what the researcher had in mind when he conceived his research. As Becker and Horowitz write:

> But some techniques, indeed, require sociologists to leave out things they *know* might be important. Thus, it is difficult, though not altogether impossible, to study certain kinds of power relationships and many kinds of historical changes by the use of survey research techniques. (Becker & Horowitz, 1972, p. 62)

Becker and Horowitz do not get more explicit. But knowing their proximity to Blumer, it seems clear that the elements that sociologists *know* that might be important for the attribution of causality are linked to the interpretation process that Blumer highlighted. Here, Becker and Horowitz reuse Blumer's argument about surveys—including Blumer's precautions and lack of radical condemnation.

(2) Sociologists, similar to other members of society, tend to believe more in the versions of the elite than those of other people, because the elite runs the organizations. That is, they tend to believe "official versions and analyses of most social problems", and thus they "find it hard to free [themselves] from official analysis, sufficiently to consider causes not credited in those versions" (Becker & Horowitz, 1972, p. 63). Becker and Horowitz here refer not only to Blumer but also to Cicourel's argument about the performing effect of official statistics, producing the causes of social problems. We believe official statistics, because their "version" is that of the elite.

(3) Finally, agency sponsorship might put conservative limits on a radical search of causes. It is not necessarily the case that they are politically biased, but when they fund research it is to solve an operational difficulty, so that they, too, limit the range of the answers that are worth giving. In particular, they tend not to see their own operations as being the cause of the problem. Although Becker and Horowitz do not discuss this directly, but one of the main "organizations" at stake here was the military itself. Even though the 1960s were the decade when the Army began to lose its near exclusivity in financing public research, it remained the main finance provider (Moore, 2008, p. 34). Again, the authors refer

to a contribution of Blumer published in a book edited by Horowitz entitled *The Rise and Fall of Project Camelot* (Horowitz, 1974) about the amazing story of a project financed by the army to use social sciences in the goal of predicting (and thus controlling) revolutionary upsurge in South America (Camelot is mentioned on the caricature Mo-Jan system, right above the *Mad* face on the rocket).

Thus, "the remedy for that is to travel light, to avoid acquiring the obligations and inclinations that make large scale funds necessary" (Becker & Horowitz, 1972, p. 64). It is obvious that, here, to "travel light", that is without the money of the Army, is also to renounce the surveys that were among the most expensive research techniques of the times. But it is not against *any* quantities, on the contrary. As stated above, collecting figures on the field or using any available figures is not shocking to them at all. In this, the 1970s radicals act towards numbers as gleaners towards ears of corns abandoned in the field. They are not cultivated; they are simply used when found here and there. Radical figures are gleanings.

This argument made a much bigger splash than the other one about the fascist character of quantification. For example, Alvin Gouldner, who was himself a core figure of radical sociology, especially since the publication of his *The Coming Crisis of Western Sociology* (Gouldner, 1970) acclaimed both authors in a later comment of the special issue, writing that "their effort to characterize radical sociology is one of the more probing I have seen" (Gouldner, 1973, p. 1079). Articulating general arguments against quantification did lead to contradiction or unrealism. So, the best was simply to ignore them, or maybe be ironic or sarcastic about them.

In conclusion, radicalism changed the relationship of sociology towards quantification. On the one hand, there was indeed a definitive condemnation of any use of quantities, bearing on Cicourel's "measurement by fiat" argument and expanding it to the point of calling "fascist" any process of quantification. This was a radical rejection of quantification, but it was paid for by an expulsion from the sociological academic field. On the other hand, Blumer's heirs, represented here by Becker and Horowitz, built the "travelling light" argument. For them statistics and quantification can be useful, and often are, both when produced by the researcher and when collected in the field. But most of the time statistics and quantification force the researcher to cope with the "demons" of power (the Welfare State, the Army, large companies), because they require a large infrastructure and funds. Thus, the safest, for radicals who did not want to compromise with these demons, was simply not to use such method-ological tools. The argument wound up not exactly *against* quantification but only *without* surveys. Even though not really to the taste of the most powerful sociologists, this one could still be swallowed by the academic field.

One question that remains is that if the radical sociologists wanted to stay away from the liberals in charge of most of the power institutions, for whom was their knowledge produced? As Jack Katz has argued (Katz, 2015), the public of the radical sociologists was the youth that at this time that was flowing in the universities, and especially in the sociology departments (Turner & Turner, 1990). Radical sociology was oriented towards the students—and professors who saw themselves primarily as teachers. The actual institutionalization of a "qualitative sociology", that as we have seen was seldom mentioned before the 1970s, came out of this movement.

Institutionalization of a "Qualitative Sociology"

We have described the criticisms which had been expressed towards quan-tification by sociologists. Their arguments were defensive, against the wave of quantities that washed over their discipline. But, from the 1970s onwards, the strategy of those opposed to surveys changed: they began to make the category fit to their own work. We will see that they would address themselves to the large number of students that were flocking to sociology departments by publishing textbooks and the creation of a new journal. Finally, we will use the Jstor database to measure the success of the enterprise.

Common Ground

One of the very first books to use the word "qualitative" on the cover, actually in the subtitle, was Glaser & Strauss's, 1967 *The Discovery of Grounded Theory, Strategies for Qualitative Research* (Glaser & Strauss, 1967) which immediately received a lot of attention and success world-wide. The subtitle would have the public think that it would be a fierce engagement against quantitative analysis. But actually, those who read it discovered that this was not the case. The book performed splendidly as a classic, albeit difficult, rhetorical tour de force: it consolidated the divide between the categories of "qualitative" and "quantitative", but only to show simultaneously the authors' exceptional ability to overcome it. The authors dug a ditch, so that everyone could see how well they were able to jump over it.

Indeed, contrary to Blumer and the ethnomethodologists, Glaser and Strauss *accepted* Stouffer and Lazarsfeld's reading of the development of sociology. They accepted the dichotomy between qualitative and quantitative and observed that, since the 1930s, quantitative research "swept over American sociology" because quantitative methods had developed "systematic canons and rules of evidence on such issues as sampling, coding, reliability, validity" etc. which were much more "rigorous" than the equivalent canons used by empirical qualitativists remaining "too impressionistic". And thus, "qualitative research was to provide quantitative research with a few substantive categories and hypotheses" (Glaser & Strauss, 1967, pp. 15–16). Qualitative sociology had come to be dominated.

But they also argued that the fundamental function of sociology was the discovery of theories based on data. Their book was supposed to be a handbook for *abstraction*, and thus an attack against those logico-deductive theorists who promoted the *verification* of theories through quantitative data. The authors called this opposition *"generation vs. verification"* of theory (Glaser & Strauss, 1967, p. 12) and proposed "strategies" to perform the former. In this, they were once again very close to the American pragmatist tradition and indeed referred often to C. Wright Mills and Blumer. In particular, they worked on the categories established by the former and they opposed the kind of sociology that C. Wright Mills had baptized "Grand Theory" (Glaser & Strauss, 1967, p. 10), meaning a theory severed from any empirical ground. There is obviously a pun between "ground" and "grand" theory.

But they also argued that, although there had been a historical connection between the quantitative and verification theories, this connection was only contingent. There was no epistemological necessity to it. On the contrary, abstraction could be performed on both kinds of data, qualitative or quantitative:

> Our position in this book is as follows: there is no fundamental clash between the purposes and capacities of qualitative and quantitative methods or data. What clash there is concerns the primacy of emphasis on verification of generation of theory – to which heated discussions on qualitative versus quantitative data have been linked historically. We believe that

each form of data is useful for both verification and generation of theory, whatever the primacy of emphasis. [...] In many instances, both forms of data are necessary. (Glaser & Strauss, 1967, p. 17)

According to Glaser and Strauss, both qualitative and quantitative data constituted a common ground on the basis of which theories could be built. With both, the researcher had to use or establish sampling methods, move from substantive to formal theory, and proceed to comparisons among sets of data. The main difference was simply that when using quantitative data for the development of theory, the researcher had to "relax the usual rigor of quantitative analysis so as to facilitate the generation of theory" (Glaser & Strauss, 1967, p. 187). She had to simply use "freedom and flexibility" with her data (Glaser & Strauss, 1967, p. 186). Any kind of data could thus support the generation of theories, if wisely utilized. But, if theory building could be achieved with both, why did they then nevertheless insist on qualitative research? They argued the following:

We focus on qualitative data for a number of other reasons: because the crucial elements of sociological theory are often found best with a qualitative method, that is from data on structural conditions, consequences, deviances, norms, processes, patterns, and systems; because qualitative research is more often than not, the end product of research within a substantive area beyond which few research sociologists are motivated to move; and because qualitative research is often the most "adequate" and "efficient" way to obtain the type of information required and to contend with the difficulties of an empirical situation. (Glaser & Strauss, 1967, p. 18)

The argument amounts finally to a question of different emphasis, not of opposition between the two. Glaser and Strauss accepted a dichotomy between qualitative and quantitative research where the latter was constructed to be dominating the former. Quantitative sociology was supposed to be more "scientific", more "rigorous", more "accomplished" than qualitative sociology. Yet, Glaser and Strauss reversed the stigma (to use the title of one of Goffman's books) highlighting qualitative research's particular suitability for the generation of theory from data. Through their work, qualitative became "better", even though "relaxed quantities" could do a comparable job.

Glaser and Strauss's book enjoyed an impressive success and participated importantly in establishing methodological guidelines that would

Table 13.1
Reproduced from
Schwartz and Jacobs
(1979, p. 5)

Data	Goals of sociology	
	Positive science	*Actor's point of view*
Use of numbers Use of natural language		

"travel light", that is guidelines that would not involve quantification and yet at the same time be considered scientific. The book was followed by a series of other methodological books on qualitative methods that would give the same argument. The very first book published under the title *Qualitative Sociology* (Schwartz & Jacobs, 1979) is particularly striking. It presents symbolic interactionism and ethnomethodology, next to quantitative research, in a fourfold empty table (Schwartz & Jacobs, 1979, p. 5) (see also Table 13.1):

But the authors don't explore the table at all. They are interested only in data based on natural language and the actors' point of view, i.e. the right hand lower cell. They don't discuss any of the other categories, or try to fill the cells out. This strategy is the one that would generally be adopted by the many textbooks on qualitative sociology that would be published in the 1970s, such as Filstead (1970), Lofland (1971), Bogdan and Taylor (1975) or, later, Taylor and Bogdan (1984).

Another academic innovation important for the institutionalization of "qualitative sociology" was the creation of the eponymous journal *Qualitative Sociology*. The first issue came out in May 1978. In this issue, the journal's title is neither explained nor justified. It is only stated on page 2, along with the list of editorial board members, that the journal is dedicated to "qualitative interpretation of social life" and that "manuscripts dealing with the qualitative analysis of social life" are welcomed. Noticeable is a letter to the editor where the author expresses his happiness to witness the birth of the journal, because he feels "disenchanted with indiscriminate number-crunching and the attending tendency for the process to become an end in itself". Nonetheless, "the editors discussed their own reaction to this letter and concluded that in fact do not see [their] project as an attack on quantitative sociology" (*Qualitative Sociology*, 1978, Vol. 1, No. 1, p. 163).

References to the label "qualitative sociology" by those who ignored, or sought to oppose, quantification became important first and foremost in sociological *textbooks*. The label was addressed primarily to the young students flocking the university. It was intended to hawk the good word to students and help newly hired undergraduates. It also created a legitimate spot in a department curriculum and provided positions for professors entering the job market.

A Bipolar Category

To measure the students and professors' role in institutionalizing "qualitative sociology", we will now use easy quantitative methods ourselves, since, as Gabriel Tarde has argued, they can help us follow the "imitation trends" of an innovation (Didier, 2010). Once the "tribe" of "qualitative sociology" was knotted together, we might ask who got interested in it and reused the label. Here, statistics are not used, as they often are, to set up the "context" of a social event, but on the contrary to follow the social effects of this event. To this end, JStor helps us conveniently. The interface "Data for Research" makes it fairly easy to track quantitatively the use of any expression (association of words) in JStor's entire database.[5]

My searches resulted in the following. The word "qualitative", as far as it is related to the words "research", "method" or "sociology", takes off right after the war. "Qualitative research" and "qualitative method" raised much faster and higher than "qualitative sociology", which actually began to rise later, in the 1950s (see Fig. 13.2).

But sociology was not the only discipline experiencing a consolidation of the dichotomy between quantitative and qualitative. A search by disciplines shows that social work, on the one hand, and several biology specialties (such as developmental and ecology), on the other, are among the most important ones driving the results for "qualitative research" and "qualitative method" presented in Fig. 13.2.

Now, let's zoom in to study "qualitative sociology" itself. I excluded publications before WWII, when they were mainly noise, and I cut off my search after 1985, when many of the actors had changed, and the publication rate had generally grown and results were hence no longer as informative. This being done, it appears that the 20 authors that used most often the expression "qualitative sociology" (names are followed by the number of articles using these words) were not only those who belonged to the "tribe" as defined by *Qualitative Sociology* (see Table

13.1). On the contrary, many among them (Alexander, Duncan, Blau, Goodman) (see Table 13.2) were scholars who were famous for their *use* of statistics. In fact, it appears that Lazarsfeld's definition of qualitative sociology had been as powerful as his advances in qualitative research, so that the quantitativists participated themselves in establishing a second school of qualitative research, as defined initially by their famous predecessor—and thus obviously making also massive use of quantities.

What were the topics addressed by these qualitativists? The distribution of the keywords of the papers allows the hypothesis that those who used the expression "qualitative sociology" did so in two different contexts. Table 13.3 below shows the first cluster of keywords, used in 800 to 1500 papers, which are words associated with the "abstracted empiricism" kind of sociology: variable, model, population, per cent, table, class.

Used in only 380 to 400 papers, we find a different semantic group comprising: member, field, pattern, and person (see Table 13.4). The fact that the amount of papers using this second set of words is so different from the amount using the first set of words leads us to think that they

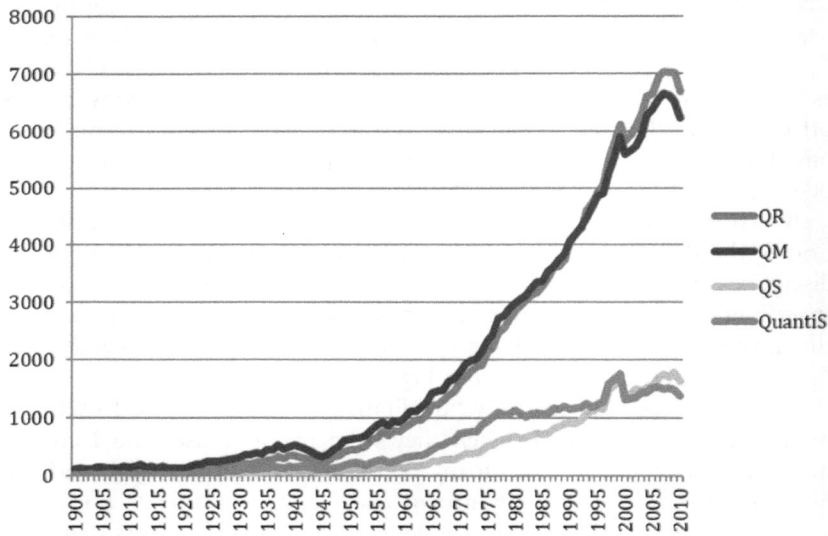

Fig. 13.2 Yearly distribution of the expressions "qualitative research", "qualitative method", "qualitative sociology" and "quantitative sociology" in the Jstor full database

Table 13.2 Name of author and number of their articles containing the expression "qualitative sociology", 1945–1985

Karl L. Alexander	16	John Hagan	12
Kenneth C. Land	16	Aaron M. Pallas	11
Otis Dudley Duncan	16	Glendon Schubert	11
Peter M. Blau	16	J. David Singer	11
Helen M. Robinson	15	James S. Coleman	11
Leo A. Goodman	15	Michel Vale	11
Helen K. Smith	13	Peter H. Rossi	11
Seymour Martin Lipset	13	Samuel Weintraub	11
Charles Tilly	12	David Knoke	10
David Riesman	12	David Snyder	10

Source JStor database, author's own compilation

Table 13.3 Twenty most used keywords of papers using the expression "qualitative sociology" and their number of appearance, 1945–1985

Variable	1559	Population	971
Theory	1353	Percent	967
Study	1305	Table	958
System	1207	School	951
Student	1154	Problem	937
Model	1136	Class	918
Behaviour	1105	Education	907
Political	1105	Change	892
Science	1078	Child	837
Analysis	1010	Family	788

Source JStor database, author's own compilation

Table 13.4 Twenty successive keywords beginning at the row 75 associated with the expression "qualitative sociology", 1945–1985

Member	446	Pattern	404
Unite	439	Empirical	401
Organizational	429	Activity	399
Field	427	Approach	398
Specie	426	Person	396
Rural	419	Power	394
Hypothesis	407	Interaction	392
Politics	407	Teaching	390
Historian	405	College	389
Number	405	Occupational	389

Source JStor database, author's own compilation

might also be related to two different sets of papers. If this hypothesis were true, then the data would prove also that the "interpretative" tribe remained less productive—probably because they were much smaller in number.

These data about the papers listed in the JStor database let us lead to think that the label "qualitative sociology" did indeed take shape consequent to the conceptual innovations that we have described. But "qualitative sociology" is a category where two sets of "good examples" of papers are in opposition. On the one hand, there are those which belong to the "Lazarsfeldian" cluster where qualitative and quantitative are in a hierarchy. On the other hand, there is a set of papers pertaining to an "interpretative" definition of the qualitative influenced by Blumer, Cicourel and their intellectual descendants which seeks to set itself apart from such "quantitativist" uses of qualitative information.

CONCLUSION

The story of "qualitative sociology" begins right after WWII in a paradox. It was imported from German-speaking Europe, defined and used, first and foremost, not by opponents to quantitative methods, but, on the contrary, by Lazarsfeld in an inherent—but dominated—relation to quantitative analysis. Qualitative analysis was for him a propaedeutic to quantitative sociology or the use of "qualitative" statistical variables. At the same time, many sociologists started to oppose the apparently unstoppable rise of polls and survey analysis and expressed strong arguments against these methods. Blumer raised the problem of the neglect of members' ability to interpret situations, and Cicourel furbished the measurement by fiat argument. These arguments were reused and pushed to their limit a decade later by the "radical sociologists" in their denunciation of the drawbacks of the Welfare State, seen as closely associated to quantitative surveys. But these sociologists did not explicitly ban quantification, they simply did without surveys, they "travelled light". And it was only at the very end of the 1960s that the category "qualitative sociology" became institutionalized, especially through textbooks and curricula.

What's more, it should be remembered that the sociologists studied here criticized *a method* of quantification, not the general use of quantities. The social spheres that pretended to be free from numbers had in fact been *purified only* from *a certain method of* quantification. De-quantification is the result of an activity aiming to suppress certain kinds of quantities. Society is everywhere quantitative, only some spheres have banished certain methods (of quantification). All sociologists referred to here, still used quantities in one way or another. It is thus apparently not possible (nor desirable) to completely wipe out quantities from an epistemic system. All that actors have been doing is rearranging quantities, reorganizing them with or without one another, reshaping their relations in new and innovative fashions. But they never completely quantified nor qualified society; rather, they rearranged the quantities that they found already within.

In this respect, in a 1984 special issue of *Qualitative Sociology* entitled "Computer and qualitative data", the editors insisted that "large mainframe computers" had changed sociology since 1946, but that they were expensive and owned by third parties who could control and influence the research (Conrad & Reinharz, 1984). According to the editors, since the war, computers had been in the hands of either the (Welfare) State or big (capitalist) companies. But they also remarked that very recently, microcomputers had appeared and had become so cheap that every single researcher could now have his or her own. Thus they raised a new question: "How can the personal computer aid that group of sociologists who do not rely on mathematical analysis of data but who search their data for patterns and meanings"? (Conrad & Reinharz, 1984, p. 4). Stated differently, microcomputers are the material tool of knowledge making compatible to "travelling light", and at the same time they are dealing with something close to mathematical analysis.

Contemporary radical sociologists might notice that the conditions that justified the rejection of quantitative reasoning in the 1970s have nowadays lost their relevance. Today, the baby boomers are old, the Welfare State is weak, and everybody has a personal computer and an internet provider through which one can access a number of fascinating databases. A wealth of new methods independent from those "demonized" by the radicals in the 1970s is available. At this point, it seems to me that the dichotomy qualitative/quantitative barely has teeth anymore and could diligently be forgotten. As we have argued elsewhere, today, radical sociologists should all be also "statactivists" (Bruno et al., 2014).

Acknowledgements I would like to thank warmly those who read this paper in advance and whose comments were extremely helpful: Stefan Bargheer, Zachary Griffen, Jack Katz, Nadine Levin, Andrea Mennicken, Robert Salais, the UCLA Sociology Theory Working Group and Epidapo Seminar.

NOTES

1. The opposition to quantification was in the 1930s the feature of a conservative ethos, criticizing standardization, state centralization and progressivism, associated to numbers (Boltanski, 2014). It became clearly progressive after WWII.
2. There were debates about quantification in other disciplines, especially in anthropology, but here we will concentrate only on sociology.
3. It is important to keep in mind that at the time Blumer was losing ground in sociology on two sides. On the one hand, he was much less empiricist than the Lazarsfeldians. He was proposing philosophical-like arguments against the data used by the quantitative researchers. Nonetheless, empiricism was then, indeed, exciting. And, on the other hand, he was also missing important innovations in philosophy itself—especially the developments of phenomenology showing that individuals are always embedded in relations to others. So, even on the "qualitative" side, he was seen as being slightly outdated.
4. Other arguments have been advanced concerning numbers, but they are ecumenists in that they seek a wise articulation of the relationship between the quantitative and qualitative, not an opposition. For instance, Erving Goffman never published about the relationship between the quantitative and qualitative. He was apparently simply not interested in the question.
5. The web address is http://dfr.jstor.org/. I want to thank warmly Erik Gjesfjeld for introducing me to this very useful resource.

REFERENCES

Abbott, A. (1999). *Department and discipline: Chicago sociology at one hundred.* Chicago University Press.

Adorno, T. W. (1998). *Critical models: Interventions and catchwords.* Columbia University Press.

Adorno, T. W., Frenkel-Brunswik, E., Levinson, D. J., & Sanford, R. N. (1950). *The authoritarian personality.* Harper and Row.

Austrian, G. (1982). *Herman Hollerith: Forgotten giant of information processing.* Columbia University Press.

Becker, H. S. (1958). Problems of inference and proof in participant observation. *American Sociological Review, 23*(6), 652–660.

Becker, H. S. (1970). *Sociological work: Method and substance.* Aldine Publishing Company.

Becker, H. S. (1988). Herbert Blumer's conceptual impact. *Symbolic Interaction, 11*(1), 13–21.

Becker, H. S. (1996). The epistemology of qualitative research. In R. Jessor, A. Colby, & R. A. Shweder (Eds.), *Ethnography and human development: Context and meaning in social inquiry* (pp. 53–71). University of Chicago Press.

Becker, H. S., & Horowitz, I. L. (1972). Radical politics and sociological research: Observations on methodology and ideology. *American Journal of Sociology, 78*(1), 48–66.

Blumer, H. G. (1933). *Movies and conduct.* Macmillan & Company.

Blumer, H. G. (1943). Morale. In W. F. Ogburn (Ed.), *American society in wartime* (pp. 207–231). University of Chicago Press.

Blumer, H. G. (1951). Morale: Certain theoretical implications of data in the American Soldier: Comment. *American Sociological Review, 16*(3), 308–309.

Blumer, H. G. (1969). *Symbolic interactionism: Perspective and method.* Prentice Hall.

Blumer, H. G. (1990). *Industrialization as an agent of social change: A critical analysis* (edited with an Introduction by David R. Maines and Thomas J. Morrione). Aldine de Gruyter.

Blumer, H. G., & Hauser, P. M. (1970 [1933]). *Movies, delinquency, and crime: Motion pictures and youth.* Arno Press.

Bogdan, R., & Taylor, S. J. (1975). *Introduction to qualitative research methods: A phenomenological approach to the social sciences.* Wiley.

Boltanski, L. (2014). *Mysteries and conspiracies: Detective stories, spy novels and the making of modern societies.* Polity.

Brain, R. M. (2001). The ontology of the questionnaire: Max Weber on measurement and mass investigation. *Studies in History and Philosophy of Science, 32 Part A*(4), 647–684.

Bruno, I., Didier, E., & Prévieux, J. (Eds.). (2014). *Statactivisme: Comment lutter avec les nombres.* Zones.

Cantril, H. et al. (1939). *Industrial conflict: A psychological interpretation* (First Yearbook of the Society for the Psychological Study of Social Issues). Cordon.

Chapoulie, J.-M. (2001). *La tradition sociologique de Chicago: 1892–1961.* Seuil.

Cicourel, A. V. (1964). *Method and measurement in sociology.* Free Press.

Cicourel, A. V. (1974). *Theory and method in a study of Argentine fertility.* Wiley.

Conrad, P., & Reinharz, S. (1984). Computers and qualitative data: Editor's introductory essay. *Qualitative Sociology, 7*(1–2), 3–15.

Daston, L. (1988). *Classical probability in the enlightenment.* Princeton University Press.

Daston, L. (1992). Objectivity and the escape from perspective. *Social Studies of Science, 22*(4), 597–618.

Desrosières, A. (1998 [1993]). *The politics of large numbers: A history of statistical reasoning.* Harvard University Press.

Desrosières, A. (2003). Managing the economy: The state, the market and statistics. In T. M. Porter & D. Ross (Eds.), *The Cambridge history of science* (pp. 553–564). Cambridge University Press.

Desrosières, A. (2008). *Pour une sociologie historique de la quantification*. Presses de l'Ecole des Mines de Paris.

Didier, E. (2009). *En quoi consiste l'Amérique ? Les statistiques, le New Deal et la Démocratie*. La Découverte.

Didier, E. (2010). Gabriel tarde and statistical movement. In M. Candea (Ed.), *The social after Gabriel Tarde: Debates and assessments* (Vol. 4, pp. 299–325, Vol. Culture, Economy and the Social). Routledge.

Didier, E. (2020). *America by the numbers: Quantification, democracy, and the birth of national statistics*. Cambridge, MA: The MIT Press.

Dupont, F. (2015). BigData : Entre régulation et architecture - Introduction. *Statistique Et Société, 2*(4), 9–12.

Durkheim, E. (1986 [1897]). *Le Suicide: Étude de Sociologie* (Collection "Quadrige" 19). PUF.

Espeland, W. N., & Sauder, M. (2007). Rankings and reactivity: How public measures recreate social worlds. *American Journal of Sociology, 113*(1), 1–40.

Filstead, W. J. (1970). *Qualitative methodology: Firsthand involvement with the social world (Markham Sociology Series)*. Markham Pub. Co.

Garfinkel, H. (1999 [1967]). *Studies in ethnomethodology*. Polity Press.

Genel, K. (2013). *Autorité et émancipation: Horkheimer et la théorie critique*. Payot.

Gigerenzer, G., Swijtink, Z., Porter, T. M., Daston, L., Beatty, J., & Krüger, L. (1989). *The empire of chance*. Cambridge University Press.

Glaser, B. G., & Strauss, A. L. (1967). *The discovery of grounded theory: Strategies for qualitative research*. Aldine Publishing Company.

Gouldner, A. W. (1968). The sociologist as partisan: Sociology and the welfare state. *The American Sociologist, 3*(2), 103–116.

Gouldner, A. W. (1970). *The coming crisis of western sociology*. Basic Books.

Gouldner, A. W. (1973). For sociology: 'Varieties of political expression' revisited. *American Journal of Sociology, 78*(5), 1063–1093.

Hacking, I. (1982). Bio-power and the avalanche of printed numbers. *Humanities in Society, 5*(3–4), 279–295.

Heritage, J. (1984). *Garfinkel and ethnomethodology*. Polity Press.

Hoffman, A. (1968). *Revolution for the hell of it*. Dial Press.

Horowitz, I. L. (Ed.). (1974). *The rise and fall of project Camelot: Studies in the relationship between social science and practical politics* (Rev. ed.). MIT Press.

Joas, H. (1996). *The creativity of action*. University of Chicago Press.

Katz, J. (2015). Foreword. In R. M. Emerson (Ed.), *Everyday troubles: The micro-politics of interpersonal conflict*. Chicago University Press.

Kitsuse, J. I. (1980). Coming out all over: Deviants and the politics of social problems. *Social Problems, 28*(1), 1–13.

Kitsuse, J. I., & Cicourel, A. V. (1963). A note on the uses of official statistics. *Social Problems, 11*(2), 131–139.

Krüger, L., Daston, L., & Heidelberger, M. (Eds.). (1987). *The probabilistic revolution, vol. 1: Ideas in history*. MIT Press.

Lazarsfeld, P. F., & Kendall, P., L. (1982). *The varied sociology of Paul F. Lazarsfeld: Writings*. Columbia University Press.

Lazarsfeld, P. F., & Rosenberg, M. (1955). *The language of social research: A reader in the methodology of the social sciences*. Free Press.

Lerner, D., & Lasswell, H. D. (1951). *The policy sciences: Recent developments in scope and method (Hoover Institution Studies)*. Stanford University Press.

Lipset, M. S., & Ladd, E. C. (1972). The politics of American sociologists. *American Journal of Sociology, 78*(1), 67–104.

Lofland, J. (1971). *Analyzing social settings: A guide to qualitative observation and analysis* (The Wadsworth Series in Analytic Ethnography). Wadsworth Pub. Co.

Maynard, D. W., Houtkoop-Steenstra, H., Schaeffer, N. C., & van der Zouwen, J. (Eds.). (2002). *Standardization and tacit knowledge: Interaction and practice in the survey interview*. Wiley.

McNall, S. G., & Johnson, J. C. M. (1975). The new conservatives: Ethnomethodologists, phenomenologists, and symbolic interactionists. *Critical Sociology, 5*(4), 49–65.

Merton, R. K. (1972). Insiders and outsiders: A chapter in the sociology of knowledge. *American Journal of Sociology, 78*(1), 9–47.

Merton, R. K., & Lazarsfeld, P. F. (1950). *Continuities in social research: Studies in the scope and method of "The American Soldier."* Free Press.

Mills, C. W. (1959). *The sociological imagination*. Oxford University Press.

Molotch, H., & Lester, M. (1973). Accidents, scandals, and routines: Resources for insurgent methodology. *Critical Sociology, 3*(4), 1–11.

Molotch, H., & Lester, M. (1974). News as purposive behavior: On the strategic use of routine events, accidents, and scandals. *American Sociological Review, 39*(1), 101–112.

Moore, K. (2008). *Disrupting science: Social movements, American scientists, and the politics of the military, 1945–1975* (Vol. Princeton Studies in Cultural Sociology). Princeton University Press.

Pollak, M. (1986). Un texte dans son contexte: L'enquête de Max Weber sur les ouvriers agricoles. *Actes De La Recherche En Sciences Sociales, 65*(1), 69–75.

Porter, T. M. (1995). *Trust in numbers: The pursuit of objectivity in science and public life*. Princeton University Press.

Rainwater, L., & Yancey, W. L. (1967). *The Moynihan report and the politics of controversy*. MIT Press.

Reitman, R. (1978). Review. *Contemporary Sociology, 7*(4), 512.

Roach, J. L. (1970). The radical sociology movement: A short history and commentary. *The American Sociologist, 5*(3), 224–233.

Ryan, W. (1971). *Blaming the victim*. Vintage Books.

Schwartz, H. B., & Jacobs, J. (1979). *Qualitative sociology: A method to the madness*. Free Press.

Schweber, L. (2002). Wartime research and the quantification of American sociology: The view from "The American Soldier". *Revue d'Histoire des Sciences Humaines, 6*(1), 65–94.

Social Science Research Council (U.S.). (Ed.). (1949). *Studies in social psychology in World War II*. Princeton University Press.

Sternberg, D. J. (1977). *Radical sociology: An introduction to American behavioral science*. Exposition Press.

Tarde, G. d., & Parsons, E. W. C. (1903). *The laws of imitation*. H. Holt and Company.

Taylor, S. J., & Bogdan, R. (1984). *Introduction to qualitative research methods: The search for meanings* (2nd ed.). Wiley.

Torgerson, W. S. (1958). *Theory and methods of scaling*. Wiley.

Turner, S. P., & Turner, J. H. (1990). *The impossible science: An institutional analysis of American sociology*. Sage.

Witzel, A., & Mey, G. (2004). Aaron V. Cicourel: I Am NOT opposed to quantification or formalization or modeling, but I do not want to pursue quantitative methods that are not commensurate with the research phenomena addressed. *Forum Qualitative Sozialforschung Forum: Qualitative Social Research, 5*(3), http://www.qualitative-research.net/index.php/fqs/article/view/549.

Zeisel, H. (1950). *Say it with figures* (Publications of the Bureau of Applied Social Research, 3rd ed.). Harper.

Zentner, H. (1951). Morale: Certain theoretical implications of data in the American Soldier. *American Sociological Review, 16*(3), 297–307.

CHAPTER 14

Afterword: Quantifying, Mediating and Intervening: The R Number and the Politics of Health in the Twenty-First Century

Peter Miller

In 1985, Anthony Hopwood remarked as follows: "A world of the seemingly precise, specific and quantitative can in this way emerge out of that of the contentious and the uncertain" (Hopwood, 1988 [1985], p. 262). This was more than a decade before the "performative turn" in economic sociology and several years before the academic explosion of "New Public Management" studies. Hopwood was speaking here about accounting, and how costs, consequences and benefits come to be divided into the defined and the seemingly known, and the imprecise and the intangible, and how this can give a calculative priority to the economic rather than the social.

P. Miller (✉)

Department of Accounting and Centre for Analysis of Risk and Regulation, London School of Economics and Political Science, London, UK
e-mail: p.b.miller@lse.ac.uk

© The Author(s) 2022
A. Mennicken and R. Salais (eds.), *The New Politics of Numbers*, Executive Politics and Governance,
https://doi.org/10.1007/978-3-030-78201-6_14

But, as he and many others have shown since, and as the contributors to this volume show, the point is more general. Numbers have acquired an unassailable power in modern political life. Political authority and the stewardship of people's lives are today inseparable from the vast range of different sorts of numbers that are deployed in the governing of advanced liberal democratic capitalist societies. Debates about the health of "the economy" are inconceivable without numerical measures of various kinds. The same applies to the quantification of the social economy, whether this be a matter of transforming poverty into the number of people claiming benefits, public order into the crime rate, the state of family life into the divorce rate or the governing of sexual conduct into the rate of spread of AIDS. And, just as political decisions come to depend increasingly on quantification, there is a simultaneous "de-politicization" of politics. The boundaries between politics and objectivity are redrawn, by proclaiming that political decisions are little more than automated technical mechanisms that tell us what to do and when, and what to prioritize (Rose, 1991).

This much will be familiar to many readers of this volume. But even for those well aware of such issues, the phrase "follow the science", and its numerical counterpart the "R" number, has attained an ascendancy that none of us could have imagined only a few months ago.[1] This affirmation of scientific expertise is all the more remarkable, given its contrast with the statement by Michael Gove, the then Secretary of State for Justice, in the context of debates about Brexit in 2016, that "people in this country have had enough of experts" (*The Financial Times*, 3 June 2016).

At its simplest, in an epidemic the R number—the reproduction number—is one of the most important numbers. As almost every citizen now knows, if the R number is below one, then that is good news. For if it is below one, the number of new infections will fall over time. But if R is above one, that is definitely not good news. It means that the number of new infections is accelerating; the higher the number, the faster the virus spreads through the population.

The R number can be used as a device for shutting schools, shops, restaurants, hotels, gyms, factories, university campuses, international travel and indeed most forms of social life. In the other direction, it can also be used as a device for opening some or all such venues and interactions. Two newspaper headlines illustrate this well. The first was printed on 1 May 2020 in the *Financial Times*, and stated as follows: "R number: the figure that will determine when lockdown lifts". Describing the R

number as the average number of new cases generated by an infected individual, the article went on to say that politicians viewed it as a key indicator for lifting lockdowns enough for significant social and economic activity to resume, without allowing a resurgence of the virus. Also, and crucially, it was viewed as a relatively simple number to convey success or otherwise to the general public. However, it also went on to say that unfortunately, and despite the repeated appeals to it in statements by politicians, things are more complicated. Not only is it incredibly difficult to measure, as an aggregate number for a large geographical area it is also potentially misleading or at least uninformative, because it does not tell you what is happening in your local area. Further, while it is widely described as if it were the actual number of new cases generated by an infected individual, in most countries the R number is in fact an estimate generated from mathematical models and simulations, with different modelling teams even in the same country arriving at different results.

The second headline, printed two weeks later in *The Financial Times*, was more cautious: "R numbers offer no easy answers for UK to lift lockdown". The starting point of this piece was the significant regional differences in the number of officially recorded new infections per day. In London, which previously had one of the highest number of new infections, the number of new infections was just 24 per day according to data from Public Health England (PHE) and Cambridge University, whereas the comparable figure in Yorkshire and North-East England was 4,320. This was cited as evidence of the difficulty of having uniform policies even across England. The modelling conducted by PHE and Cambridge suggested a median R of 0.75 for England as a whole, but varying from 0.4 in London to 0.8 in the North-East and Yorkshire. Other modelling groups, such as the London School of Hygiene and Tropical Medicine and Imperial College, London, gave higher figures for R in London, but all showed the number significantly below 1 in the capital. And things became even more difficult once differences in approach to the relaxation of restrictions between England and the devolved governments in Scotland, Wales and Northern Ireland began to appear. In rejecting the lockdown easing in England, all cited worries about the R number in their regions.

But the different R number estimates only partially explain the divergence in policy across the different parts of the UK. Statisticians reiterated that the R number was extraordinary difficult to calculate, the Scottish government's chief statistician commenting that the official R number in

Scotland takes 56 hours for an Edinburgh university supercomputer to calculate. Further, he emphasized that the number, even when checked against the numbers produced by the other models, should only be expressed as a range, rather than a single figure, as there is a roughly 50% risk that the R level is higher than any specific point estimate.

These differences in results arising from different modelling assumptions, when combined with the different approaches to the relaxation of restrictions in the devolved administrations, began to undermine the appeal of the slogan to "follow the science". An article in the publication *Wired* went as far as to adopt the headline: "Boris Johnson's brief love affair with science is well and truly over" (Matt Reynolds, 6 June 2020, Wired).[2] Moreover, it soon became clear that, while the R number needed to be below 1 to ease restrictions, no politician was willing to say how far below 1 it needed to be in order to ease restrictions. In addition, prominent scientific advisors began to distance themselves from specific government policies. For instance, on 3 June the Chief Scientific Adviser to the UK government refused to explicitly endorse the government's decision to impose quarantine on new arrivals to the UK with effect from the following week (*The Guardian*, Andrew Sparrow, "Evening Summary" 3 June 2020, updated 4 June 2020).

The R number is a key part of what one might call a conditional "trust in numbers" (Porter, 1995), albeit one where the authority of the number is tempered not only by political judgement but also by an array of other numbers, including GDP and unemployment, together with reports by official bodies predicting either a V-shaped or a U-shaped recovery, a further spike in infections, and much else besides. It thus stands at the heart of the politics of health in the twenty-first century, a perfect "mediating instrument" (Miller & O'Leary, 2007) linking the health and well-being of the population with the health of the economy. A calculative assemblage that facilitates a level of intervention in the lives and activities of citizens in advanced liberal democracies that is not only unprecedented but fundamentally at odds with so much that is at the heart of our political culture. As Foucault remarked of the politics of health in the eighteenth century, the biological characteristics of the population become relevant factors for economic management. It becomes necessary to organize around the population an apparatus that will ensure its subjection and even its enforced idleness so as to (hopefully) increase or at least maintain its utility as and when the pandemic subsides (Foucault,

1980 [1976]). Epidemiology, virology and statistical science thus assume an increasingly important place in the machinery of power.

Medico-administrative knowledge, albeit tempered by political expediency, has achieved a political hold on a population at the mercy of a virus that in just over a few months has certainly killed more than 50,000 people in the UK as of the time of writing, and may well have killed more than 60,000. And this display of medical knowledge has taken place in the most public manner. Daily briefings (initially), 92 in total, ending on 23 June,[3] saw government ministers flanked on most occasions by their most senior scientific advisers, and in most instances diplomatically endorsing the actions of the government. Perhaps the most notable exception to this being the critical comments made by the Deputy Chief Medical Officer Jonathan Van-Tam on 30 May, when asked about the behaviour of Boris Johnson's most senior special adviser Dominic Cummings. As the journalist John Crace remarked, the relationship between the government and the scientists never really recovered thereafter,[4] and Van-Tam did not appear again at the daily briefings.[5]

At every briefing there would be slides showing graphs of the number of new cases, the total number of cases, the 7-day rolling average, the number of patients on mechanical ventilators, the number of people in hospital with COVID-19, and, most depressingly, the number of deaths in the previous 24 hours, as well as the total number of deaths. This unprecedented public display of medical knowledge was backed up by various government websites that provided access to the materials displayed, while the Office for National Statistics, the National Records of Scotland, and the Northern Ireland Statistics and Research Agency provided further data. This included three different ways of measuring the number of deaths: those with a positive COVID-19 test result; those where the death certificate mentions COVID-19; and the third being the number of "excess" deaths for the time of year. As of mid-/late June, the three numbers for the UK as a whole stood at 43,414, 53,009, and 65,138, respectively.[6]

The political ascendancy of the beguilingly simple R number is all the more remarkable, as it is a relative newcomer to epidemiology. Now regarded as arguably the most important quantity in the study and control of epidemics (Heesterbeek, 2002), it was only clearly defined for the first time in 1975 by the German mathematician Klaus Dietz, as follows:

> The quantity R is called the reproduction rate, since it represents the number of secondary cases that one case can produce if introduced to a susceptible population. (Dietz, 1975, p. 106)

Yet, despite the existence of this clear definition, it still took a number of years for epidemiologists to fully embrace the R number. Two events in 1982 provided the stimulus for the R number to become central to the analysis of epidemics by epidemiologists. The first was an article published in February that year in the journal *Science*, which made extensive use of R_0, calling it "the intrinsic reproduction rate" (Anderson & May, 1982a, p. 1055). The second was an influential workshop held in the Berlin suburb of Dahlem in March of that year (Anderson & May, 1982b), with almost all contributors using R_0 as if the concept had been used in epidemiology for decades, which was certainly not the case (Heesterbeek, 2002, p. 200).

There had of course been earlier attempts to model the spread of epidemics, most notably through the work of Ronald Ross (1857–1932), a medical doctor, a colonel in the British army, a minor poet and a self-taught mathematician, and the first Briton to be awarded a Nobel Prize (Heesterbeek, 2002, p. 192). He led several anti-malaria campaigns, dissected many mosquitoes, and discovered in 1898 that (bird) malaria was transmitted by mosquitoes, rather than by "bad air" from marshes as was previously believed. He received a Nobel Prize for this discovery in 1902.

His work in modelling epidemics started with showing that trying to control malaria by fighting mosquitoes was a real possibility. This was in contrast to general opinion at the time that fighting mosquitoes was not viable because it would be impossible to kill all mosquitoes locally and therefore impossible to stop transmission of malaria. Ross identified the main factors in malaria transmission and calculated the number of new infections arising per month as the product of these factors. He referred to his discovery as the "Mosquito Theorem". His conclusion was that instead of having to eradicate all mosquitoes in a given area, it was sufficient to depress the ratio of mosquitoes to man below a particular threshold. There was, he argued, a "critical density of mosquitoes" below which the malaria parasite could not be sustained.

While the notion of a critical threshold (critical community size) was helpful for the study and control of malaria, it was not conducive for the development of the notion of a reproduction threshold or rate. As Ross himself had discovered, malaria is a vector-transmitted infection, rather than a directly horizontally (i.e. person to person) transmitted infection. Ross published a series of three papers (two co-authored with Hilda Hudson) (see Ross, 1916; Ross & Hudson 1917a, 1917b) in an attempt to develop a general theory of epidemic phenomena, a "theory of happenings". He referred to his approach as "a priori pathometry" (Heesterbeek, 2002, pp. 192–193; see also Kucharski, 2020). This led Heesterbeek (2002, p. 193) to comment that Ross was the first to try and develop a general theory of epidemic phenomena using prior assumptions about mechanisms that could be acting in the spread of infections, rather than trying to obtain insight a posteriori by studying real epidemics. Heesterbeek concluded that this work represents the first development in abstract or modern epidemic theory, even if it did not result in the formulation of R_0.

But it was to be more than 50 years before the notion of the reproduction rate was to be formulated in epidemiology. This, despite Ross's aspiration to "establish the general law of epidemics", and his encouragement to McKendrick, a medical doctor who served in the British army under his command in Sierra Leone in 1901 during one of the anti-malaria campaigns, to continue his work further. As Ross remarked rather ambitiously to McKendrick: "We shall end by establishing a new science. But first let you and me unlock the door and then anybody can go in who likes" (Ross, in a letter to McKendrick in 1911, cited in Heesterbeek, 2002, p. 195).

Meanwhile, in 1925, and within demography rather than epidemiology, the concept of R_0 or the reproduction rate was formulated in a paper titled "On the true rate of natural increase", published in the *Journal of the American Statistical Association* (Dublin & Lotka, 1925). One of the authors was Alfred Lotka, who worked for the Metropolitan Life Insurance Company in New York. Lotka had started the chain of reasoning with a short note in *Science* in 1907 on the "rate of natural increase per head", which he called r, of a population with constant birth and death rate.

The 1925 paper was published just one year after President Coolidge had signed into law the Immigration Act of 1924, the most stringent US immigration policy up till then in the nation's history. The paper began by remarking that "The present policy of restricting immigration into the United States lends a particular interest to inquiries into the powers of natural increase of our population" (Dublin & Lotka, 1925, p. 305). The paper went on to comment that the excess of birth-rate over death rate may appear to provide a measure of natural increase. However, that would be misleading because it fails to take into account the age distribution of the population. If one factors in reduced immigration, which would over time result in a reduction of productive and reproductive members of the population, combined with a falling birth-rate, then sooner or later the birth-rate would become stationary or nearly so. Numerically, this would mean that the excess of the birth-rate over the death rate would fall from 11 per thousand per annum to 5.5 per thousand per annum, that is, it would be reduced by one half (Dublin & Lotka, 1925, p. 307). Having considered fecundity, mortality (i.e. a life table), together with the age schedule for fecundity of females in the United States in 1920, the authors conclude as follows:

> The net result is that if we follow the history of 100,000 females at the current rate of fecundity we find that throughout their life they give birth to 116,700 daughters; or, on average, one female gives birth to 1.168 daughters in the course of her life. This, then, is the ratio of the total births (of daughters) in two successive generations. *It will be convenient for future reference to denote this ratio by the symbol R_0.* (Dublin & Lotka, 1925, p. 310, emphasis added)

This way of expressing things enabled the authors to speak of a "standardized" or "stable natural rate of increase" under specified conditions of maternity and mortality. While our interest here is primarily in terms of this early formulation of R_0 within demography, it is difficult in our current socio-political circumstances to avoid remarking on this linking of the positive impact of immigration on the productive and reproductive health of the population. Once the impact of reduced levels of immigration, combined with a rapidly declining birth-rate, have had time to manifest themselves, the authors remarked that the country would no longer have a disproportionately high population in the productive and reproductive age group, something that is rarely remarked on publicly in current debates concerning the age profile of the UK (Dublin & Lotka, 1925, p. 328).

The 50-year gap between Dublin & Lotka's formulation of R_0 in demography, and the formulation with regard to epidemics by the German mathematician Klaus Dietz in 1975, is even more remarkable as Lotka worked in the fields of both demography and epidemiology. Despite this:

> It took a long time for modellers in epidemiology to realise that the formulation in terms of reproduction potential is a much clearer and more powerful concept for infectious diseases as well, which is moreover much more amenable to generalization to heterogeneous populations, and can be tied much more easily to data and hence applications. (Heesterbeek, 2002, p. 190)

Heesterbeek attributes this to the much closer link to data in the field of demography, than was the case in the early development of epidemiology. Researchers working in the field of epidemiology were "much more interested in presenting a mathematically coherent theory" than in engaging with data (Heesterbeek, 2002, p. 191). This was compounded, he suggested, when a large number of mathematicians "took over" the field of epidemiology in the early 1950s (Heesterbeek, 2002, p. 197). Unfortunately, many of these depended on a review of the field by Norman Bailey published in 1957, devoted entirely to the mathematical study of epidemic phenomena. It was unfortunate because, although it opened up the subject for mathematicians, it neglected to extrapolate from a paper by George Macdonald published in 1952 in the *Tropical Diseases Bulletin*. Macdonald was the Director of the Ross Institute at the London School of Hygiene and Tropical Medicine. He devoted his paper entirely to malaria, but also in the appendix took a more general view of epidemic phenomena, which included the "basic reproduction rate". Although Bailey had, apparently, read the paper by Macdonald, he did not recognize the potential of the definition for a much more general class of infections. It is no wonder, Heesterbeek remarks, that none of the mathematicians was enticed to read the original Macdonald papers for a number of years to come, for "mathematicians would not easily be led to read a paper in the Tropical Diseases Bulletin unless they would be told that it contained a mathematically interesting idea" (Heesterbeek, 2002, p. 197).

So, although the theory of epidemics blossomed for a number of years, by the end of the nineteen-sixties the field had come no closer to defining R_0. As already noted above, it was not until 1975 that the concept was finally formulated clearly within epidemiology, fifty years after it had been formulated within demography, and twenty-five years after its potential had been registered within epidemiology, even if the symbol Z_0 rather than R_0 had been used. At last, the use of the concept R_0 in examining the spread and control of infectious diseases with epidemiological models could start to grow.

CONCLUSIONS

The emergence of the R number in multiple and dispersed sites will be no surprise to those sociologists of science who have long demonstrated the non-linear nature of scientific discovery. That said, the bifurcation or compartmentalization of demography and epidemiology in this instance is quite remarkable, not least given the existence of key figures who worked in both disciplines. However, it is perhaps reassuring that the challenges of interdisciplinary work are not limited to the social sciences. Also, it is possibly unsurprising to see the close links between a particular calculative instrument within epidemiology and the politics of health in the twenty-first century. As noted above, this linkage between medicine and governing was already established in the eighteenth century, if not before. As for the almost totemic significance of the R number, a number which turns out in fact to be a range rather than a single number; again, researchers studying accounting, management, macro-economics and no doubt many other domains have demonstrated the power of the single figure. What is somewhat unusual though, in the case of COVID-19, is the prominence such a number has rapidly achieved in popular social and political discourse.

The current crisis also reminds us more generally of the fraught relationship between expertise and government, whether in the UK or beyond. In the UK, and in the current pandemic, "Following the science" has turned out to be more a slogan than a description of policy formulation. The R number has acted here as a crucial mediating instrument, linking the health and well-being of the population with the health of the economy, and supporting arguments both in favour of and against restrictions of various kinds. As this volume demonstrates, the triptych of quantification, administrative capacity and democracy is far from harmonious. For now, perhaps the best we can hope is that if and when the pandemic finally subsides, responsibility for key decisions will be laid at the door of those who made the decisions, rather than those medics who sought to offer advice, however difficult that would have been in light of the data available. Also, and even more importantly, let us hope that not too many more people will suffer and die, or lose their jobs, before a vaccine is discovered for COVID-19.

NOTES

1. This piece was written in July 2020, when the pandemic was only a few months old.
2. Matt Reynolds, 6 June 2020, Wired. Wired is a monthly magazine based in San Francisco and focusing on emerging technologies and how they affect culture, the economy and politics.
3. These resumed briefly on 2 July (and intermittently thereafter), to address the issue of schools reopening in England with attendance becoming mandatory, quite possibly in light of the low attendance until then among those eligible to return to school. The following day the government announced that later in the year there would be White House-style daily televised press briefings.
4. John Crace, "A daily dose of world-beating waffle ends", *The Guardian*, 24 June 2020.
5. As of the time of writing.
6. See the following link for the slides and datasets displayed in these briefings: https://www.gov.uk/government/collections/slides-and-datasets-to-accompany-coronavirus-press-conferences, accessed 14 July 2020. See also https://www.statisticsauthority.gov.uk/news/covid-19-and-the-uk-statistics-system, accessed 14 July 2020.

REFERENCES

Anderson, R. M., & May, R. M. (1982a). Directly transmitted infectious diseases: Control by vaccination. *Science, 215*(4536), 1053–1060.

Anderson, R. M., & May, R. M. (Eds.). (1982b, March 14–19). *Population biology of infectious diseases. Report of the Dahlem workshop on population of infectious disease agents, Berlin*. Springer-Verlag Berlin.

Dietz, K. (1975). Transmission and control of arboviruses. In D. Ludwig & K. L. Cooke (Eds.), *Epidemiology* (pp. 104–121). Society for Industrial and Applied Mathematics.

Dublin, L. I., & Lotka, A. J. (1925). On the true rate of natural increase. *Journal of the American Statistical Association, 20*(151), 305–339.

Foucault, M. (1980 [1976]). The politics of health in the eighteenth century. In C. Gordon (Ed.), *Michel Foucault: Power/ knowledge—Selected interviews and other writings 1972–1977* (pp. 166–182). Harvester Press.

Heesterbeek, J. A. P. (2002). A brief history of R0 and a recipe for its calculation. *Acta Biotheoretica, 50*, 189–204.

Hopwood, A. G. (1988 [1985]). Accounting and the domain of the public: Some observations on current developments (The Price Waterhouse Public

Lecture on Accounting, University of Leeds, 1985). In *Accounting from the outside: The collected papers of Anthony G. Hopwood*. Garland Publishing.

Kucharski, A. (2020). *The rules of contagion: Why things spread and why they stop*. Profile Books Limited.

Miller, P., & O'Leary, T. (2007). Mediating instruments and making markets: Capital budgeting, science and the economy. *Accounting, Organizations and Society, 32*(7–8), 701–734.

Porter, T. M. (1995). *Trust in numbers: The pursuit of objectivity in science and public life*. Princeton University Press.

Rose, N. (1991). Governing by numbers: Figuring out democracy. *AccOunting, Organizations and Society, 16*(7), 673–692.

Ross, R. (1916). An application of the theory of probabilities to the study of a priori pathometry – Part I. *Proceedings of the Royal Society, London A, 42*, 204–230.

Ross, R., & Hudson, H. P. (1917a). An application of the theory of probabilities to the study of a priori pathometry—Part II. In *Proceedings of the Royal Society, London* (pp. 212–225)

Ross, R., & Hudson, H. P. (1917b). An application of the theory of probabilities to the study of a priori pathometry—Part III. In *Proceedings of the Royal Society, London* (pp. 225–240).

Index